LIBRARY
University of Glasgow

ALL ITEMS ARE ISSUED SUBJECT TO RECALL

GUL 18-08

Progress in Mathematical Physics

Volume 56

Editors-in-Chief

Anne Boutet de Monvel (Université Paris VII Denis Diderot, France)
Gerald Kaiser (Center for Signals and Waves, Austin, TX, USA)

Pankaj Sharan

Spacetime, Geometry and Gravitation

Birkhäuser
Basel · Boston · Berlin

HINDUSTAN
BOOK AGENCY

Author:

Pankaj Sharan
Physics Department
Jamia Millia Islamia, Jamia Nagar
New Delhi, 110 025
India
e-mail: pankajsharan@gmail.com

2000 Mathematics Subject Classification: 83-01, 83Cxx

Library of Congress Control Number: 2009924925

Bibliographic information published by Die Deutsche Bibliothek.
Die Deutsche Bibliothek lists this publication in the Deutsche Nationalbibliografie;
detailed bibliographic data is available in the Internet at http://dnb.ddb.de

ISBN 978-3-7643-9970-2 Birkhäuser Verlag AG, Basel · Boston · Berlin

© 2009 Hindustan Book Agency (HBA)
Authorized edition by
    Birkhäuser Verlag, P.O. Box 133, CH-4010 Basel, Switzerland
    Part of Springer Science+Business Media
for exclusive distribution worldwide except India.
The distribution rights for print copies of the book for India remain with
Hindustan Book Agency (HBA).
Printed in India, bookbinding made in Germany.

ISBN 978-3-7643-9970-2   e-ISBN 978-3-7643-9971-9
9 8 7 6 5 4 3 2 1   www.birkhauser.ch

Contents

III Gravitation   209

Preface

This is an introductory book on the general theory of relativity based partly on lectures given to students of M.Sc. Physics at my university.

The book is divided into three parts. The first part is a preliminary course on general relativity with minimum preparation. The second part builds the mathematical background and the third part deals with topics where mathematics developed in the second part is needed.

The first chapter gives a general background and introduction. This is followed by an introduction to curvature through Gauss' Theorema Egregium. This theorem expresses the curvature of a two-dimensional surface in terms of intrinsic quantities related to the infinitesimal distance function on the surface. The student is introduced to the metric tensor, Christoffel symbols and Riemann curvature tensor by elementary methods in the familiar and visualizable case of two dimensions. This early introduction to geometric quantities equips a student to learn simpler topics in general relativity like the Newtonian limit, red shift, the Schwarzschild solution, precession of the perihelion and bending of light in a gravitational field.

Part II (chapters 5 to 10) is an introduction to Riemannian geometry as required by general relativity. This is done from the beginning, starting with vectors and tensors. I believe that students of physics grasp physical concepts better if they are not shaky about the mathematics involved.

There is perhaps more mathematics in Part II than strictly required for Part III of this introductory book. My aim has been that, after reading the book, a student should not feel discouraged when she opens advanced texts on general relativity for further reading. The advanced books introduce mathematical concepts far too briefly to be really useful to a student. And the student feels lost in the pure mathematical textbooks on differential geometry. In that sense, this book offers to fill a gap.

The final part is devoted to topics that include the action principle, weak gravitational fields, gravitational waves, Schwarzschild and Kerr solutions and the Friedman equation in cosmology. A few special topics are touched upon in the final chapter.

Many exercises are provided with hints and very often complete solutions in the last section of chapters. These exercises contain material which cannot be ignored and has been put in this format purposely to help students learn on their own.

Note

I have generally used the female gender for the imagined student reader of the book, but occasionally, the male pronouns 'he' or 'his' are also slipped in for political correctness.

Acknowledgements

I am grateful to H.S. Mani, teacher and friend, for suggesting that I transform the notes I had written down for classroom lectures into a book and to R. Ramaswamy for useful advice and help during the final stages of the book.

I thank Pravabati Chingangbam for discussions on many points in geometry.

A preliminary version of Part I of the book was used in '2006 Enrichment Course in Physics' at the Indian Institute of Astrophysics Observatory, Kodaikanal in June 2006. I am grateful to the Organizers of the Course, particularly Professor Vinod Gaur and Dr. K. Sundara Raman for hospitality at the observatory. Richa Kulshreshtha assisted with many tutorials in that course.

I am grateful to my colleagues M. Sami, Sanjay Jhingan, Anjan Ananda Sen, Rathin Adhikari and Somasri Sen at the Center for Theoretical Physics, and Tabish Qureshi and Sharf Alam in the Physics Department, Jamia Millia Islamia, New Delhi, for helping in various ways. L. K. Pande and Patrick Dasgupta made helpful suggestions on the manuscript.

I am grateful to Basabi Bhaumik for help during the preparation of the book in many ways, too numerous to mention.

Finally, I would like to remember my science teacher in school Mr. M. C. Verma who gave me a book on Relativity to read in 1963.

Pankaj Sharan
New Delhi, 2009

Notation and Conventions

We use super- and sub-scripts ('upper' and 'lower' indices) aplenty. In Part II of the book indices $i, j, k, l, m, \ldots$ range from 1 to n where n is the dimension of the vector space or the manifold. In rare cases n too is pressed into service when indices are running short in supply, but I hope no confusion arises when n is summed from 1 to n! Indices $i, j, k, \ldots$ take value $1, 2$ while discussing two-dimensional surfaces in chapter 2.

For physical four-dimensional spacetime (in Parts I and III) we use Greek indices $\mu, \nu, \sigma, \tau, \alpha, \beta, \ldots$ for components in the coordinate bases. These indices take on values $0, 1, 2, 3$. The index 0 refers to the time-related coordinate (usually) and in this context indices $i, j, ..$ etc. take values $1, 2, 3$ for space-like coordinates. For example

$$A^\mu = (A^0, A^i) = (A^0, A^1, A^2, A^3)$$

On one or two occasions alphabets $a, b, c, \ldots$ running from 0 to 3 are employed when a basis other than a coordinate basis is used. For example, when an ortho-normal basis for vector fields is chosen to calculate Ricci coefficients.

We use Einstein's summation convention which assumes a sum over the full range of values of any repeated index (called "dummy index") in a term without explicitly writing the summation sign Σ. In most cases one of the summed indices is lower, the other upper. For example, $A^i B_i = \sum_{i=1}^{n} A^i B_i$ in mathematical chapters, and $A^\mu B_\mu = \sum_{\mu=0}^{3} A^\mu B_\mu$ in physics chapters. Any departure from the summation convention is explicitly pointed out in the text.

We use a comma followed by an index to denote partial differentiation with respect to some coordinate system, the name symbol of the coordinate (usually x) is to be understood from the context:

$$f_{,i} = \frac{\partial f}{\partial x^i}, \qquad F_{,jk} = \frac{\partial^2 F}{\partial x^j \partial x^k}$$

When specific coordinates are used, a comma followed by a coordinate symbol denotes differentiation, for example $f_{,r} = \partial f / \partial r$.

We use physicists' convention of multiplying a vector (or tensor) by a scalar number on the left or right according to convenience.

The signature of the metric tensor is $+2$. The Minkowski metric $\eta_{\mu\nu}$ has diagonal components $\eta_{00} = -1, \eta_{11} = 1, \eta_{22} = 1, \eta_{33} = 1$.

We have used the physicists' notation ds^2 to denote the metric tensor most of the time and $\mathbf{g}$ or $\underline{\underline{g}}$ at one or two places. It is written $ds^2 = g_{ij}dx^i dx^j$ instead of $\mathbf{g} = g_{ij}dx^i \otimes dx^j$. Since g_{ij} is symmetric, the terms like $g_{12}dx^1dx^2 + g_{21}dx^2dx^1$ are written in combined form as $2g_{12}dx^1dx^2$ instead of $g_{12}(dx^1dx^2 + dx^2dx^1)$.

In raising and lowering of indices (the isomorphism induced by the metric tensor between a tangent space and its dual space of one-forms) we follow the usual practice of not changing the name symbol of a vector or tensor field for important tensors like the Riemann tensor $R_{ijkl} = g_{im}R^m{}_{jkl}$, the Ricci tensor $R^{ij} = g^{ik}g^{jl}R_{kl}$, the Einstein tensor $G^{ij} = g^{ik}g^{jl}G_{kl}$ or the stress energy tensor $T_{ij} = g_{ik}g_{jl}T^{kl}$. But we write $\delta^i_j = g^{ik}g_{kj}$ and not g^i_j. We avoid this convention when it serves no special purpose.

The covariant derivative is denoted by a semicolon (;), for example $\phi_{;i}, A_{i;j}$ etc. Repeated covariant derivatives are sometimes written without an additional semicolon: $(A_{i;j})_{;k} = A_{i;jk}$. Unlike ordinary derivatives where $A_{i,jk} = A_{i,kj}$, covariant derivatives do not commute, $A_{i;jk} \neq A_{i;kj}$ in general and one has to be careful.

The components of the Riemann curvature tensor are given by

$$-R^i{}_{jkl} = \Gamma^i{}_{jk,l} - \Gamma^i{}_{jl,k} + \Gamma^m{}_{jk}\Gamma^i{}_{lm} - \Gamma^m{}_{jl}\Gamma^i{}_{km}$$

This is the same as the convention of Hawking-Ellis, Misner-Thorne-Wheeler and Hartle but differs by a minus sign from Weinberg's

$$(R^i{}_{jkl})_{\text{our}} = -(R^i{}_{jkl})_{\text{Weinberg's}}$$

and by a rearragement of indices with Wald's

$$(R^i{}_{jkl})_{\text{our}} = (R_{lkj}{}^i)_{\text{Wald's}}$$

Part I

Spacetime

Chapter 1

Introduction

General Theory of Relativity (or General Relativity) is Einstein's theory of gravitation given by him in 1915. The name also applies to its later developments. According to the theory spacetime is a Riemannian space whose metric $g_{\mu\nu}$ determines the gravitational field.[1] The **Einstein equation**

$$R_{\mu\nu} - \frac{1}{2}g_{\mu\nu}R = \frac{8\pi G}{c^4}T_{\mu\nu} \tag{1.1}$$

governs the gravitational field. In this equation the quantities $R_{\mu\nu}$, R are functions of the metric $g_{\mu\nu}$ and its various derivatives and $T_{\mu\nu}$ on the right-hand side are determined by distribution of matter.

Bodies move along the straightest possible curves if no forces other than gravity act on them.

Understanding the general theory of relativity means understanding:

1. the left-hand side of this equation, which relates to curvature of spacetime,

2. the right-hand side which contains $T_{\mu\nu}$, the stress-energy tensor,

3. the nature of solutions to the equation in various situations.

In cosmological applications the Einstein equation is written with an additional term,

$$R_{\mu\nu} - \frac{1}{2}g_{\mu\nu}R + \Lambda g_{\mu\nu} = \frac{8\pi G}{c^4}T_{\mu\nu}$$

where Λ is called the 'cosmological constant'. We can include this term by adding to $T_{\mu\nu}$ an additional $T_{\mu\nu}^{\Lambda} \equiv -c^4 g_{\mu\nu}\Lambda/8\pi G$. Such a form for a stress energy tensor is unusual in the sense that it corresponds to a perfect fluid with negative pressure. This term is of little consequence except in cosmology. That is where we will discuss it.

---

[1]The explanation of these mathematical concepts will come as we go along in the book. For the purposes of initial chapters, quantities with indices should be treated as a set of physical quantities which appear together in an equation, much like the components of a vector.

1.1 Inertial and Non-Inertial Frames

1.1.1 Inertial Frames

In the Newtonian mechanical view the world is made up of 'mass-points' and the motion of mass-points is determined by the three laws of motion.

To describe the motion of a mass-point, we need a **frame of reference**. A frame of reference is (i) a set of orthogonal axes from which the position of mass-points can be determined, and (ii) a system of measuring time accurately. One can imagine a frame as a set of orthogonal coordinate axes made of light and thin but rigid rods with markings of length, and an accurate clock. The position of a mass-point at any time t shown by the clock can be described by its space coordinates $\mathbf{r}(t) = (x^1(t), x^2(t), x^3(t))$. Here t is the Newtonian **absolute time** which has the same numerical value in all frames of reference.

The functions $\mathbf{r}(t)$ describe the trajectory of the mass-point. The velocity and acceleration are defined as the first and second derivatives with respect to time t of these functions.

The first law of motion (law of inertia) states that a mass-point either stays put at a fixed position or moves with uniform velocity if there are no forces acting on it. If the motion is accelerated, then there must be forces acting on the mass-point.

It is assumed that we can identify the forces as real physical agents independently of their capacity to produce accelerations. So, if there are no forces, and we find mass-points accelerated, then it is because we have chosen a wrong frame of reference. The acceleration we see is due to the acceleration of the frame of reference itself.

The 'correct' frame of reference is one in which this first law holds good. Such a frame is called an **inertial frame of reference**.

It is necessary to make sure there exist inertial frames. Isaac Newton gave the following recipe for a close approximation to one inertial frame. Choose a frame in which the distant stars are stationary. The distant stars thus define a reference frame at **absolute rest**.

Given this one inertial frame we can define an additional infinite number of them. Any frame whose origin moves with a constant velocity in any direction with respect to this given inertial frame and whose axes move parallel to themselves is another inertial frame. Moreover, there are inertial frames which can be obtained by a shifting of the origin of coordinates by a fixed amount or those obtained by a rotation of the axes with respect to the original inertial frame by a fixed angle. And furthermore, there are frames that can be obtained by constant relative velocities given to frames which have been translated and rotated in this manner.

The second law of motion equates the rate of change of momentum $p^i = mdx^i/dt$ in the i-direction to the component of force F^i acting on the particle in that direction. Here the **inertial mass** m is a measure of the 'quantitiy of matter' in the particle and is a constant. The second law shows that the force is equal to inertial mass times acceleration d^2x^i/dt^2,

$$\frac{dp^i}{dt} = m\frac{d^2x^i}{dt^2} = F^i.$$

Note that these equations are formulated in rectangular cartesian coordinates.

This fundamental law of physics is *not* a definition of force. The numbers F^i on the right-hand side have to be measured separately (for example by comparing with some standard spring pulled to a given distance) and must be equal (up to experimental accuracy) to mass times the accelerations produced in the particle by observing the particle trajectory.

The third law states that, in an inertial frame, when two mass-points interact with each other, the force produced by one on the other is equal and opposite to the force produced by the other on the first. One should remember here that the two forces act at the same instant of time and at the locations of the two mass-points, which may be quite far. The Newtonian "action at a distance" works with a cartesian coordinate system which is defined everywhere and in which it is possible to "parallel transport" a force vector located at one point to the other point without ambiguity and compare it with the force vector at that point to see that it is indeed equal and opposite.

1.1.2 Inertial or Psuedo Forces

Newton's laws hold in inertial frames. Bodies on which no real forces act will move with uniform velocities as seen in inertial frames but will seem to be accelerated in a non-inertial frame. If we insist that Newton's laws hold for such frames too, we have to invent fictitious forces acting on bodies to account for the accelerations.

For example, if there is a frame of reference rotating with a constant angular velocity ω with respect to an inertial frame, then in this rotating frame all bodies seem to have an acceleration $r|\omega|^2$ radially away from the axis of rotation where r is the distance from the axis. This is the **centrifugal** acceleration. Similarly there is the **Coriolis** acceleration equal to $2\mathbf{v} \times \omega$ on any body moving with a velocity $\mathbf{v}$ in the rotating frame.

In the rotating frame *all* bodies have these accelerations. We are accustomed to explaining acceleration in material bodies caused by the presence of forces. Therefore, in such a frame it seems as if there is a universal field of force acting on all bodies with the peculiarity that the *force is exactly proportional to their inertial masses*. Thus we have the centrifugal *force* or the Coriolis *force* in a rotating frame. These forces are called **inertial** or **psuedo forces**. The proportionality of these forces on the inertial masses of bodies on which they act is trivial because we multiply acceleration by the mass to get the force.

The common characteristics of these inertial forces are, as noted above:

(i) They are universal, that is they act on all bodies,

(ii) they are proportional to mass, and

(iii) they can be "transformed away" completely by changing to a suitable frame of reference. That is, by going to an inertial frame.

1.1.3   Absolute Space and Mach's Principle

Newton's classical mechanics rests on the existence of inertial frames of reference. Is it sheer luck that there are 'distant stars' which provide a definition of inertial frames? Newton invented the concept of **absolute space** simply to avoid having to depend on the existence of distant stars to provide a foundation of mechanics. According to Newton, the distinction between an inertial and an accelerated frame is absolute and not dependent on existence of matter elsewhere in the universe. Suppose there were nothing in the universe except two huge spherical balls of matter, far away from each other and one of them rotating relative to the other about the axis joining the centres of the two balls. Then, according to Newton, one can find out how much both are rotating with respect to absolute space by measuring the equatorial bulge due to the centrifugal forces. If one of the balls is at rest with respect to the absolute space, it will show no bulging while the other will.

The concept of absolute space was criticized by Ernst Mach in the late 19th Century. Mach's view is that there is no experimental way to establish the existence of absolute space because we cannot remove the distribution of matter in the universe and compare it with the situation when it is present. Mach assumed that the inertial forces like the centrifugal force are *actually caused by distribution of matter in the universe*. The distant stars do, in fact, determine the inertial frame here in the vicinity of earth.

Although Mach did not give any mechanism or formula for calculating the acceleration of a body from the knowledge of matter distribution, Einstein was influenced by Mach's ideas in formulating the general theory of relativity.

1.1.4   Inertial and Gravitational Mass

Gravity is a force very different from other forces. Newton not only discovered the laws of motion but also the way the most important of the known forces, gravity, acts.

Newton's formula for gravitational force *on* a mass m *due to* a mass M is usually written as

$$\mathbf{F}_{m \to M} = -\frac{GMm\mathbf{r}}{r^3}. \tag{1.2}$$

Here G is a universal constant and $\mathbf{r}$ the position vector of the point-mass m with respect to M. The force is attractive, and acts along the straight line joining the two masses.

Gravity, which is as real as a force can be, curiously, seems to share the first two, and a-little-of-the-third, of properties of pseudo forces listed above.

(i) Gravity is universal. It acts on *all* bodies, unlike electric or magnetic forces which act only on charged or magnetized matter.

(ii) Secondly, the gravitational force on a body is proportional to its inertial mass m. Which means that all bodies accelerate by the same amount in a gravitational field. (See section 1.6.3 in this chapter for further remarks.)

(iii) And lastly, gravitation has a bit of the third property of psuedo forces. If you were to fall freely in a gravitational field, you would not feel it. If you surrender to gravity, gravity surrenders. It vanishes! This is only partly true though because only *static* and *uniform* gravitational fields can be so eliminated everywhere by going to a **freely falling frame**. The real gravitational field can be eliminated in this manner in a very small region of space and for not too long a time. This is the essence of the "Equivalence Principle" which is the starting point of Einstein's general theory of relativity.

1.1.5 Special Theory of Relativity

Relativity was born out of attempts to relate the descriptions of electrodynamic phenomena in different inertial frames. Einstein's fundamental paper on special theory of relativity, published in 1905, is titled "Electrodynamics of Moving Bodies".

Maxwell had showed in 1861 that electric and magnetic fields will travel in empty space as waves with a speed determined by a constant c appearing in formulas as the ratio of (the then prevalent) systems of electrostatic and electromagnetic units.(See historical note later in this chapter.) That this constant (which has dimension of velocity) was found to be very close to the measured speed of light was a surprise and the first indication that light was composed of electromagnetic waves. It was presumed that electromagnetic waves propagated as vibrations of a medium called **aether**. The frame of reference in which the aether is at rest is therefore especially distinguished. In this frame the velocity of waves is equal to the constant c of the electromagnetic theory. The question then was: what is the velocity of earth with respect to aether? And, can we determine how fast we are going with respect to aether by carefully measuring the light velocity?

Very accurate optical experiments were done for over twenty years by A. A. Michelson and later by Michelson and Morley and others. There was no direct or indirect evidence of the supposed "light-medium" aether, and light was seen to travel in vacuum always with the same speed in all directions.

Transformation formulas for electromagnetic fields as seen in different inertial frames were obtained by Lorentz and Poincare. These formulas suggest that one must associate with each inertial frame not only a system of cartesian coordinates but also a separate time coordinate. Thus for one frame there are coordinates x, y, z, t and for another x', y', z', t'. The specification of a physical place and a definite time, given in each frame by the coordinates (x, y, z, t) is called an 'event'. The same event whose coordinates are (x, y, z, t) in one frame has coordinates (x', y', z', t') in another. These coordinates are related to each other by formulas which depend on how the two frames are related. Once these formulas for events are established, quantities like electric and magnetic fields as observed in the two frames also get related.

The Lorentz transformations for a frame (x', y', z', t') which moves with velocity v with respect to frame (x, y, z, t) along the x-axis such that the two origins and axes coincide at time $t = t' = 0$ are given by

$$x' = \frac{x - vt}{\sqrt{1 - v^2/c^2}}, \qquad t' = \frac{t - vx/c^2}{\sqrt{1 - v^2/c^2}}$$

with $y' = y$ and $z' = z$. The time variable t' is different for different frames. The Newtonian 'absolute time' is supposed to have the same value for each frame. Lorentz and others regarded a separate time variable attached to each frame to be a mere mathematical convenience without any physical importance. Instead, they tried to explain the negative results of experiments of Michelson and Morley by supposing that all physical lengths shorten by a factor $\sqrt{1 - v^2/c^2}$ in the direction of motion of the aether, thereby making the change in velocity of light with respect to aether undetectable.

Around the same time, in 1904, Poincare had formulated the Principle of Relativity, as the requirement that laws of physics should look the same in all inertial frames and that there should be no distinguished frame of reference. There was therefore a very genuine problem of reconciling the principle of relativity with the properties of the light medium aether.

It was Albert Einstein who solved the problem in 1905, quite independently.

He accepted both the experimental fact of constancy of velocity of light as well as the principle of relativity.

He critically examined the concept of time and of **simultaneity** of two events or happenings. He found that two events which happen at different places but at the same time in one frame will not happen at the same time in another frame moving with respect to the first one.

He rederived the Lorentz transformation formulas from these two principles.

The **relativity of simultaneity** leads to a revision of all the basic concepts of Newtonian physics. Values of length and time intervals depend on the frame in which they are measured. The transformation formulas for velocities are changed. No object can move with a speed greater than that of light. The expressions of momentum and kinetic energy in terms of velocity have to be altered. The mass m of a body in a frame in which it is at rest determines a "rest-energy" equal to mc^2. Energy is conserved only if we include this rest energy for each particle in the expression for total energy. The law of conservation of mass no longer holds, and every form of energy E has an inertial mass equivalent to E/c^2.

Einstein's theory was called the theory of relativity. The reason why most of the Newtonian mechanics had been doing fine till Einstein's time is that these 'relativistic effects' are of the order of (v^2/c^2), which are very small for objects moving with velocity v much smaller than the velocity of light.

Einstein required the Lorentz transformations to hold for all inertial frames of reference and for *all* physical phenomena, and not just the electromagnetic ones where they were first discovered.

Curiously, gravitational force could not be brought in line with this theory of relativity, although attempts were made the next few years.

Theory of relativity implies that the set of all Lorentz transformations (which form a transformation group) classify all physical quantities as vectors or tensors transforming as representations of this group. This means that physical quantities occur in subsets or 'multiplets'. This unification of spacetime and consequent ordering of all physical quantities as four-dimensional vectors or tensors (and later, spinors, in quantum theory) was the single most important conceptual advance in physics after Newton's mechanics and the Faraday-Maxwell theory of fields. The theory of relativity as given by Einstein in 1905 is restricted to inertial frames of reference and is called the **special theory of relativity**. The **general theory of relativity** is Einstein's generalization of the special theory in order to include gravity. But then it becomes a fundamental theory, not just a relativistic theory of gravity. See remarks in section 1.6.1 of this chapter on the suitability of the name 'general relativity'.

1.1.6 Equivalence Principle

The special theory of relativity deals with physical phenomena as seen by observers in inertial frames of reference. In these frames bodies which are not under the influence of any force move with constant velocity. It seems natural to extend the basic principle of relativity to all observers or frames of reference and assume they are all equally good for describing physics.

The simplest non-inertial frames of reference are those which move with a constant linear acceleration or those which rotate at a constant angular velocity with respect to an inertial frame of reference. In these non-inertial frames one sees the appearance of inertial or pseudo forces like centrifugal or Coriolis force. As discussed above, these forces are universal, they are proportional to the mass of the body on which they act, and they can be eliminated entirely by changing to an inertial frame.

Gravitation is similar: it is universal, proportional to mass (that is why all bodies fall with the same acceleration in a gravitational field), and if a person falls freely, there is no gravitational field in the falling frame.

It must have been thoughts in this direction, coupled with the failure of all attempts to generalise Newton's theory of gravitation to make it compatible with special theory of relativity, that led Einstein to see the equivalence of gravitation and the inertial forces that are produced in an accelerated frame.

The electric force on a body in an electric field is proportional to its charge. Given the electric force on the body its acceleration is determined by its mass in accordance with Newton's second law.

The physical fact that in a gravitational field the gravitational force on a body is proportional to the same mass which occurs in the equation force = mass × acceleration is called the equality of the gravitational and the inertial mass. Because of this equality all bodies fall with the same acceleration in a gravitational field.

1.1.7   Falling Elevator

Imagine a box the size of a small room with an observer inside it resting on Earth's surface. The observer cannot see out of the box. Sitting inside she can infer that there is a gravitational field by observing that bodies in the box fall with the same acceleration **g**.

Now take this box (along with the observer) to a place far away from gravitating bodies like the Sun or Earth and (with respect to any inertial frame) give a constant acceleration −**g** to the box.

To the observer inside everything would again be seen to fall with acceleration **g** and there would be no way of knowing that she is not, in fact, in a constant gravitational field.

This equivalence of constant gravitational field with the physical effects in an accelerated frame was called by Einstein the **Equivalence Principle**. He made an even stronger assumption that one would not be able to distinguish between the two situations *by any physical experiments* (and not merely the mechanical experiments used for measuring the accelerations of bodies). Sometimes this is called the Strong Principle of Equivalence.

An immediate consequence of the principle is that an observer falling freely in a constant gravitational field would be in an inertial frame. The freely falling observer experiences *no gravitational field!* Every object around her will fall with the same acceleration under gravity and there would be no motion ascribable to gravity.

Using special relativity it was possible to find physical effects in an instantaneous frame "comoving" with an accelerated frame, and by using the Equivalence Principle one could therefore guess what would happen in a gravitational field. For example, Einstein proved in 1907 that clocks near a massive body (where gravitational potential is lower) run slowly compared to clocks farther away.

We come to Einstein's ingenious argument a little later in this chapter.

Strictly speaking only a *uniform* gravitational field can be so removed by using a freely falling frame. Any realistic gravitational field will have **tidal forces** associated with it. The tidal forces depend on second derivatives of the gravitational potential, and in principle can be measured in the falling elevator. This does not reduce the importance of the equivalence principle as a basic input in the formulation of general relativity. See tutorial problem on tidal forces at the end of the chapter.

1.2 Space and Time

1.2.1 Galilean Relativity

Einstein critically examined the concept of time and of **simultaneity** of two events. He found that two events which happen at different places but at the same time in one frame will not happen at the same time in a frame moving with respect to the first one.

The idea to represent motion of a particle as a curve in a diagram with one of the coordinate axes representing the time is natural. For a particle trajectory one can show it as in Figure 1.1, a curve in the graph of distance and time.

The equivalence of all inertial frames in Newtonian physics is called "Galilean relativity". This means, in particular, that frames moving uniformly with respect to each other are physically equivalent.

A frame of reference which moves with respect to a stationary frame with constant velocity v along the positive x-axis, will have its assignment of position coordinate x' of any object at time t related to those of the stationary frame as $x' = x - vt$. The time coordinate t' of the moving frame is, of course, the Newtonian absolute time $t' = t$. What is important to see is that we can view the two frames as two coordinate systems on an underlying spacetime. This is shown in Figure 1.2.

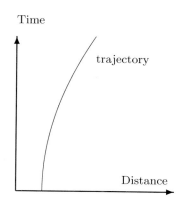

Fig. 1.1: Space and Time.

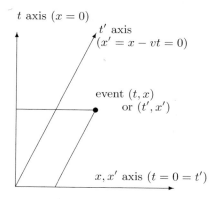

Fig. 1.2: Galilean Relativity.

$$(x' = x - vt, \qquad t' = t)$$

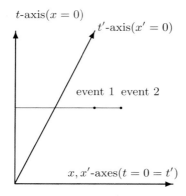

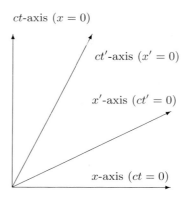

Fig. 1.3: Absolute Simultaneity.   **Fig. 1.4: Special Relativity.**

$$x' = \gamma(x - \beta ct), \ \ ct' = \gamma(ct - \beta x)$$

$$\beta = v/c, \ \gamma = (1 - \beta^2)^{-1/2}$$

Note that there is a new time axis (corresponding to $x' = x - vt = 0$) for deciding the the coordinates x', t' for the moving frame. However due to $t' = t$ the new x'-axis that is $t' = 0$ coincides with the x-axis.

In Galilean relativity two events which are simultaneous in one frame are simultaneous in any other frame. This is because there is only one absolute time associated with an event. (Figure 1.3)

1.2.2   Space and Time in Special Relativity

The Lorentz transformations between the same two frames can be written in terms of convenient variables ct and ct' as

$$x' = \gamma(x - \beta ct), \ \ ct' = \gamma(ct - \beta x),$$

$$\beta = \frac{v}{c} \qquad \gamma = (1 - \beta^2)^{-\frac{1}{2}}.$$

The scaling of the time axes by a factor c has the effect of showing the relativistic effects very clearly on a diagram by magnifying by this large factor.

When the axes are plotted for the space and time for the two frames we see that in the spacetime plane the axes corresponding to x and x' (that is sets of points corresponding to $t = 0$ for x-axis, and $t' = 0$ for x'-axis) are separated. See Figure 1.4.

We notice the **relativity of simultaneity**. Two events in spacetime which are simultaneous for (x, t) frame (that is they have the same value for the time co-ordinate) are not so in the other (x', t') frame which moves with respect to this frame. (Figure 1.5).

Consider the trajectory of a particle moving in spacetime. For two neighbouring events (x_1, ct_1) and (x_2, ct_2) on its trajectory in the frame (x, ct) define

$$s^2 = (x_2 - x_1)^2 - c^2(t_2 - t_1)^2.$$

It is a consequence of Lorentz transformations that the similar quantity has the *same* value for the (x', t') frame as well,

$$s^2 = (x'_2 - x'_1)^2 - c^2(t'_2 - t'_1)^2.$$

Thus this quantity is the **invariant distance** or **invariant interval** between the two events independent of the inertial frame of reference in which it is calculated.

Because the speed of all objects is less than that of light, s^2 is actually negative. Define $\tau = \sqrt{-s^2/c^2}$ or,

$$\tau_{21} = (t_2 - t_1)\sqrt{1 - v^2/c^2} = (t'_2 - t'_1)\sqrt{1 - v'^2/c^2}$$

where $v = (x_2 - x_1)/(t_2 - t_1)$ and $v' = (x'_2 - x'_1)/(t'_2 - t'_1)$ are the velocities of the particle as determined by these two infinitesimally close events. Because τ_{21} has the same value in all frames, it is equal to its value in the frame in which the particle is at rest. Thus τ_{21} *is the time recorded by a clock carried along with the particle.*

A particle moves with a non-uniform velocity in general. We define the proper time along the trajectory of the particle as

$$\int d\tau = \int \sqrt{1 - v^2/c^2}\, dt.$$

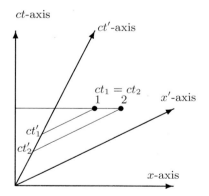

Fig. 1.5: Relative Simultaneity.

Events 1 and 2 are simultaneous for one frame $(t_1 = t_2)$ and are not so for another frame $(t'_1 \neq t'_2)$.

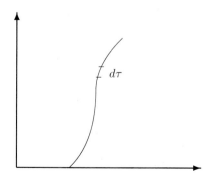

Fig. 1.6: Proper time.

$\int d\tau = \int \sqrt{1 - v^2/c^2}\, dt$ is the time shown by a clock travelling with the particle.

1.2.3   Minkowski Space

A point in Minkowski spacetime is called an **event**. It is specified by the spatial coordinates $\mathbf{x} = (x^1, x^2, x^3)$ of the point and the time t. We denote by $x^\mu = (x^0 = ct, x^1, x^2, x^3)$ the coordinates of an event where the time coordinate t is multiplied by the light velocity c in order to get greater symmetry in the formulas. It also makes all the four quantities have the same physical dimension of length.

This four-dimensional space, called the **Minkowski space**, differs from the usual three-dimensional Euclidean space in one crucial way. The length squared between two points in three-dimensional Euclidean space is given by the positive definite expression

$$(\mathbf{x} - \mathbf{y}).(\mathbf{x} - \mathbf{y}) = (x^1 - y^1)^2 + (x^2 - y^2)^2 + (x^3 - y^3)^2$$

which is always positive. In the Minkowski space the square of the interval is given by

$$
\begin{aligned}
(x - y).(x - y) &= -(x^0 - y^0)^2 + (x^1 - y^1)^2 + (x^2 - y^2)^2 + (x^3 - y^3)^2 \\
&= -(x^0 - y^0)^2 + (\mathbf{x} - \mathbf{y}).(\mathbf{x} - \mathbf{y}) \\
&= \eta_{\mu\nu}(x^\mu - y^\mu)(x^\nu - y^\nu)
\end{aligned}
$$

where we use summation over indices $\mu, \nu = 0, .., 3$ as explained in the note on notation (page xiii). This expression can be positive, negative or zero. The numbers $\eta_{\mu\nu}$ are elements of a matrix η which are equal to $-1, +1, +1, +1$ on the diagonal and zero elsewhere.

The important point is that this **interval** between two spacetime events x and y is *invariant*, that is a number which has the same value in all coordinate frames.

Each inertial frame of reference is represented by a choice of axes in this space. Lorentz transformations between inertial frames are 4×4 matrices Λ which connect the coordinates x^0, x^1, x^2, x^3 and x'^0, x'^1, x'^2, x'^3 corresponding to the *same physical event* (written as column matrices): $x' = \Lambda x$ which must satisfy the condition that the interval has the same invariant value $s^2 = (x-y).(x-y) = (x'-y').(x'-y')$ for a pair of events whose coordinates in one frame are x and y and in the other x' and y'.

Between any two infinitesimally close spacetime events with coordinates x^μ and $x^\mu + dx^\mu$ the infinitesimal interval can be written

$$(ds)^2 = \eta_{\mu\nu}dx^\mu dx^\nu.$$

When the interval is positive the events are called **space-like separated**, when it is negative they are **time-like separated events**, and when it is zero they are **light-like separated**. If a point particle is present at x and also at $x + dx$ then (assuming $dt > 0$) its velocity $\mathbf{v} = d\mathbf{x}/dt$ where $\mathbf{x} = (x^1, x^2, x^3)$, is related to the invariant interval as

$$(ds)^2 = -c^2(dt)^2 + (d\mathbf{x})^2 = -c^2(dt)^2 + \mathbf{v}^2(dt)^2 = -(c^2 - \mathbf{v}^2)dt^2.$$

As the particle cannot have velocity greater than that of light the interval is time-like. We call

$$d\tau = \frac{1}{c}\sqrt{-\eta_{\mu\nu}dx^{\mu}dx^{\nu}} = dt\left(1 - \frac{\mathbf{v}^2}{c^2}\right)^{\frac{1}{2}}$$

the **proper time** between the particle's two succesive spacetime positions. The invariant number $d\tau$ is the time as seen by an observer in the rest frame of the particle, because for a clock traveling with the particle, the velocity $\mathbf{v}$ is zero.

For an arbitrary motion of the particle, the sum of infinitesimal proper times along its four-dimensional trajectory will be the time shown by a clock we can imagine travelling along with the particle. Two clocks starting from the same initial event and showing the same time at that event can follow seperate trajectories and may meet again at a common spacetime point later on. But on camparison they will show different times in general. The integrated proper-times along the two trajectories need not be equal. This is no more surprising than the fact that arc-lengths of two arbitrary curves between two fixed points in the Euclidean plane are different in general. In fact the proper time plays the same role as the length of a curve in Euclidean space.

The motion of a mass-point can be described by a curve $x(\tau)$ in spacetime, where τ is the proper time labelling points on the trajectory curve. For a free particle without any force acting on it the trajectory is a time-like straight line in spacetime:

$$x^{\mu}(\tau) = x^{\mu}(0) + \tau U^{\mu}, \qquad \frac{dx^{\mu}}{d\tau} = U^{\mu}, \qquad \frac{d^2 x^{\mu}}{d\tau^2} = 0.$$

A time-like straight line in spacetime between two fixed points is a **geodesic**, that is, a path for which the integral of ds along the path is extremal. A variational principle can be written with action A,

$$\delta A = -mc^2 \delta \int d\tau = -mc\delta \int \sqrt{-\eta_{\mu\nu}dx^{\mu}dx^{\nu}} = 0.$$

This looks a little more familiar if we rewrite it in a non-relativistic limit of small velocities $|\mathbf{v}| \ll c$,

$$A = -mc^2 \int d\tau = -mc^2 \int dt\left(1 - \frac{\mathbf{v}^2}{c^2}\right)^{\frac{1}{2}} \approx -mc^2(t_2 - t_1) + \frac{1}{2}\int m\mathbf{v}^2 dt$$

where the first term depends only on the endpoints and the second term shows the kinetic energy as the Lagrangian of a free non-relativistic particle.

In four dimensions, *dynamics* becomes *geometry*. The force-free motion becomes straight lines or extremal paths given by $\delta \int d\tau = 0$.

Light signals also move along straight lines but the proper time along the line is zero.

The main lesson of Minkowski's contribution to special theory of relativity is that physical quantities have a four-dimensional character. Quantities which were described by three-vectors (that is vectors with three components) before the theory of relativity are seen to have a fourth partner corresponding to the time-axis. Such quantities were called **four-vectors**.

For example the "four-velocity" of the particle U is defined as the derivative with respect to the proper time: it is the tangent vector to the trajectory,

$$U^\mu = \frac{dx^\mu}{d\tau} = (c\gamma(|\mathbf{v}|), v^i\gamma(|\mathbf{v}|)) \qquad (1.3)$$

where $v^i \equiv dx^i/dt$, the usual three-velocity of the particle. Note that the three-velocity is not just the "space" part of the velocity four-vector. Note also, that all four components of the velocity four-vector are not independent

$$\langle U, U \rangle \equiv \eta_{\mu\nu} \frac{dx^\mu}{d\tau} \frac{dx^\nu}{d\tau} = -c^2. \qquad (1.4)$$

We can say that the *four-velocity of a material particle is a time-like vector of constant magnitude.*

At any (proper) time τ there exists a coordinate system in which the particle is momentarily at rest, that is, its velocity four-vector has components $(c, 0, 0, 0)$.

1.3   Linearly Accelerated Frame

In this section we follow Einstein's argument of 1907 to show that clocks run slowly where gravitational potential is low compared to clocks at higher value of potential.

Let there be an inertial frame S with respect to which another frame S_1 is at rest at $t = 0$ with coinciding axes and origin. Let S_1 start accelerating in the x-direction at $t = 0$ with acceleration g.

Let there be two clocks C_1 and C_2 in the accelerating frame S_1. They are at events O_1 and O_2 at $t = 0$ as seen by S. As shown in Figure 1.7 the trajectories of the clocks C_1 and C_2 according to frame S are given by $x_1(t)$ and $x_2(t)$ where

$$x_1(t) = \frac{1}{2}gt^2, \qquad x_2(t) = L + \frac{1}{2}gt^2.$$

These equations describe the accelerated frame at low enough velocities, that is, at small values of t.

After a small time Δt by the clock of S, the frame S_1 is moving with velocity $g\Delta t$, and the clocks in S_1 are located respectively at events $A : (x = g(\Delta t)^2/2, t = \Delta t)$ and $B : (x = L + g(\Delta t)^2/2, t = \Delta t)$. The proper time shown by these clocks is the same

$$\int_0^{\Delta t} dt \sqrt{1 - g^2 t^2/c^2} = \Delta t + O(\Delta t^3).$$

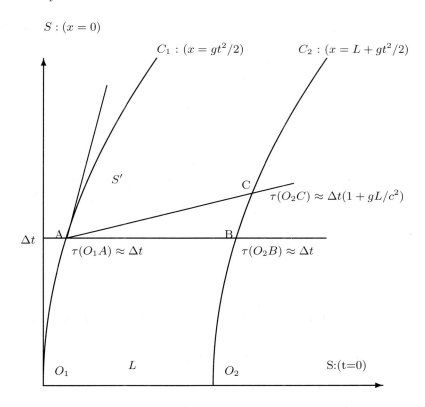

Fig. 1.7: Einstein's argument of 1907.
Clocks run slowly at lower gravitational potential.

Events A and B are simultaneous in frame S but not simultaneous in the accelerated frame S_1 which (at time $t = \Delta t$) has gained a velocity $g\Delta t$ with respect to S.

Let S' be a third frame which moves with constant velocity $v = g\Delta t$ with respect to S. Einstein chooses this frame just so that it is "co-moving" with the accelerated frame S_1 at exactly the S time $t = \Delta t$.

It is simultaneity in S' (which is co-moving with S_1), that should be used to decide which two events are simultaneous in the accelerated frame S_1 at this instant. The $t' =$ constant line (which determines the simultaneity in S') passes through event A on the trajectory of C_1 and cuts the trajectory of the second clock C_2 a little further at C. The time shown by clock C_2 kept at L in the accelerated frame S_1 will be the proper time *up to event C while at the same instant (according to S') the time shown by the clock C_1 will be the proper time up to A.*

It is clear from the diagram that the clock C_2 (at event C) will show more time than the clock C_1 at A. We calculate the proper times $\tau(O_1A)$ and $\tau(O_2C)$. $\tau(O_1A)$ is equal up to lowest order to Δt. To find proper time up to C, we need the time coordinate t_C of event C in frame S. The coordinates of these events are as follows in the two frames:

	S	S'
A	$x_A = g(\Delta t)^2/2, \ t = \Delta t$	$t' = t'_A$
C	$x_C = gt_C^2/2, \ t = t_C$	$t' = t'_C = t'_A$

where we have not written those coordinates we do not need.

The basic idea is to calculate t'_A from the first line and then relate $t'_C = t'_A$ to (x_C, t_C). Use Lorentz transformation between S and S' with relative velocity $v = g\Delta t$:

$$t'_A = \gamma(v)\left[\Delta t - \frac{v}{c^2}\frac{g(\Delta t)^2}{2}\right], \qquad (v = g\Delta t),$$

$$t'_C = \gamma(v)\left[t_C - \frac{v}{c^2}\left(L + \frac{g(t_C)^2}{2}\right)\right].$$

Neglecting order $(\Delta t)^3$ terms the first equation gives $t'_A \approx \Delta t$ and so the second gives

$$t'_C = t'_A = \Delta t = t_C - \frac{g\Delta t}{c^2}\left[L + \frac{g(t_C)^2}{2}\right].$$

Solving this equation for t_C and keeping to lowest order we get

$$t_C = \Delta t(1 + gL/c^2),$$

which is also equal to the proper time $\tau(O_2 C)$ in this approximation.

Therefore, in the accelerated frame, which experiences a constant gravitational field in the $-x$ direction , the clock at 'height' L shows a reading $\Delta t(1+\Phi/c^2)$ when at the same instant (according to this frame) the clock at the origin shows a reading Δt. Here $\Phi = gL$ is the gravitational potential difference between the locations of the clocks.

Similarly one can prove that light rays moving in a gravitational field bend from a rectilinear path in the direction of the gravitational field.

These are preliminary results of an incomplete theory, but were crucial for progress.

Gravitational fields can be appproximated to be uniform only in very small regions of spacetime. The equivalence principle therefore holds only in such regions. Fortunately, laws of physics are expressed in terms of local differential equations for quantities like fields. Therefore the priciple can be used to generalise laws from their formulations in a 'freely falling local inertial frame' to arbitrary gravitational fields.

The equivalence principle proved to be the guiding principle for finding the general equations. One was required to write (*in a suitable form*) the equations in the inertial freely falling frame where we knew the special theory of relativity holds, and then declare them to be valid for all frames of reference.

1.4 Need for the Riemannian Geometry

There are three physical questions to be answered for a relativistic theory of gravity: (1) which quantities describe the gravitation field? (2) what are the equations that relate the gravitational field to matter-energy distribution? and (3) what is the equation of motion of a particle in the gravitational field.

In the Newtonian theory the gravitational potential $\Phi(x)$ describes gravity. At any instant the matter distribution is given by the mass density ρ. This density determines the potential at the same instant of time by the Poisson equation

$$\nabla^2 \Phi = 4\pi\rho.$$

In this gravitational field a mass point moves along a trajectory determined by

$$\frac{d^2\mathbf{x}}{dt^2} = -\nabla\Phi.$$

It looks natural to generalise Φ into a scalar field to describe gravity in the relativistic case. But such a theory proposed by Einstein and others between 1908-1912 does not work.

The first clues came from the equivalence principle.

According to the Equivalence Principle, there must exist in a small region of spacetime a coordinate system X^0, X^1, X^2, X^3 in which the equation of the particle is a straight line, i.e., force free,

$$\frac{d^2 X^\mu}{d\tau^2} = 0.$$

As seen in the section on relativity theory this equation can be derived from the variational priciple $\delta \int d\tau = 0$. In a general coordinate system, the same equation holds because $d\tau$ is invariant:

$$c^2(d\tau)^2 = -(ds)^2 = (dX^0)^2 - (dX^1)^2 - (dX^2)^2 - (dX^3)^2 = -g_{\mu\nu}dx^\mu dx^\nu$$

where $g_{\mu\nu}$ are determined by the relation between the coordinates X^μ and x^μ,

$$g_{\mu\nu} = \eta_{\alpha\beta} \frac{\partial X^\alpha}{\partial x^\mu} \frac{\partial X^\beta}{\partial x^\nu}.$$

In a general frame the particle will be seen moving under the influence of gravitational forces, along paths determined by $\delta \int d\tau = 0$ where proper time $d\tau$

is given by the expression in terms of $g_{\mu\nu}$. Therefore $g_{\mu\nu}$ must be intimately related to the gravitational field. The first question (about which quantities represent the gravitational field) was thus resolved.

This expression for $(d\tau)^2$ is the familiar 'line element' in the Gauss-Riemann geometry. It became clear that the Riemannian geometry is the proper tool for gravitational theory.

Working along these lines, Einstein collaborated with his mathematician friend Marcel Grossmann. The equation $\delta \int d\tau = 0$ was seen to lead to the equation for the 'straightest curve' or the geodesic

$$\frac{d^2 x^\mu}{d\tau^2} = -\Gamma^\mu_{\nu\sigma} \frac{dx^\nu}{d\tau} \frac{dx^\sigma}{d\tau}$$

where $\Gamma^\mu_{\nu\sigma}$ are expressions in terms of derivatives of $g_{\mu\nu}$. These quantities will be defined in later chapters.

In the non-relativistic limit, this equation reduces to the Newtonian equation if we take, for $(x^0 = ct)$,

$$g_{00} = -\left(1 + \frac{2\Phi}{c^2}\right).$$

From the form of the equations for the particle, the Γ appear to be related to gravitational force and the $g_{\mu\nu}$ play the role of gravitational potential. This takes care of the third question above.

The search for the *field equations* which determine $g_{\mu\nu}$, from a knowledge of the energy-matter distribution, (that is, the second question in our list) took the greatest effort. After many false steps Einstein finally arrived at the correct theory in November 1915.

1.5   General Theory of Relativity

As we have seen, it is the geometry of space *and* time which departs from the Eulidean. The Minkowski space of special relativity has a non-Euclidean metric or line element.

Einstein's final version of the theory regards spacetime as a four-dimensional Riemannian space or manifold.

Let the neighbouring points have coordinates $x = (x^0, x^1, x^2, x^3)$ and $x + dx = (x^0 + dx^0, x^1 + dx^1, x^2 + dx^2, x^3 + dx^3)$ where x^0 is a 'time' coordinate and $x^i, i = 1, \ldots, 3$ refer to space. The infinitesimal distance squared is given by an expression

$$ds^2 = g_{\mu\nu} dx^\mu dx^\nu$$

where indices $\mu, \nu = 0, \ldots, 3$. Different coordinate systems can be used to specify the same spacetime points, but the quantity ds^2 remains the same. This invariance of ds^2 determines how the metric tensor components change from one coordinate system to another.

The special theory of relativity as interpreted by H. Minkowski in 1908 uses a spacetime in which the metric tensor

$$g_{\mu\nu} = \eta_{\mu\nu}$$

has non-zero components

$$\eta_{00} = -1, \qquad \eta_{11} = \eta_{22} = \eta_{33} = 1.$$

Lorentz transformations are precisely those coordinate changes in which these constant values of $\eta_{\mu\nu}$ remain the same. A spacetime continuum in which it is possible to choose a coordinate system which makes $g_{\mu\nu}$ constant everywhere such as here is called **flat** or **Minkowskian**.

In the presence of gravitating matter, $g_{\mu\nu}$ become modified, and it is not possible to choose coordinate systems in which they can take constant values. The space becomes **curved**, and the measure of curvature is the Riemann-Christoffel **curvature tensor**. Its components $R_{\mu\nu\sigma\tau}$ depend on $g_{\mu\nu}$ and their derivatives up to second-order.

The curvature tensor is the true measure of the 'real' gravitational field that cannot be transformed away by a choice of coordinates. Even so, in a very small neighbourhood of a spacetime point one can choose a coordinate system such that the derivatives of $g_{\mu\nu}$ vanish at the point. Therefore those effects which depend on the first derivatives of the $g_{\mu\nu}$ (like the acceleration of a particle falling in gravity) are indistinguishable from that of a particle moving in a gravity-free region. This is the physical content of the Equivalence Principle.

The $g_{\mu\nu}$'s are determined by energy and matter distribution by the **Einstein Equation**

$$G_{\mu\nu} \equiv R_{\mu\nu} - \frac{1}{2}g_{\mu\nu}R = \frac{8\pi G}{c^4}T_{\mu\nu}$$

where the **Einstein tensor** $G_{\mu\nu}$ is a simple combination of the so-called **Ricci tensor** given by $R_{\mu\nu} = R^{\sigma}{}_{\mu\sigma\nu}$ and the **scalar curvature** $R = g^{\mu\nu}R_{\mu\nu}$.

c and G are the velocity of light and Newton's gravitational constant. The constant multiplying $T_{\mu\nu}$ on the right-hand side is chosen so that in the non-relativistic limit the theory gives Newton's law of gravitation.

The right-hand side contains $T_{\mu\nu}$, the **stress-energy tensor** of matter which contains information about momentum and energy densities and pressure of matter and radiation.

Material particles move in the gravitational field determined by $g_{\mu\nu}$ along **geodesics**, the "straightest possible" curves. The equation for such curves $x^\mu(\tau)$ is

$$\frac{d^2 x^\mu}{d\tau^2} + \Gamma^\mu_{\nu\sigma} \frac{dx^\nu}{d\tau} \frac{dx^\sigma}{d\tau} = 0$$

where $\Gamma^\mu_{\nu\sigma}$, called Christoffel symbols, are functions of $g_{\mu\nu}$ and their derivatives.

Light rays also move along geodesics except that along the path the interval $d\tau$ between any two neighbouring points is zero.

It must be remembered that curves corresponding to motion of a body in a gravitational field are the straightest possible in *spacetime* and not in three-dimensional space. Thus a particle thrown vertically upwards in Earth's gravity may seem to have a highly curved path at its turning point in the three dimensions of space, but it is as straight a path as can be in the four-dimensional spacetime. This spectacular 'straightening' happens when we go from a three-dimensional projection of the trajectory to the actual curve in four dimensions because the speed of light has a very large value compared to ordinary velocities.

General theory of relativity is a theory of gravitation in the narrow sense that if there was no gravity then special theory of relativity would suffice.

In the broad sense it is a fundamental theory which includes all physical phenomena because gravity acts on everything. As gravity is a weak force at ordinary distances and mass densities, the corrections to Newton's law of gravitation are small. The typical corrections to Newtonian theory are of the order of the dimensionless number GM/rc^2 where M is the mass of the gravitating body, r the typical distance scale of the problem.

For example, a clock at a distance r from a spherical body of mass M runs in a ratio $\sqrt{1 - 2GM/rc^2}$ slower than a clock very far away from the body. This leads to frequencies of spectral lines emitted by atoms, (atoms act like a clock), at a distance r away from the massive body to be reduced by a fraction GM/rc^2 when seen by a far away observer. This is called gravitational red-shift (because in the visible spectrum lower frequencies occur near the red). For an atom sitting on the surface of the Sun this turns out to be of the order of 10^{-6}.

The angle of bending of light from a distant star grazing past the surface of the Sun is $4GM/Rc^2$ where R is the radius of the Sun.

Similarly, the angle by which the elliptical orbit of a planet fails to close in one revolution turns out in Einstein's theory to be $6\pi GM/Lc^2$ where M is the mass of the Sun, and L, the latus rectum of the orbit. For the innermost planet Mercury (which has the smallest value of L) this angle turns out to be equal to $43''$ of arc in a hundred years (415 revolutions of the planet). This "precession of the perihelion" was the observed, but unexplained, leftover discrepancy in the orbit of Mercury after all known causes for this phenomenon had been taken into account. This was a great triumph for the general theory of relativity.

Gravitational disturbances, which mean perturbations in the values of $g_{\mu\nu}$ caused by changing matter distribution, travel out as gravitational waves, taking away energy. However the direct evidence of gravitational waves is yet to come. An ambitious world-wide program for the detection of gravitational waves is currently underway and the first results are expected soon.

One of the most dramatic application of the general theory of relativity is in the **gravitational collapse** and formation of **black holes**. One finds that an idealised, extremely heavy mass concentrated in a very small region so that this region is inside its **Schwarzschild radius** $R_S = 2GM/c^2$ will give rise to a gravitational field so strong that even light cannot escape from this region. Such a region is called a black hole. It is a consequence of the general theory of relativity that if a sufficiently large mass distribution begins to collapse under its own weight, there may be no known forces of nature to stop it. In that case the collapse goes on unchecked and the mass densities may reach extremely high values whose physics we do not understand. The spacetime is so highly curved around the cores of such objects that light gets bent and 'sucked back in', unable to escape. There are believed to be black holes of all sizes existing in Nature, ranging in total mass from a few solar masses to black holes of a billion solar masses residing in the centers of most galaxies.

An application of the general theory of relativity is to the behaviour of the universe itself. Assuming the universe at the largest distance scales (of a few billion light years) to be uniform and isotropic, it is possible to infer from Einstein's equations that the universe must have started from a hot fiery ball which has been expanding ever since. The field of **cosmology** has received a great boost in recent years due to fantastic progress in observational astrophysics.

All the same the general theory of relativity is a *classical* theory. Physicists believe that all phenomena, at the appropriate level, must be described by a quantum theory. All attempts to quantize the classical gravitational field described by Einstein's theory have been unsuccessful so far. Physicists hope to learn something deep and fundamental once they are ultimately successful in doing that.

In this book we concentrate on just the basics of Einstein's theory. A guide to the literature dealing with advanced topics and applications of the theory is given in a section at the end of this chapter.

1.6   Tutorial

1.6.1   The Name 'General Relativity'

The special theory of relativity was called the 'theory of relativity' from 1905 to 1915 (and even till much later) because it refers to a *relativity* of describing the physical phenomena in different inertial frames. Two positions of observers are *relative*, two orientations are *relative* and two uniform motions are *relative* because they are equally good for description. The Lorentz transformations which connect inertial frames are linear mappings. Therefore, when theory of gravitation required accelerated frames and *general coordinate transformations* it was called general theory of relativity by Einstein, and the former theory of relativity became the special theory.

But it is surely possible to formulate the special theory of relativity in general coordinates. It may even be convenient sometimes. What is special about special relativity is that the metric tensor $g_{\mu\nu}$ (which will appear in such a formulation) is *maximally symmetric*. The spacetime of special theory is completely homogeneous and isotropic leading to Poincare invariance. Realistic gravitational fields in general relativity may not have any symmetry at all in $g_{\mu\nu}$. Therefore general theory of relativity is only general in the sense that it uses spacetimes which are not special like the flat Minkowski space of special relativity. There is no relativity of position or orientation. So it is a general theory alright, but not a general theory of *relativity* there being no relativity in the general theory.

But what is there in a name and everyone uses it! See the lecture by V. Fock reprinted in the collection edited by C. W. Kilmister for a discussion on this point.

1.6.2   Historical Note on c

RATIO OF ELECTROSTATIC AND ELECTROMAGNETIC SYSTEMS OF UNITS

One can define the electrostatic unit (e.s.u.) of charge as the charge which repels an equal charge by a unit force at a unit distance. This amounts to choosing the constant of proportionality in Coulomb's formula as 1:

$$F = \frac{q_1 q_2}{r_{12}^2} \qquad \left(F = \frac{1}{4\pi\epsilon_0} \frac{q_1 q_2}{r_{12}^2} \ (SI) \ \right).$$

Therefore q(in e.s.u.) = q(in SI)/$\sqrt{4\pi\epsilon_0}$. On the other hand the electromagnetic unit of current was defined similarly using Ampere's formula for force between current elements

$$F = \frac{i_1 i_2 \mathbf{dl_1} \times (\mathbf{dl_2} \times \mathbf{r_{12}})}{r_{12}^3} \qquad \left(F = \frac{\mu_0}{4\pi} \frac{i_1 i_2 \mathbf{dl_1} \times (\mathbf{dl_2} \times \mathbf{r_{12}})}{r_{12}^3} \ (SI) \ \right).$$

One e.m.u. of current is the current i which makes two current elements idl placed a unit distance apart, parallel to each other and perpendicular to the line joining them, attract by $(dl)^2$ units of force. (The elements dl should be chosen small enough in comparison with unit distance so that Ampere's formula remains valid.) From the above I(in e.m.u.)=I(in SI)$\sqrt{\mu_0/4\pi}$.

One e.m.u. of charge is then the amount of charge which flows past a cross-section of wire having one e.m.u. of current in unit time. The two unit systems of charge differ by a constant of dimension of velocity because the physical dimensions of these quantities are (we denote the physical dimension by the symbol [ ])

$$[\text{(e.m.u.) of charge}] = [T]\sqrt{[\text{Force}]}, \qquad [\text{(e.s.u.) of charge}] = \sqrt{[\text{Force}]}[L].$$

Therefore [(e.s.u.)]=c [(e.m.u.)] where c has dimensions of velocity. From the comparison with our SI units,

$$Q(\text{in e.s.u.}) = Q(\text{in e.m.u.})/\sqrt{\mu_0 \epsilon_0} = cQ(\text{in e.m.u.}).$$

This constant was measured carefully by Weber and Kohlrausch in 1856 by measuring the charge on a Leyden Jar (a capacitor) by using electrostatic repulsion as well as by electromagnetic effects of the current produced when discharging the jar. Their value was 3.1×10^{10} cm/sec. The coincidence of its equality with the velocity of light was noticed immediately by Kirchoff, who related it to his theory of waves of electric disturbance propagating in a wire made of a perfect conductor. Five years later, in 1861, Maxwell gave the concept of 'displacement current' and showed that electric *and* magnetic fields will propagate with velocity $c/\epsilon^{1/2}$ where ϵ is the dielectric constant (relative permittivity) of the medium. See E.T.Whittaker[1951],*History of theories of Aether and Electricity* ,Vol I, p.232

1.6.3 Remarks on Inertial and Gravitational Masses

We have put the inertial masses M and m in the formula for the gravitational force but conceptually the role of M and m is quite different here. We should have written the formula as

$$\mathbf{F}_{m \to M} = -\frac{GM_{active}m_{passive}\mathbf{r}}{r^3}.$$

M_{active} is the "active" gravitational mass which denotes the capacity of M to produce the force of gravity and $m_{passive}$ is the constant which occurs in the formula for the force just as charge q of a body occurs in the formula for electric force $q\mathbf{E}$ acting on it. The inertial mass m in contrast has nothing to do with gravity in particular. It just determines the acceleration produced in a body due to a force acting on it.

Active and passive gravitational masses for the same body can be chosen to be equal, because there is an equal and opposite force due to m on M,

$$\mathbf{F}_{M \to m} = \frac{Gm_{active}M_{passive}\mathbf{r}}{r^3} = -\mathbf{F}_{m \to M}.$$

Therefore

$$\frac{M_{active}}{M_{passive}} = \frac{m_{active}}{m_{passive}} = \text{const.}$$

By absorbing this constant in the definition of G we can write all equations with only the passive masses to be henceforth called **gravitational mass**. Therefore we write

$$\mathbf{F}_{m \to M} = -\frac{GM_{grav.}m_{grav.}\mathbf{r}}{r^3}$$

and the formula for acceleration becomes

$$\mathbf{a} = \frac{\mathbf{F}_{m \to M}}{m} = -GM_{grav.} \left(\frac{m_{grav.}}{m} \right) \frac{\mathbf{r}}{r^3}.$$

It just *happens to be a great coincidence*, that the inertial and the gravitational masses are always exactly proportional. In other words $m_{grav.}/m$ is a universal constant. The proof of this fact is that if the body with mass m is replaced by another with a different mass, we find its acceleration exactly the same. The proportionality constant of $m_{grav.}/m$ can again be absorbed in the redefinition of G and then the formula can be written as originally.

The question is, why there is this equality of gravitational and inertial mass? Why do different bodies have the same acceleration?

Galileo and Newton were aware of this coincidence, and all measurements have only confirmed the equality of gravitational and inertial masses.

1.6.4   Newtonian Gravity

Exercise 1. Derive expressions for the centrifugal and Coriolis forces in a frame rotating with constant angular velocity ω about the z-axis of an inertial frame.

Answer 1. Let the coordinates of any point with respect to the rotating frame be x', y', z'. Assume at $t = 0$ the axes coincide. Then at time t the frame has rotated by angle ωt and

$$\begin{aligned} x' &= x\cos(\omega t) + y\sin(\omega t), \\ y' &= y\cos(\omega t) - x\sin(\omega t), \\ z' &= z. \end{aligned}$$

These give

$$\begin{aligned} \ddot{x}' &= 2\omega \dot{y}' + \omega^2 x', \\ \ddot{y}' &= -2\omega \dot{x}' + \omega^2 y', \\ \ddot{z}' &= 0. \end{aligned}$$

Put these in vector form: $\omega = (0, 0, \omega)$, $\mathbf{v}' = (\dot{x}', \dot{y}', \dot{z}')$,

$$\mathbf{r}'_\perp = (x', y', 0) = \mathbf{r}' - \omega(\omega . \mathbf{r}')/|\omega|^2 = \omega \times (\mathbf{r}' \times \omega)/|\omega|^2$$

and multiply by the mass of the particle

$$\mathbf{F}' = m\ddot{\mathbf{r}}' = 2m\mathbf{v}' \times \omega + m\omega \times (\mathbf{r}' \times \omega).$$

The first term depending on the velocity is called the Coriolis force and the second the centrifugal force.

Exercise 2. Show that motion of a particle under gravitational force takes place in a fixed plane.

Answer 2. The plane is orthogonal to the angular momentum vector.

Exercise 3. Use radial coordinates r, ϕ in the plane of the motion and show that the equations of motion are

$$\frac{d^2 r}{dt^2} = -\frac{GM}{r^2} + r \left(\frac{d\phi}{dt} \right)^2, \tag{1.5}$$

$$\frac{d}{dt} \left(r^2 \frac{d\phi}{dt} \right) = 0. \tag{1.6}$$

Answer 3. Change from Cartesian to polar.

Exercise 4. Identify the constants of motion in the above equations of motion.

Answer 4. Angular momentum: related to Kepler's law of equal area in equal time,

$$h = r^2 \frac{d\phi}{dt}, \tag{1.7}$$

and energy per unit mass

$$
\begin{aligned}
E &= \frac{1}{2} \left(\frac{d\mathbf{r}}{dt} \right)^2 + \Phi(\mathbf{r}) \\
&= \frac{1}{2} \left(\frac{dr}{dt} \right)^2 + \frac{h^2}{2r^2} - \frac{GM}{r}.
\end{aligned} \tag{1.8}
$$

Exercise 5. Determine the equation for the orbit as a relation between r and ϕ.

Answer 5. When $h = 0$ the angle ϕ is constant and the motion is one-dimensional along the radial direction. For $h \neq 0$ we can use h to switch differentiation between t and ϕ,

$$\frac{d}{dt} = \frac{d\phi}{dt} \frac{d}{d\phi} = \frac{h}{r^2} \frac{d}{d\phi}. \tag{1.9}$$

Define $u = 1/r$ in terms of which the equation for the orbit is

$$\left(\frac{du}{d\phi} \right)^2 = \left(\frac{2E}{h^2} + \frac{2GMu}{h^2} - u^2 \right). \tag{1.10}$$

Another traditional form of this equation is obtained by differentiating with respect to ϕ and cancelling $du/d\phi$,

$$\frac{d^2 u}{d\phi^2} + u = \frac{GM}{h^2}. \tag{1.11}$$

Exercise 6. Solve the equation for the orbit and discuss the nature of the curve.

Answer 6. The solution is

$$u = \frac{1}{r} = \frac{1}{L} \left(1 + e \cos(\phi - \phi_0) \right). \tag{1.12}$$

This is an equation of a conic section with the origin $r = 0$ as the **focus**,

$$L = \frac{h^2}{GM} \tag{1.13}$$

the **latus rectum** and

$$e = \left(1 + \frac{2Eh^2}{G^2M^2}\right)^{1/2} \tag{1.14}$$

the **eccentricity**.

Exercise 7. Write the equation for energy as

$$\frac{1}{2}\left(\frac{dr}{dt}\right)^2 = E - V(r), \tag{1.15}$$

plot the effective potential

$$V(r) \equiv \frac{h^2}{2r^2} - \frac{GM}{r}, \tag{1.16}$$

and discuss the qualitative features of orbits.

Answer 7. The qualitative features of the orbits can be read off from the plot of the "effective potential"

$$V(r) \equiv \frac{h^2}{2r^2} - \frac{GM}{r} \tag{1.17}$$

as a function of r.

Then the energy equation for a fixed value of E determines the radial velocity from

$$\frac{1}{2}\left(\frac{dr}{dt}\right)^2 = E - V(r). \tag{1.18}$$

1. $h = 0, E \geq 0$:
   $h = 0$ implies $d\phi/dt = 0$ so that ϕ is a constant. The motion is along the radial direction. If the the velocity is towards the center, the 'radial plunge' occurs. If away from the center, and less than the escape velocity $v_{\text{esc.}} = \sqrt{2GM/r}$, the particles goes away, stops and then falls to the center. If more, escape to $r = \infty$ occurs.

2. $h \neq 0, E \geq 0$:
   The "repulsive core" in V due to non-zero angular momentum prevents a particle from falling into the gravitating center. The trajectory is a hyperbola or a parabola. The particle comes from infinity, goes round the center and back to infinity again. The distance of minimum approach can be found when the radial velocity goes to zero.

3. $h \neq 0, E < 0$:
   In this case the motion is an ellipse. The perihelion and apehelion correspond to the two points at which the line $E = $ constant cuts the graph of $V(r)$. At the minimum of V, the two extreme values of r coincide and for exactly that combination of h and E we get a circular orbit.

Exercise 8. Derive the Poisson equation for the gravitational potential.

Answer 8. For a continuous distribution of matter with mass-density $\rho(\mathbf{r})$ the potential becomes

$$\Phi(\mathbf{r}) = -G \int d^3\mathbf{r}' \frac{\rho(\mathbf{r}')}{|\mathbf{r}' - \mathbf{r}|}. \tag{1.19}$$

Using the well-known identity

$$\nabla^2 \left(-\frac{1}{4\pi|\mathbf{r}|} \right) = \delta^3(\mathbf{r}) \tag{1.20}$$

we can calculate

$$\nabla^2 \Phi(\mathbf{r}) = 4\pi G \rho(\mathbf{r}). \tag{1.21}$$

Exercise 9. TIDAL FORCES
Show that in the infinitesimal neighbourhood of an observer falling freely in a gravitational field given by potential Φ, a particle (also falling freely) experiences a tidal force given by

$$\frac{d^2\xi^i}{dt^2} = -\frac{\partial^2\Phi}{\partial x^j \partial x^i} \xi^j$$

where ξ^i are the components of the particle's position with respect to the falling observer and the axes of the cartesian frame of the observer are parallel to the axes of the frame in which the observer is falling.

Answer 9. Let the position vector of the observer be $x^i = X^i$, then the particle is at $x^i = X^i + \xi^i$. Newton's law requires

$$\frac{d^2X^i}{dt^2} = -\left. \frac{\partial\Phi}{\partial x^i} \right|_X, \qquad \frac{d^2(X^i + \xi^i)}{dt^2} = -\left. \frac{\partial\Phi}{\partial x^i} \right|_{X+\xi}.$$

Expanding the right-hand side of the second equation and using the first we get the equation.

The tidal forces are the non-relativistic counterparts of the Riemann curvature tensor. In fact when we calculate the Newtonian limit of Einstein's theory in the weak field static case, we get

$$R_{0i0j} = \frac{1}{c^2} \frac{\partial^2\Phi}{\partial x^i \partial x^j}.$$

As an example, if the oberver is in a circular orbit of radius $r = R$ in the x-y plane in a field of a mass M at the origin, then the planar components of tidal forces in the vicinity of the observer when the observer is located at a point $x^1 = R, x^2 = 0 = x^3$ are

$$T_1 = \frac{2GM}{R^3} \xi^1, \qquad T_2 = -\frac{GM}{R^3} \xi^2.$$

A diagram showing the direction of the tidal forces is given below.

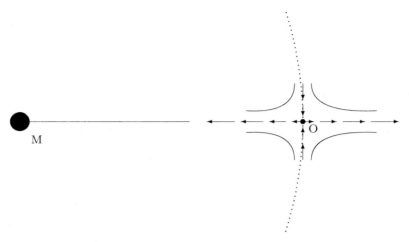

Fig. 1.8: Tidal forces in the neighbourhood of an observer O falling in a circular orbit (shown by the dotted line) in the field of a mass M. The curves near the observer are the lines of the tidal force field.

1.6.5   Minkowski Space

A trajectory can be described by functions $x^\mu(\tau)$, τ being the proper time. The proper time τ is the time shown by a clock which is carried along by the particle.

Exercise 10. Find the four-velocity of the particle U defined as the tangent vector to the trajectory, that is, derivative with respect to the proper time

$$U^\mu = \frac{dx^\mu}{d\tau}$$

in terms of the usual three-velocities as coordinate time derivatives.

Answer 10.

$$U^\mu = \frac{dx^\mu}{d\tau} = (c\gamma(|\mathbf{v}|)), v^i \gamma(|\mathbf{v}|)) \tag{1.22}$$

where $v^i \equiv dx^i/dt$ the usual three-velocity of the particle and

$$\gamma(|\mathbf{v}|)) = \left(1 - \frac{\mathbf{v}^2}{c^2}\right)^{-\frac{1}{2}}.$$

Exercise 11. Show that not all four components of the velocity four-vector are independent.

Answer 11.

$$\begin{aligned}
\langle U, U \rangle &= \eta_{\mu\nu} \frac{dx^\mu}{d\tau} \frac{dx^\nu}{d\tau} \\
&= -(U^0)^2 + (U^1)^2 + (U^2)^2 + (U^3)^2 \\
&= -c^2.
\end{aligned}$$

We can say that the *four-velocity of a material particle is a time-like four-vector of constant magnitude.*

Exercise 12. A particle is described by a spacetime trajectory $x^\mu(\tau)$ with velocity four-vector $U^\mu(\tau) = dx^\mu/d\tau$. Show that there exists a coordinate system in which the particle is momentarily at rest, that is, its velocity four-vector has components $(c, 0, 0, 0)$. Find a Lorentz transformation which connects this rest-frame to the given frame.

Answer 12. We write down one Lorentz transformation, called **direct boost** (out of an infinite number of them) which connects the "co-moving frame" in which the particle is at rest to the one in which the particle moves with three-velocity v^i (or four-velocity U^μ)

$$\Lambda(\mathbf{v}) = \begin{pmatrix} \gamma & \gamma v^i/c \\ \gamma v^i/c & \delta^{ij} + v^i v^j(\gamma-1)/|\mathbf{v}|^2 \end{pmatrix} \tag{1.23}$$

$$= \frac{1}{c}\begin{pmatrix} U^0 & U^i \\ U^i & c\delta^{ij} + U^i U^j/(c+U^0) \end{pmatrix} \tag{1.24}$$

where γ is $\gamma(|\mathbf{v}|)$.

Exercise 13. Prove that the four-acceleration $A = dU/d\tau = d^2x/d\tau^2$ is orthogonal to the four-velocity in the Minkowski sense and therefore is space-like.

Answer 13.

$$\frac{d}{d\tau}(\langle U, U \rangle) = 0 \tag{1.25}$$

implies

$$\left\langle \frac{dU}{d\tau}, U \right\rangle = 0. \tag{1.26}$$

As U is a time-like four-vector, this Minkowski-orthogonality implies that the *acceleration four-vector is space-like.* It has components

$$A^\mu = \frac{dU^\mu}{d\tau} = \frac{d^2x^\mu}{d\tau^2} = \left(\gamma^4\frac{\mathbf{v}.\mathbf{a}}{c}, \gamma^2\mathbf{a} + \gamma^4\frac{\mathbf{v}(\mathbf{v}.\mathbf{a})}{c^2}\right) \tag{1.27}$$

where $a^i = d^2x^i/dt^2$.

Exercise 14. A PARTICLE WITH CONSTANT ACCELERATION
Find the trajectory $x^\mu(\tau)$ of a particle moving along the x^1-axis which has constant acceleration $A^\mu_{\text{rest}} = (0, g, 0, 0)$ in its (instantaneous) rest-frame all along. We are given that the particle passes through the spacetime point $x^\mu(0) = (0, L, 0, 0)$ at $\tau = 0$ with spatial velocity zero $(dx^\mu/d\tau = U^\mu(0) = (c, 0, 0, 0))$. Plot the trajectory on an $x^0 - x^1$ plane.

Answer 14. We can solve this as a two-dimensional problem omitting x^2, x^3 which are zero for all τ.

Let U^0, U^1 be $U^\mu = dx^\mu/d\tau$ components and A^0, A^1 the acceleration components $A^\mu = d^2 x^\mu/d\tau^2$. The Lorentz transformation which connects the rest-frame to the frame with velocity U^μ is (Exercise 12)

$$\Lambda = \frac{1}{c} \begin{pmatrix} U^0 & U^1 \\ U^1 & U^0 \end{pmatrix}$$

where we use the relation $(U^0)^2 - (U^1)^2 = c^2$. This Lorentz 'boost' transforms the acceleration vector $A^\mu = dU^\mu/d\tau \equiv \dot{U}^\mu$ from $A^\mu_{\text{rest}} = (0, g)$ to $(A^0, A^1) = (\dot{U}^0, \dot{U}^1)$,

$$\begin{pmatrix} \dot{U}^0 \\ \dot{U}^1 \end{pmatrix} = \frac{1}{c} \begin{pmatrix} U^0 & U^1 \\ U^1 & U^0 \end{pmatrix} \begin{pmatrix} 0 \\ g \end{pmatrix},$$

the solution of this equation for U^1 (with the initial condition $U^1 = 0$ at $\tau = 0$) is $U^1 = 2A \sinh(g\tau/c)$. Then $U^0 = (c/g)\dot{U}^1$ is found to be $U^0 = 2A \cosh(g\tau/c)$. The condition $(U^0)^2 - (U^1)^2 = c^2$ forces $2A = c$. The solution for $x^\mu(\tau)$ can then be obtained. It is

$$x^0(\tau) = \frac{c^2}{g} \sinh\left(\frac{g\tau}{c}\right),$$

$$x^1(\tau) = L + \frac{c^2}{g}\left[\cosh\left(\frac{g\tau}{c}\right) - 1\right],$$

$$x^2(\tau) = 0 \quad, \quad x^3(\tau) = 0.$$

Let $g > 0$. At $\tau \to -\infty$ the particle is at $x^1 \to \infty$ moving with velocity close to c towards the origin. It keeps moving with decreasing speed due to accleration g till it comes to a stop at $x^1 = L$ when $\tau = 0$. It then goes back to $x^1 \to \infty$ with increasing velocity eventually moving with asymptotic velocity c. This type of motion is described sometimes as **hyperbolic motion**.

Exercise 15. (i) Show that two non-null orthogonal vectors are linearly independent. (ii) Show that two non-orthogonal null vectors are linearly independent.

Answer 15. (i) Let

$$a\mathbf{u} + b\mathbf{v} = \mathbf{0}$$

where $\langle \mathbf{u}, \mathbf{u} \rangle \neq 0$, $\langle \mathbf{v}, \mathbf{v} \rangle \neq 0$ and $\langle \mathbf{u}, \mathbf{v} \rangle = 0$. Take the inner product in the above equation by $\mathbf{u}$,

$$a\langle \mathbf{u}, \mathbf{u} \rangle = 0$$

or $a = 0$. Similarly taking the product by $\mathbf{v}$ gives $b = 0$. This proves linear independence of $\mathbf{v}$ and $\mathbf{u}$.
(ii)

$$a\mathbf{u} + b\mathbf{v} = \mathbf{0}$$

where $\langle \mathbf{u}, \mathbf{u} \rangle = 0$, $\langle \mathbf{v}, \mathbf{v} \rangle = 0$ and $\langle \mathbf{u}, \mathbf{v} \rangle \neq 0$. Take the inner product in the above equation by $\mathbf{v}$,

$$a\langle \mathbf{u}, \mathbf{v} \rangle = 0$$

or $a = 0$. Similarly taking the product by $\mathbf{u}$ gives $b = 0$. This proves linear independence of $\mathbf{v}$ and $\mathbf{u}$.

1.7 Literature

The literature on relativity is vast. The following is just an indication of easily traceable books and other sources. The choice of this material is guided by students' needs (particularly easy accessibility) and *the list is grossly incomplete.*

Original Papers

Lorentz, **Einstein**, **Minkowski** and **Weyl**, *Principle of Relativity*, Dover books, 1952
A collection of original papers on special and general theory of relativity in English translation with notes by A. Sommerfeld.
    C. W. Kilmister (Ed.), *Special Theory of Relativity*, Pergamon Press, 1970, *General Theory of Relativity*, Pergamon Press, 1973
Collection of almost all the basic papers on relativity. These volumes have introductory chapters and commentary by C. W. Kilmister.

Historical Matter

Abraham Pais, *'Subtle is the Lord...', The Science and the Life of Albert Einstein*, Oxford University Press, 1982
A thorough discussion of development of Einstein's thinking with an almost day-by-day account.
    Edmond T. Whittaker, *History of theories of Aether and Electricity*, Vol I and II, Thomas Nelson and Sons, London. Reprinted by Humanities Press, New York, 1973
This is another standard reference for the history of field theories of classical physics.

Texts

Albert Einstein, *Meaning of Relativity*, Indian Edition by Oxford Book Company, 1965
These are Einstein's 1921 Princeton lectures, originally published by The Princeton University Press. Every student of relativity should read these 100 odd pocket-book sized pages for the clarity and brevity of the man who discovered the theory. The available editions have appendices containing Einstein's later unified theories, none of which seem relevant today. But who knows?
    Wolfgang Pauli, *Theory of Relativity*, Pergamon Press 1958
Pauli's *Mathematical Encyclopedia* review article of 1921 written at the age of 21 by the author. The article gives a complete account of relativity theory till 1921.
    Herman Weyl, *Space-Time-Matter*, Dover, 1952
First published in German in 1918. It was already in its fourth edition in 1920. One of the earliest expositions of relativity by a great mathematician who contributed to the theory. The third edition of 1919 introduces Levi-Civita's 1917 discovery of infinitesimal parallel displacement.

Lev Landau and **E. M. Lifshitz**, *Classical Theory of Fields*, Pergamon Press, Reprint 2004
Volume 2 of the famous Course of Theoretical Physics. The first nine chapters are on Electrodynamics, last five on the General Theory of Relativity. These (less than two hundred) pages constitute an introduction which is both deep and brief.

Arther S. Eddington, *Mathematical Theory of Relativity*, Cambridge University Press, 7th reprint 1963
One of the clearest early expositions, first written as mathematical notes to his delightful and popular book *Space Time and Gravitation* published in 1920.

Peter G. Bergmann, *Introduction to the Theory of Relativity*, Prentice Hall, 1942
(Now available as a paperback in Dover, New York)
This book has a forward by Einstein. It is the complete, perhaps the first, textbook, written with students' needs in mind. It covers quite 'advanced' ideas (for those days) such as the Kaluza-Klein theory, which has made a comeback in theoretical physics recently.

Vladimir A. Fock, *The Theory of Space, Time and Gravitation*, Macmillan, 1964
The authoritative book by the great Russian physicist who took a critical look at many of the fundamental ideas of relativity.

C. Møller, *The Theory of Relativity*, Oxford, 1952
Authoritative book on all aspects of the theory.

J. L. Synge, *Relativity, the General Theory*, North Holland, 1966
This is a book written in a very independent style very unlike other standard books.

C. M. De Witt and B. S. De Witt (Eds.), *Relativity, Groups and Topology*, Blackie and Sons, 1964
Contains delightful introductory lectures on general theory of relativity given by Synge at Les Houche School in 1963. This collection also includes lectures by Wheeler, Penrose, Sachs, and Misner on various aspects of general relativity.

Charles W. Misner, Kip S. Thorne and John A. Wheeler, *Gravitation*, Freeman, 1973
The absolute darling of students and researchers in general relativity. Its *twenty fourth* reprint came out in 2002 ! 'Merely holding the book in your hand makes you think about gravity !', we used to say as students in the 1970's. The presently available paperback is lighter, a little above two kilograms. This large sized (20 cm× 25 cm), 1272 page book begins at the beginning and has everything on gravity (up to 1973). There are hundreds of diagrams and special boxes for additional explanations, exercises, historical and biographical asides and bibliographical details. It must have converted a fair number of people into research in general relativity. And conversion is a good word here because S. Chandrasekhar, while reviewing the book, is supposed to have commented on its missionary spirit! What makes it a pleasure to read is that no idea is introduced without its motivation. A student

is told why the idea is natural to expect, and, if the natural expectation is wrong, why it is so. It has a cheery delightful style throughout.

Steven Weinberg, *Gravitation and Cosmology*, John Wiley, 1973
A modern classic which reduces the emphasis on geometry and reinforces the power of the equivalence principle. Extremely readable with a discussion of experimental data (up to 1973).

Paul M. Dirac, *General theory of Relativity*, Princeton University Press, 1996, Reprinted by Prentice-Hall of India, 2001
Dirac's 1975 Lectures at Florida State University. As concise, to-the-point as only Dirac could be. This slim 35-section, 70 page booklet is written for a beginner. The book has significantly five sections on the action principle.

S. Chandrasekhar, *The Mathematical Theory of Black Holes*, Oxford University Press, 1983
The exhaustive treatise on black hole solutions and their properties. If you need anything, anything at all, on the Schwarzschild or the Kerr spacetime it is here.

Robert M. Wald, *General Relativity*, University of Chicago Press, 1984
The deservedly famous advanced textbook includes extremely readable introductory parts in the first six chapters and advanced topics on researches done in the 1960's and 1970's in chapters 7 to 14. A lot of mathematical background is condensed and relegated to Appendices at the end of the book which one cannot do without.

S. W. Hawking and G. F. R. Ellis, *The large scale structure of space-time*, Cambridge University Press, 1973
A classic on spacetime structure in general relativity, known for its clarity and rigour. All proofs are complete, every concept well defined, most details included. But it requires considerable mathematical maturity to follow the line of thought. The mathematical apparatus used is indispensable for research in the area but the introduction to differential geometry is too brief (forty pages) to be of any actual help to a beginner.

R. Adler, M. Bazin, M. Schiffer, *Introduction to General Relativity*, Second Edition, McGraw Hill, 1975
A very good textbook although not widely available. Its derivation of the Kerr metric is particularly good.

W. Rindler, *Essential Relativity*, Springer-Verlag, 1977
This book, written specially for the advanced undergraduate student, is known for conceptual clarity and style. Written in extremely lucid style it is an enjoyable but deep book.

E. F. Taylor and **J. A. Wheeler**, *Spacetime Physics*, W. H. Freeman, 1963
A delightful undergraduate book on basics.

Bernard F. Schutz, *A first course in general relativity*, Cambridge University Press, 1985
A good textbook from a beginner's point of view. It develops the mathematical background of tensor calculus through hundreds of exercises.

Ray d'Inverno, *Introducing Einstein's Relativity*, Clarendon Press Oxford, 1992
A textbook with several advanced topics as well.

H. C. O'Hanian and R. Ruffini, *Gravitation and Spacetime* W. W. Norton and Co., 1994
An introductory book which looks at gravity in the most logical and straightforward way. In spirit it is closer to Weinberg's book. There is a good collection of problems in each chapter. And the book contains *a very detailed* guide to further reading in each chapter.

James B. Hartle, *Gravity*, Pearson Education, 2002
Written in the spirit of Misner, Thorne, Wheeler, this is the best recent treatment of general relativity available to the advanced undergraduate student. The book is complete with all the exciting experimental data up to the end of the 20th century. This book is a must for every beginner.

S. M. Carroll, *Spacetime and Geometry: An Introduction to General Relativity*, Addison Wesley, 2004
This is a well-written recent textbook which gives plenty of space to geometry as needed in general relativity. Carroll's lecture notes on general theory of relativity are also available on the arxiv.org as gr-qc/9712019.

N. Straumann, *General Relativity – With Applications to Astrophysics*, Springer Verlag, 2004
A thorough recent book. It has a condensed mathematical introduction in Part III, used throughout the book.

Eric Poisson, *A Relativist's Toolkit*, Cambridge University Press, 2004
A recent advanced book on selected topics in general relativity. Although the topics are advanced, the author has taken pains to provide details and explanation.

Important Reviews and Internet Sources

S. W. Hawking and W. Israel, *Einstein Centenary Survey*, Cambridge University Press, 1979
S. W. Hawking and W. Israel, *300 Years of Gravitation*, Cambridge University Press, 1987

Living Reviews on Relativity: (http://relativity.livingreviews.org/)
An internet source of reviews which are periodically updated. In addition there are research articles and reviews available on (http://arxiv.org/) in the "gr-qc" (general relativity and quantum cosmology) section.

Books on Mathematics

For the convenience of a physics student all texts on the general theory of relativity do try to give an introduction to Riemannian geometry with varying degrees of pedagogical attention or success.

B. F. Schutz, *Geometrical Methods in Mathematical Physics*, Cambridge University Press, 1980
An introduction to geometry and topology used in gravity and gauge theories.

Y. Choquet-Bruhat, C. De Witt-Morette and **M. Dillard-Bleick**, *Analysis, Manifolds and Physics*, North Holland., 1989
A thorough introduction to modern differential geometry as needed by physicists. It has rigourous approach illustrated by examples from physics.

C. Isham, *Modern Differential Geometry for Physicists*, World Scientific, 1989
Another thorough introduction to differential geometry as used in gravity and gauge theories.

H. K. Nickerson, D. C. Spencer and **N. E. Steenrod**, *Advanced Calculus*, Von Nostrand, 1959
An undergraduate textbook for introduction to vectors, tensors, forms and differentiable manifolds based on lectures at Princeton University.

I. M. Singer and **J. A. Thorpe**, *Lecture Notes on Elementary Topology and Geometry*, Undegraduate Texts in Mathematics, Springer-Verlag, 1976
Another classic with undergraduate students in mind.

F. W. Warner, *Foundations of Differentiable Manifolds and Lie Groups*, Springer-Verlag, 1983
A more advanced introduction.

B. A. Dubrovin, A. T. Fomenko and **S. P. Novikov**, *Modern Geometry – Methods and Applications*, Springer-Verlag, 1992
A good relaxed introduction to geometry in three volumes.

S. S. Chern, W. H. Chen and **K. S. Lam**, *Lectures on Differential Geometry*, World Scientific, 1999
The Peking University lectures in 1980 by the great mathematician S. S. Chern.

N. J. Hicks, *Notes on Differential Geometry*, Von Nostrand, 1965
This is an all-time favourite. A slim, 175 page book which introduces Riemannian geometry through hypersurfaces and theorema egregium.

Books on Astrophysics and Cosmology

Astrophysics and Cosmology are two important branches of physics where general theory of relativity is applied. The subject of cosmology has been in very rapid growth in the last ten years.

T. Padmanabhan, *Theoretical Astrophysics*, Cambridge University Press, Vol. I, 2000, vol. II, 2001, Vol. III, 2002
This is a thorough introduction in three volumes.

J. V. Narlikar, *An Introduction to Cosmology*, Cambridge University Press, 2002
For cosmology there are many recent texts but as introduction, this is perhaps the best. It gives a clear, detailed, account of all concepts with their historical background.

P. Coles and **F. Lucchin**, *Cosmology*, John Wiley and Sons, 2002

V. Mukhanov, *Physical Foundations of Cosmology*, Cambridge University Press, 2005

These are two of the many recent textbooks on cosmology.

S. Weinberg, *Cosmology*, Oxford, 2008.

A very recent advanced book on Cosmology.

Chapter 2

What is Curvature?

In this chapter we discuss Gauss' work to explain the concept of curvature tensor and the geodesic in the familiar case of a two-dimensional curved surface.

Differential geometry as used in the general theory of relativity, and in much of gauge theories of elementary particles, was the result of Riemann's generalization of Gauss' work on curvature of two-dimensional surfaces to any number of dimensions.

It is instructive to see how these ideas developed.

2.1 Concept of Curvature

The curvature of a curve in a plane is determined by how fast its unit normal vector $\mathbf{n}$ (or the tangent vector for that matter) changes as we move along the curve. A measure of curvature is the ratio of the small change $|d\mathbf{n}|$ in the unit normal vector to the distance ds moved by the point on the curve.

A straight line has zero curvature because the unit normals are all parallel and do not change. A circle of radius R has curvature $1/R$ because for the distance ds that the point P moves, the unit normal vector changes by an angle ds/R so $|d\mathbf{n}| = ds/R$.

Gauss defined the curvature of a surface analogously.

Let S be a two-dimensional surface whose points $\mathbf{r}(u)$ are labelled by two independent parameters u^1, u^2. Let the unit normal be denoted by $\mathbf{n}(u)$ at the point determined by the parameters u.

When parameters u^1, u^2 change by small amounts du^1, du^2 the point on the surface traces a small displacement

$$d\mathbf{r} = \frac{\partial \mathbf{r}}{\partial u^1} du^1 + \frac{\partial \mathbf{r}}{\partial u^2} du^2 \equiv \mathbf{r}_{,i} du^i$$

tangential to the surface.

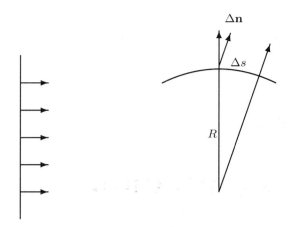

Fig. 2.1: Curvature of planar curves.
The straight line on the left has zero curvature because the normal
does not change, whereas for the arc of a circle (of radius R) the ratio
$\Delta\mathbf{n}/\Delta s$ is $1/R$.

Here we are using the convenient notation $\mathbf{r}_{,i}$ for the derivative with respect to
u^i and have omitted the summation sign, it being understood that whenever there
is a *repeated* index, there will be a sum. See page xiii for notational conventions.

Vectors $\mathbf{r}_{,1}$ and $\mathbf{r}_{,2}$ are linearly independent because u^1, u^2 are independent
parameters, so that changes in $\mathbf{r}$ along the u^1 and u^2 directions cannot be collinear.
These derivatives $\mathbf{r}_{,1}$ and $\mathbf{r}_{,2}$ form a basis for vectors tangent to the surface.

Because the unit normal $\mathbf{n}$ is a vector of constant length $\mathbf{n}.\mathbf{n} = 1$, infinitesimal
changes in it due to changes in the parameters u are orthogonal to it. That is,
$d(\mathbf{n}.\mathbf{n}) = 2\mathbf{n}.d\mathbf{n} = 0$. Therefore $d\mathbf{n} = \mathbf{n}_{,i}du^i$ is tangential to the surface.

The normal vector changes by $(d\mathbf{n})_1 = \mathbf{n}_{,1}du^1$ when the parameter u^1 is
changed, but u^2 is kept fixed while it changes by $(d\mathbf{n})_2 = \mathbf{n}_{,2}du^2$ when the opposite
is true. Note that in order to find the change in the normal vector the normal at
the displaced point is brought *parallel to itself* so that its base point coincides with
that of the normal at the original point. The change $d\mathbf{n}$ is then the infinitesimal
vector which joins the tip of the original normal to that of the parallel normal
from the neighbouring point.

As $(d\mathbf{n})_1, (d\mathbf{n})_2$ are tangential, (we have omitted the numerical coefficients
du^1 and du^2), we can expand them in basis vectors $\mathbf{r}_{,i}, i = 1, 2$,

$$\mathbf{n}_{,i} = L_i{}^j\mathbf{r}_{,j}. \tag{2.1}$$

The coefficients $L_i{}^j$ depend on u and they determine the way the tips of the unit
normal vectors move when their base is carried on the curving surface. It defines
a mapping of the tangent vector $d\mathbf{r}$ into another tangent vector $d\mathbf{n}$. This mapping
is called the **Weingarten map** and matrix $L_i{}^j$ which determines it the **Weingarten**

matrix. Generalizing the notion of curvature of a plane curve, Gauss defined the curvature of the surface at a point P to be the ratio of the **area** spanned by the increments $(d\mathbf{n})_1, (d\mathbf{n})_2$ of the normal, to the area spanned by the tangent vectors $(d\mathbf{r})_1 = \mathbf{r}_{,1}du^1$ and $(d\mathbf{r})_2 = \mathbf{r}_{,2}du^2$ on the surface.

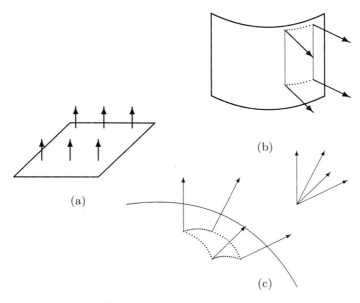

Fig. 2.2: Curvature of surfaces.

The curvature is defined to be the ratio of the surface area spanned by infinitesimal changes in the normal vector $d\mathbf{n}_1$ and $d\mathbf{n}_2$ in two independent directions to the surface area of the infinitesimal area element along whose sides the change in normal unit vector is calculated.

The curvature of a plane **(a)** is zero because the normal vector does not change at all. For a cylindrical surface **(b)** the curvature is zero again because, even if the normal changes along the circular side of the surface, it does not change in a direction parallel to the axis. For the surface of a sphere of radius R, the curvature **(c)** is equal to $1/R^2$ because the solid angle that dS subtends at the center of the sphere is the same as that covered by the normal vectors at the end of the area element.

The area of the parallelogram spanned by vectors $\mathbf{a}$ and $\mathbf{b}$ is $|\mathbf{a} \times \mathbf{b}|$. Therefore, the measure of curvature is given by

$$
\begin{aligned}
(d\mathbf{n})_1 \times (d\mathbf{n})_2 &= (\mathbf{n}_{,1} \times \mathbf{n}_{,2})\, du^1 du^2 \\
&= (L_1^1 \mathbf{r}_{,1} + L_1^2 \mathbf{r}_{,2}) \times (L_2^1 \mathbf{r}_{,1} + L_2^2 \mathbf{r}_{,2})\, du^1 du^2 \\
&= (L_1{}^1 L_2{}^2 - L_1{}^2 L_2{}^1)((d\mathbf{r})_1 \times (d\mathbf{r})_2).
\end{aligned}
$$

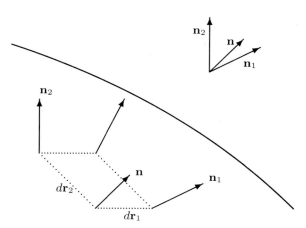

Fig. 2.3: Gaussian Curvature of a surface.

Curvature of the surface at a point is the ratio of the area spanned
by the increments $(d\mathbf{n})_1 = \mathbf{n}_1 - \mathbf{n}$ and $(d\mathbf{n})_2 = \mathbf{n}_2 - \mathbf{n}$ of the normal,
to the area spanned by the tangential displacement vectors $(d\mathbf{r})_1$ and
$(d\mathbf{r})_2$ on the surface.

The quantity $\det L \equiv L_1{}^1 L_2{}^2 - L_1{}^2 L_2{}^1$ is called the **Gauss curvature** or the
total curvature of the surface.

Thus, a plane has zero curvature because the normals are all in the same
direction. If we imagine a sheet of paper in place of the plane, and roll it in the
form of a cylinder, the normals at points lying on the circular direction point in
different directions, but those along a line parallel to the axis of the cylinder are
parallel. Thus the area spanned by the tips of the normal when the base point is
moved in a small rectangle on the paper with sides along the axis and along the
circle is stll zero and the cylinder has zero total curvature. The spherical surface
of radius R however has curvature $1/R^2$ as can be seen easily.

2.2   "Theorema Egregium" of Gauss

Gauss, considered to be one of the three greatest mathematicians of all times, the
other two being Archimedes and Newton, was not known for overrating his own (or
anybody else's) work. But he must have felt reasonably pleased with himself to call
the following result as standing out in a flock of theorems ("theorema egregium",
1827):

The curvature of a surface can be defined entirely in terms of quantities
intrinsic to the surface, without any reference to how the surface is located or
embedded in the three-dimensional surrounding space.

The intrinsic quantities in question are the coefficients g_{ij} of the quadratic
form called the **metric form** or the **line element** which determines the distance
between two infinitesimally close points on the surface.

Expressed in terms of the parameters (u^1, u^2) and $(u^1 + du^1, u^2 + du^2)$ the distance can be calculated as follows. As $d\mathbf{r} = \mathbf{r}_{,i} du^i$, the infinitesimal distance squared is

$$ds^2 = d\mathbf{r}.d\mathbf{r} = \mathbf{r}_{,i}.\mathbf{r}_{,j} du^i du^j \equiv g_{ij} du^i du^j, \tag{2.2}$$

$$g_{ij} = \mathbf{r}_{,i}.\mathbf{r}_{,j} = g_{ji}. \tag{2.3}$$

The quadratic form of the metric is positive definite, symmetric and non-singular (that is, $\det g \neq 0$).

Measurements of small distances made by two-dimensional creatures living on the surface are sufficient to determine these coefficients. Functions and derivatives of g_{ij} with respect to u^i are again intrinsic.

What Gauss showed, specifically, was that the combination of $L_i{}^j$ which determines the total curvature is given in terms of intrinsic quantities by the following equation known as the **Gauss equation**:

$$L_j{}^m L_l{}^k - L_l{}^m L_j{}^k = -g^{mi} R^k{}_{ijl} \tag{2.4}$$

where g^{ij} is the matrix inverse to g_{ij} and the quantities $R^k{}_{ijl}$, called the **Riemann-Christoffel curvature tensor** are given by

$$-R^k{}_{ijl} = \Gamma^k_{ij,l} - \Gamma^k_{il,j} + \Gamma^m_{ij}\Gamma^k_{ml} - \Gamma^m_{il}\Gamma^k_{mj} \tag{2.5}$$

which in turn is composed from

$$\Gamma^k_{ij} = \frac{1}{2} g^{kl}(g_{il,j} + g_{lj,i} - g_{ij,l}) \tag{2.6}$$

and its derivatives.

The antisymmetry in j, l and in m, k restricts the Gauss equation to essentially one combination of indices j, l, m, k, (because there are only two values possible for any of these indices, 1 and 2) and that combination gives $\det L$, the total curvature.

Moreover, the straightest possible curves on the surface, called geodesics, are given by functions $u^i(s)$, where s is the distance measured along the curve, which satisfy

$$\frac{d^2 u^k}{ds^2} + \Gamma^k_{ij} \frac{du^i}{ds} \frac{du^j}{ds} = 0. \tag{2.7}$$

We give a proof of this revolutionary theorem later in this chapter.

Riemann generalised Gauss' formula for curvature of two-dimensional surfaces to any number of dimensions in 1854 by introducing the curvature tensor named after him.

All information of the space is contained in the line element

$$ds^2 = g_{ij}dx^i dx^j$$

and quantities defined in terms of these, like the Γ^k_{ij} and the curvature tensor $R^k{}_{lij}$. Riemann's work was extended by Christoffel, Ricci and Levi-Civita and other mathematicians.

2.3   The Gauss Equation

We have seen that the first drivatives of $\mathbf{r}$ are tangential to the surface $\mathbf{r}_{,i}.\mathbf{n} = 0$. But the second derivatives

$$\mathbf{r}_{,ij} \equiv \frac{\partial^2 \mathbf{r}}{\partial u^i \partial u^j}$$

have components both along the normal and tangent to the surface.

(1) Differentiating $\mathbf{r}_{,i}.\mathbf{n} = 0$ once, we get

$$\mathbf{r}_{,ij}.\mathbf{n} + \mathbf{r}_{,i}.\mathbf{n}_{,j} = 0$$

and using the definition of the Weingarten map,

$$\mathbf{r}_{,ij}.\mathbf{n} = -\mathbf{r}_{,i}.(L_j{}^k \mathbf{r}_{,k}) = -L_j{}^k g_{ki} \equiv -L_{ji} \tag{2.8}$$

where we have used the symmetry $g_{ki} = g_{ik}$ and defined L_{ij}. This equation shows that the L_{ij} closely related to the Weingarten map is **symmetric** $L_{ij} = L_{ji}$. The second derivatives have a component in the direction of the normal.

(2) Similarly, differentiating the defining equation $g_{ij} = \mathbf{r}_{,i}.\mathbf{r}_{,j}$ we get

$$\mathbf{r}_{,ik}.\mathbf{r}_{,j} + \mathbf{r}_{,i}.\mathbf{r}_{,jk} = g_{ij,k}.$$

We rewrite this equation twice more, but with changed names of indices

$$\mathbf{r}_{,ij}.\mathbf{r}_{,k} + \mathbf{r}_{,i}.\mathbf{r}_{,kj} = g_{ik,j},$$

$$\mathbf{r}_{,ki}.\mathbf{r}_{,j} + \mathbf{r}_{,k}.\mathbf{r}_{,ji} = g_{kj,i}$$

and add these later two and subtract the first one from these to get

$$2\mathbf{r}_{,ij}.\mathbf{r}_{,k} = g_{ik,j} + g_{kj,i} - g_{ij,k}.$$

This shows that the second drivatives have non-zero components in the tangential direction $\mathbf{r}_{,k}$ as well.

(3) Let us expand the second derivatives in the three linearly independent vectors $\mathbf{n}, \mathbf{r}_{,1}, \mathbf{r}_{,2}$ and write temporarily

$$\mathbf{r}_{,ij} = (\ )_{ij}\mathbf{n} + (\ )^k_{ij}\mathbf{r}_{,k}.$$

By taking the dot product of this equation with $\mathbf{n}$ and with $\mathbf{r}_{,l}$ we identify these unknown co-efficients from the steps (1) and (2),

$$\mathbf{r}_{,ij} = -L_{ij}\mathbf{n} + \Gamma_{ij}^{k}\mathbf{r}_{,k} \tag{2.9}$$

where L_{ij} has already been defined, and

$$\Gamma_{ij}^{k} = \frac{1}{2}g^{kl}(g_{il,j} + g_{lj,i} - g_{ij,l}) \tag{2.10}$$

where g^{ij} is the matrix inverse to g_{ij},

$$g^{ij}g_{jk} = \delta_{k}^{i}$$

(g_{ij} is non-singular, as noted above).

(4) Differentiating $\mathbf{r}_{,ij}$ above with respect to u^{l} to get the *third* derivatives,

$$\begin{aligned}
\mathbf{r}_{,ijl} &= -L_{ij,l}\mathbf{n} - L_{ij}\mathbf{n}_{,l} + \Gamma_{ij,l}^{k}\mathbf{r}_{,k} + \Gamma_{ij}^{k}\mathbf{r}_{,kl} \\
&= -L_{ij,l}\mathbf{n} - L_{ij}L_{l}{}^{k}\mathbf{r}_{,k} + \Gamma_{ij,l}^{k}\mathbf{r}_{,k} + \Gamma_{ij}^{k}(-L_{kl}\mathbf{n} + \Gamma_{kl}^{m}\mathbf{r}_{,m}).
\end{aligned}$$

Because of the symmetry of mixed partial derivatives, $\mathbf{r}_{,ijl} - \mathbf{r}_{,ilj} = 0$. This, when written out fully using the above expression, gives

$$(-R^{k}{}_{ijl} - L_{ij}L_{l}{}^{k} + L_{il}L_{j}{}^{k})\mathbf{r}_{,k} - (L_{ij,l} - L_{il,j} + \Gamma_{ij}^{k}L_{kl} - \Gamma_{il}^{k}L_{kj})\mathbf{n} = 0$$

where we have gathered coefficients of normal and tangential parts separately and interchanged index names k and m in $\Gamma\Gamma$ terms. The quantity $R^{k}{}_{ijl}$ is called Riemann-Christoffel curvature tensor and is given by

$$-R^{k}{}_{ijl} = \Gamma_{ij,l}^{k} - \Gamma_{il,j}^{k} + \Gamma_{ij}^{m}\Gamma_{ml}^{k} - \Gamma_{il}^{m}\Gamma_{mj}^{k}. \tag{2.11}$$

Equating the coefficients of linearly independent vectors $\mathbf{n}, \mathbf{r}_{,1}, \mathbf{r}_{,2}$ to zero we get the **Gauss equation**

$$-R^{k}{}_{ijl} = L_{ij}L_{l}{}^{k} - L_{il}L_{j}{}^{k} \tag{2.12}$$

and the **Codazzi** equation

$$L_{ij,l} - L_{il,j} = -\Gamma_{ij}^{k}L_{kl} + \Gamma_{il}^{k}L_{kj}. \tag{2.13}$$

The Gauss equation can be rewritten to bring it closer to total curvature by using the symmetry of L_{ij} and inverse matrix g^{ij}:

$$-R^{k}{}_{ijl} = L_{ij}L_{l}{}^{k} - L_{il}L_{j}{}^{k} = L_{ji}L_{l}{}^{k} - L_{li}L_{j}{}^{k} = g_{im}(L_{j}{}^{m}L_{l}{}^{k} - L_{l}{}^{m}L_{j}{}^{k}),$$

therefore,

$$L_{j}{}^{m}L_{l}{}^{k} - L_{l}{}^{m}L_{j}{}^{k} = -g^{mi}R^{k}{}_{ijl}. \tag{2.14}$$

2.4   The Geodesic Equation

The straightest possible curve, a **geodesic**, can be found equally easily.

For such a curve given by $\mathbf{r}(\sigma)$ where σ is the parameter defining the curve, the tangent vector $d\mathbf{r}/d\sigma$, to the curve is (i) never zero, and (ii) does not sway *sideways*. This means that the change $d^2\mathbf{r}/d\sigma^2$ has a tangential component only along $d\mathbf{r}/d\sigma$.

Now,

$$\frac{d\mathbf{r}}{d\sigma} = \mathbf{r}_{,i}\frac{du^i}{d\sigma}$$

and

$$\frac{d^2\mathbf{r}}{d\sigma^2} = \mathbf{r}_{,ij}\frac{du^j}{d\sigma}\frac{du^i}{d\sigma} + \mathbf{r}_{,i}\frac{d^2u^i}{d\sigma^2}.$$

Substituting for $\mathbf{r}_{,ij}$ and gathering terms in tangential and normal directions

$$\begin{aligned}
\frac{d^2\mathbf{r}}{d\sigma^2} &= \left(-L_{,ij}\mathbf{n} + \Gamma^k_{ij}\mathbf{r}_{,k}\right)\frac{du^j}{d\sigma}\frac{du^i}{d\sigma} + \mathbf{r}_{,i}\frac{d^2u^i}{d\sigma^2} \\
&= \left(\frac{d^2u^k}{d\sigma^2} + \Gamma^k_{ij}\frac{du^j}{d\sigma}\frac{du^i}{d\sigma}\right)\mathbf{r}_{,k} - L_{,ij}\mathbf{n}.
\end{aligned}$$

The condition for the curve being straightest possible is that the tangential part of $d^2\mathbf{r}/d\sigma^2$ is in the same direction as $d\mathbf{r}/d\sigma$ with no component in the tangential perpendicular direction.

$$\left(\frac{d^2u^k}{d\sigma^2} + \Gamma^k_{ij}\frac{du^j}{d\sigma}\frac{du^i}{d\sigma}\right)\mathbf{r}_{,k} = f(\sigma)\mathbf{r}_{,k}\frac{du^k}{d\sigma}$$

where $f(\sigma)$ is an arbitrary function of σ determining the proportionality. This gives,

$$\frac{d^2u^k}{d\sigma^2} + \Gamma^k_{ij}\frac{du^j}{d\sigma}\frac{du^i}{d\sigma} = f(\sigma)\frac{du^k}{d\sigma}.$$

For a different parametrization, $t = t(\sigma)$ we can transform

$$\frac{d^2u^k}{dt^2} + \Gamma^k_{ij}\frac{du^j}{dt}\frac{du^i}{dt} = \left[-\left(\frac{dt}{d\sigma}\right)^{-2}\frac{d^2t}{d\sigma^2} + f(\sigma)\left(\frac{dt}{d\sigma}\right)^{-1}\right]\frac{du^k}{dt}.$$

We can make the quantity on the right-hand side in square brackets equal to zero for a suitable choice of parametrization. Denoting the differentiation with respect to σ by a prime we must solve $t'' = ft'$ whose solution is

$$t = a + b\int e^{\int F(\sigma)d\sigma}\, d\sigma$$

where a and b are arbitrary constants. The equation of the geodesic can then be put in the form

$$\frac{d^2 u^k}{dt^2} + \Gamma^k_{ij} \frac{du^j}{dt} \frac{du^i}{dt} = 0.$$

Affine Parameter

Note that there is a whole family of parameters for which the geodesic takes the above simple form. For different values of constants a and b of integration we get different parameters t_1 and t_2,

$$t_1 = a_1 + b_1 \int e^{\int F(\sigma)d\sigma} d\sigma,$$

$$t_2 = a_2 + b_2 \int e^{\int F(\sigma)d\sigma} d\sigma,$$

therefore any two parameters are related by an "affine transformation"

$$t_2 = At_1 + B$$

where $A = b_2/b_1$ and $B = a_2 - a_1 A$. The parameters t which allow the simple form of the geodesic equation are called **affine parameters**.

The length of the curve

$$s = \int ds = \int \left(g_{ij} \frac{du^i}{d\sigma} \frac{du^j}{d\sigma} \right)^{\frac{1}{2}} d\sigma$$

is always an affine parameter as we show below.

The general form of the geodesic (for any parameter) is

$$\frac{d^2 u^i}{ds^2} + \Gamma^i_{jk} \frac{du^j}{ds} \frac{du^k}{ds} = f(s) \frac{du^i}{ds}.$$

We show that $f(s) = 0$.

From

$$g_{jk} \frac{du^j}{ds} \frac{du^k}{ds} = 1$$

it follows that

$$\begin{aligned}
0 &= g_{jk,l} \frac{du^j}{ds} \frac{du^k}{ds} \frac{du^l}{ds} + g_{jk} \frac{d^2 u^j}{ds^2} \frac{du^k}{ds} + g_{jk} \frac{du^j}{ds} \frac{d^2 u^k}{ds^2} \\
&= g_{jk,l} \frac{du^j}{ds} \frac{du^k}{ds} \frac{du^l}{ds} + 2 g_{jk} \frac{d^2 u^j}{ds^2} \frac{du^k}{ds}
\end{aligned}$$

as the last two terms are the same in the above equation because of the symmetry of g_{jk} in j, k and there is summation over both the indices. Substituting the second derivative d^2u^j/ds^2 from the general form of the geodesic equation and using the expression

$$\Gamma^j_{pr} = \frac{1}{2}g^{jl}[g_{lp,r} + g_{lr,p} - g_{pr,l}]$$

we see that all terms cancel and we are left with $f(s) = 0$.

The straightest possible curves are also the curves with shortest distance between two fixed points.

$$\delta \int ds = 0. \tag{2.15}$$

2.5   Historical Note on Riemann

In 1854, Riemann generalised Gauss' formula for curvature of two-dimensional surfaces to any number of dimensions by introducing the curvature tensor named after him.

Bernhard Riemann (1826-1866) presented his generalization in a lecture to the faculty of the University of Gottingen as a candidate for 'Privatdozent', an unpaid lecturership that was traditionally the starting point of an academic career. Gauss, then 77, was in the audience among those judging him. He was suitably impressed.

Riemann's contribution is remarkable because he totally abandoned the idea of thinking of curvature as a property of the way the space is embedded as a subspace in still higher dimensional Euclidean space. Riemann showed that for the existence of Euclidean coordinates, all components of the curvature tensor must be zero. He also introduced the 'normal coordinate system' around a point which makes the g_{ik} constant in a small neighbourhood of the point. (The derivatives of the g_{ik} vanish at the point, making all Γ's zero.) The existence of this coordinate system is precisely the content of the Equivalence Principle in the General Theory of Relativity. Riemann's work was extended by Christoffel, Ricci and Levi-Civita and other mathematicians.

It must have been a very bold step to think of our three-dimensional space as not being the homogeneous Euclidean space, but an intrinsically defined curved space without being a surface of some higher dimensional Euclidean space. Riemann pointed out in a later paper that if homogeneity and isotropy of space is assumed (independence of bodies from position in an older language) the space is of constant curvature and measurements of departure from Euclidean geometry (assuming them to be small) will be seen only at very large distances, possibly too far to be observable. But if the space is not homogeneous, there is no such restriction and it is possible that, at infinitely small distances, departures from Euclidean geometry may occur and still not be noticeable at commonplace distances. Riemann suggests that if required, even the quadratic form for the line element could be replaced by a more general expression. He also mentions that the physical concept of distance is based on the rigid body and light rays, and these phenomena may require revision for infinitely small scale.

Riemann's paper on the relation of physics and geometry was translated and published by W.K.Clifford in Nature, vol.183 (1873) page 14. It is reprinted in the collection edited by Kilmister [1973].

Riemann published not many papers. Before his premature death due to tuberculosis at the age of 39 years, only nine papers were published. His contribution to analytic function theory (the Riemann surfaces), integration theory (the Riemann integral), prime numbers (through the zeros of the Riemann zeta-function), Abelian functions etc. are all great breakthroughs. But the real physical world seems to follow the Riemannian geometry. That was discovered by Einstein only after sixty years of Riemann's work.

2.6 Tutorial on Surfaces

Two-Dimensional Surfaces in Euclidean Space

Exercise 16. Choose appropriate coordinates and write the metric for the following two-dimensional surfaces embedded in three-dimensional Euclidean space. **Draw diagrams too.**

1. The cylinder $x^2 + y^2 = R^2$.

2. The sphere given by $x^2 + y^2 + z^2 = R^2$.

3. The hyperboloid of one sheet $x^2 + y^2 = R^2 + z^2$.

4. One sheet of the hyperboloid of two sheets $z^2 = R^2 + x^2 + y^2, z > 0$.

Answer 16. For any two neighbouring points (x, y, z) and $(x + dx, y + dy, z + dz)$ on the surface,

$$(ds)^2 = (dx)^2 + (dy)^2 + (dz)^2.$$

We just change this to our chosen coordinates on the surface.

The cylinder

Choose $x = R\cos\phi$ and $y = R\sin\phi$ with z and ϕ giving the coordinates on the cylinder.

$$dx = -R\sin\phi d\phi, \qquad dy = R\cos\phi d\phi.$$

Therefore $(dx)^2 + (dy)^2 = R^2(d\phi)^2$ and

$$(ds)^2 = (dz)^2 + R^2(d\phi)^2.$$

Calling the coordinates $(x^1 = z, x^2 = \phi)$ the metric written as a matrix is

$$g_{ij} = \begin{pmatrix} 1 & 0 \\ 0 & R^2 \end{pmatrix}.$$

The Sphere

Use polar coordinates $x = R\sin\theta\cos\phi, y = R\sin\theta\sin\phi, z = R\cos\theta$ and so with $x^1 = \theta$ and $x^2 = \phi$,

$$g_{ij} = R^2 \begin{pmatrix} 1 & 0 \\ 0 & \sin^2\theta \end{pmatrix}.$$

The hyperboloid of one sheet

Use $x = r\cos\phi, y = r\sin\phi$; then we have $r^2 = R^2 + z^2$. Moreover $r\,dr = z\,dz$, so we can choose r and ϕ as coordinates with r ranging from R to ∞:

$$
\begin{aligned}
(ds)^2 &= (dx)^2 + (dy)^2 + (dz)^2 \\
&= (dr)^2 + r^2(d\phi)^2 + \frac{r^2}{z^2}(dr)^2 \\
&= (dr)^2 + r^2(d\phi)^2 + \frac{r^2}{r^2 - R^2}(dr)^2 \\
&= \frac{2r^2 - R^2}{r^2 - R^2}(dr)^2 + r^2(d\phi)^2.
\end{aligned}
$$

One sheet of the hyperboloid of two sheets

Use, again, $x = r\cos\phi, y = r\sin\phi$; then we have $z^2 = R^2 + r^2$. Moreover $r\,dr = z\,dz$, so we can choose r and ϕ as coordinates with r ranging from 0 to ∞:

$$
\begin{aligned}
(ds)^2 &= (dx)^2 + (dy)^2 + (dz)^2 \\
&= (dr)^2 + r^2(d\phi)^2 + \frac{r^2}{z^2}(dr)^2 \\
&= (dr)^2 + r^2(d\phi)^2 + \frac{r^2}{r^2 + R^2}(dr)^2 \\
&= \frac{2r^2 + R^2}{r^2 + R^2}(dr)^2 + r^2(d\phi)^2.
\end{aligned}
$$

Two-Dimensional Surfaces in Minkowski Space

Exercise 17. Find the metric for the two-dimensional "upper hyperboloid" in Minkowski space $(ds)^2 = -(cdt)^2 + (dx)^2 + (dy)^2 + (dz)^2$ given by the conditions

$$z = 0, \qquad (ct)^2 = R^2 + x^2 + y^2.$$

Answer 17. As in the previous exercise choose $x = r\cos\phi, y = r\sin\phi$. We have $c^2t\,dt = r\,dr$ by differentiating $(ct)^2 = R^2 + x^2 + y^2 = R^2 + r^2$.

$$
\begin{aligned}
(ds)^2 &= -(cdt)^2 + (dx)^2 + (dy)^2 \\
&= \left[1 - \frac{r^2}{c^2t^2}\right](dr)^2 + r^2(d\phi)^2 \\
&= \frac{R^2}{r^2 + R^2}(dr)^2 + r^2(d\phi)^2.
\end{aligned}
$$

Therefore the metric is

$$g_{ij} = \begin{pmatrix} R^2/(r^2 + R^2) & 0 \\ 0 & r^2 \end{pmatrix}.$$

Curvature of Two-Dimensional Surfaces

Exercise 18. Check that the curvature of the cylindrical surface is zero.

Answer 18. The metric tensor has constant components. Therefore all Γ's are zero. This makes all R-components zero.

Exercise 19. Calculate the g^{ij}, Γ^i_{jk}, $R^i{}_{jkl}$, $R_{ijkl} \equiv g_{im} R^m{}_{jkl}$, $R_{ij} \equiv R^k{}_{ikj}$ and $R_S \equiv g^{ij} R_{ij}$ for the two-dimensional spherical surface of radius R whose metric is

$$(ds)^2 = R^2 \left[(d\theta)^2 + \sin^2 \theta (d\phi)^2 \right].$$

Answer 19. We call $x^1 = \theta$ and $x^2 = \phi$.

$$g_{ij} = R^2 \begin{pmatrix} 1 & 0 \\ 0 & \sin^2 \theta \end{pmatrix},$$

$$g^{ij} = R^{-2} \begin{pmatrix} 1 & 0 \\ 0 & 1/\sin^2 \theta \end{pmatrix}.$$

Connection coefficients

There are six Γ's to calculate.

$$\begin{aligned}
\Gamma^1_{11} &= 0, \\
\Gamma^1_{12} &= 0, \\
\Gamma^1_{22} &= -\sin\theta \cos\theta, \\
\Gamma^2_{11} &= 0, \\
\Gamma^2_{12} &= \cot\theta, \\
\Gamma^2_{22} &= 0.
\end{aligned}$$

Curvature tensor

In two dimensions there is only one independent component of the curvature tensor, namely, $R_{1212} = g_{1j} R^j{}_{212}$. Since $g_{12} = 0$, we have $R_{1212} = g_{11} R^1{}_{212}$. Now,

$$\begin{aligned}
-R^1{}_{212} &= \Gamma^1_{21,2} - \Gamma^1_{22,1} + \Gamma^j_{21}\Gamma^1_{2j} - \Gamma^1_{22}\Gamma^1_{1j} \\
&= -\frac{\partial}{\partial \theta}(-\sin\theta\cos\theta) + \cot\theta(-\sin\theta\cos\theta) \\
&= -\sin^2\theta.
\end{aligned}$$

Therefore,

$$R_{1212} = R^2 \sin^2 \theta.$$

The Ricci tensor is

$$\begin{aligned}
R_{11} &= R^1{}_{111} + R^2{}_{121} = g^{22} R_{2121} = g^{22} R_{1212} = 1, \\
R_{12} &= R^1{}_{112} + R^2{}_{122} = 0, \\
R_{22} &= R^1{}_{212} + R^2{}_{222} = g^{11} R_{1212} = \sin^2 \theta.
\end{aligned}$$

The curvature scalar is

$$R_S = g^{ij} R_{ij} = g^{11} R_{11} + g^{22} R_{22} = \frac{2}{R^2}.$$

The Gauss curvature scalar is

$$
\begin{aligned}
\det L &= L_1{}^1 L_2{}^2 - L_2{}^1 L_1{}^2 = -g^{1j} R^2{}_{j12} = -g^{11} g^{22} R_{2112} \\
&= g^{11} g^{22} R_{1212} \\
&= \frac{1}{R^2}.
\end{aligned}
$$

Exercise 20. Calculate the g^{ij}, Γ^i_{jk}, $R^i{}_{jkl}$, R_{ijkl}, R_{ij}, and R_S for the two-dimensional space-like "upper hyperboloid" of the Minkowski space whose metric is (Exercise 17 above)

$$(ds)^2 = \frac{R^2}{r^2 + R^2}(dr)^2 + r^2 (d\phi)^2.$$

Answer 20. We call $x^1 = r$ and $x^2 = \phi$.

$$g_{ij} = \begin{pmatrix} R^2/(r^2 + R^2) & 0 \\ 0 & r^2 \end{pmatrix},$$

$$g^{ij} = \begin{pmatrix} (r^2 + R^2)/R^2 & 0 \\ 0 & 1/r^2 \end{pmatrix}.$$

Connection coefficients

There are just three non-zero independent Γ's,

$$\Gamma^1_{11} = -\frac{r}{r^2 + R^2}, \qquad \Gamma^1_{22} = -\frac{r(r^2 + R^2)}{R^2},$$

$$\Gamma^2_{12} = \frac{1}{r} = \Gamma^2_{21}.$$

Curvature tensor

Therefore,

$$R_{1212} = g_{11} R^1{}_{212} = -\frac{r^2}{r^2 + R^2}.$$

The Ricci tensor is

$$
\begin{aligned}
R_{11} &= R^1{}_{111} + R^2{}_{121} = g^{22} R_{2121} = g^{22} R_{1212} = -\frac{1}{r^2 + R^2}, \\
R_{12} &= R^1{}_{112} + R^2{}_{122} = 0, \\
R_{22} &= R^1{}_{212} + R^2{}_{222} = -\frac{r^2}{R^2}.
\end{aligned}
$$

The curvature scalar R_S is

$$R_S = g^{ij} R_{ij} = g^{11} R_{11} + g^{22} R_{22} = -\frac{2}{R^2}.$$

The Gauss curvature scalar is

$$
\begin{aligned}
\det L &= L_1{}^1 L_2{}^2 - L_2{}^1 L_1{}^2 = -g^{1j} R^2{}_{j12} = -g^{11} g^{22} R_{2112} \\
&= g^{11} g^{22} R_{1212} \\
&= -\frac{1}{R^2}.
\end{aligned}
$$

Chapter 3

General Relativity Basics

We begin with Einstein's theory in this chapter. Since the full mathematical background will be developed in Part II, for the present the student is advised to regard $g_{\mu\nu}, \Gamma^{\mu}_{\nu\sigma}, R_{\mu\nu}$ etc. as sets of physical quantities which are labelled by indices just as the familiar vectors and fields are. The words "tensor", "connection components" or coefficients should not bother the student.

3.1 Riemannian Space

We discussed Gauss' theorem in great detail in the previous chapter. Riemann generalised the result to any number of dimensions. The Riemannian geometry has the following basic quantities:

Coordinates

The points of the space are labelled by n coordinates $x^i, i = 1, \ldots, n$. The points of the space can be labelled by a different set of coordinates as well. Different sets of coordinates are differentiable functions of each other.

Metric Tensor

In a given set of coordinates the infinitesimal distance between neighbouring points can be written as

$$(ds)^2 = g_{ij} dx^i dx^j \tag{3.1}$$

where g_{ij} are called the components of the metric tensor. In general the g_{ij} depend on the coordinates of the point about which the distance is being calculated $g_{ij} = g_{ij}(x)$. Although written as $(ds)^2$ the quantity is not positive definite in relativity.

The metric tensor written as a matrix is (i) symmetric and (ii) non-singular. The inverse of the matrix g_{ij} is written g^{ij} and is called the contravariant form of the metric tensor. The non-zero determinant $\det |g_{ij}|$ is traditionally written as g.

Connection Components

These are quantities defined by the derivatives of the metric

$$\Gamma_{ij}^k = \frac{1}{2} g^{kl}(g_{il,j} + g_{lj,i} - g_{ij,l}). \tag{3.2}$$

Riemann-Christoffel Curvature Tensor

Defined in terms of the derivative and products of the connection coefficients,

$$-R^k{}_{mjl} = \Gamma_{mj,l}^k - \Gamma_{ml,j}^k + \Gamma_{mj}^i \Gamma_{il}^k - \Gamma_{ml}^i \Gamma_{ij}^k. \tag{3.3}$$

The covariant form of the Riemann curvature tensor

$$R_{ijkl} \equiv g_{im} R^m{}_{jkl} \tag{3.4}$$

shows many interesting symmetries in its indices. These are

1. $R_{ijkl} = -R_{ijlk}$,

2. $R_{ijkl} = -R_{jikl}$,

3. $R_{ijkl} = R_{klij}$,

4. $R_{ijkl} + R_{iklj} + R_{iljk} = 0$.

This reduces the number of independent components of R_{ijkl} to $n^2(n^2 - 1)/12$ in n-dimensions. This means there are 20 components in four dimensions, six in three and just one, R_{1212}, in two dimensions.

Ricci Tensor, scalar curvature and Einstein Tensor

The Ricci tensor is just a linear combination of components of Riemann tensor

$$R_{ij} = R^k_{ikj}. \tag{3.5}$$

It can be shown that it is symmetric, $R_{ij} = R_{ji}$. The Einstein tensor is defined by

$$G_{ij} = R_{ij} - \frac{1}{2} g_{ij} R \tag{3.6}$$

where the scalar $R = g^{ij} R_{ij}$.

With these ingredients we proceed to the application of the Riemannian geometry to physics.

3.2 General Relativity

As discussed in the first chapter the general theory of relativity makes the following assumptions about the nature of spacetime and gravitation.

Spacetime Continuum

Spacetime is a four-dimensional continuum whose points ("events") can be specified by systems of coordinates x^μ, x'^μ etc. where $\mu = 0, 1, 2, 3$ and one of these is a time-like coordinate.

The infinitesimal "interval" between two neighbouring points is given (in the coordinate system x^μ for example) by

$$ds^2 = g_{\mu\nu}(x)dx^\mu dx^\nu. \tag{3.7}$$

The quantities $g_{\mu\nu}(x) = g_{\nu\mu}(x)$ can be treated as elements of a symmetric 4×4 matrix depending on coordinates in general. These fundamental quantities are called components of the **metric tensor**. Because of symmetry there are only ten independent quantities.

The Gravitational Field

The gravitational field is determined by the $g_{\mu\nu}$ and quantities derived from these. In a certain definite sense the metric tensor represents ten gravitational "potentials" in place of the one Newtonian potential Φ, and $\Gamma^\mu_{\nu\sigma}$ represent the forty gravitational "forces".

Minkowski Space

In the absence of gravitation, coordinates x^μ can be chosen such that $g_{\mu\nu}$ are equal to $\eta_{\mu\nu}$ (with $\eta_{00} = -1, \eta_{11} = \eta_{22} = \eta_{33} = 1$, all others zero). In this case $x^0 = ct$ where t is the time coordinate, and x^1, x^2, x^3 are the rectangular cartesian coordinates:

$$(ds)^2 = -c^2(dt)^2 + (dx^1)^2 + (dx^2)^2 + (dx^3)^2. \tag{3.8}$$

One can also choose r, θ, ϕ coordinates for the spatial part $(x^1, x^2, x^3) = (r \sin\theta \cos\phi, r \sin\theta \sin\phi, r \cos\theta)$ so that

$$(ds)^2 = -c^2(dt)^2 + (dr)^2 + r^2(d\theta)^2 + r^2 \sin^2\theta(d\phi)^2. \tag{3.9}$$

The Einstein Equation

Matter distribution is given by stress-energy tensor $T_{\mu\nu}$, (we will discuss it in a later chapter). It determines the gravitational field $g_{\mu\nu}$ through the **Einstein equation**

$$R_{\mu\nu} - \frac{1}{2}g_{\mu\nu}R = \frac{8\pi G}{c^4}T_{\mu\nu}. \tag{3.10}$$

In a *matter-free region* – *and we are interested in only such a region in Part I* – the Einstein equation further simplifies to

$$R_{\mu\nu} = 0 \tag{3.11}$$

because $T_{\mu\nu} = 0$ implies $g^{\mu\nu}(R_{\mu\nu} - g_{\mu\nu}R/2) = -R = 0$.

Calculation of $R_{\mu\nu}$

The Ricci tensor $R_{\mu\nu}$ is defined through the following steps:
   First we define $g^{\mu\nu}$ which, as a matrix, is inverse to the matrix $g_{\mu\nu}$.
   Secondly, we calculate the forty coefficients $\Gamma^\alpha_{\mu\nu} = \Gamma^\alpha_{\nu\mu}$ depending on derivatives of the metric

$$\Gamma^\alpha_{\mu\nu} = \frac{1}{2}g^{\alpha\beta}\left(\frac{\partial g_{\beta\mu}}{\partial x^\nu} + \frac{\partial g_{\beta\nu}}{\partial x^\mu} - \frac{\partial g_{\mu\nu}}{\partial x^\beta}\right). \tag{3.12}$$

There are forty coefficients because Γ's are symmetric in the two lower indices so that there are ten possibilities and the upper index can take all four values.
   In terms of these we then determine the **curvature tensor**

$$-R^\mu{}_{\nu\sigma\tau} = \Gamma^\mu_{\nu\sigma,\tau} - \Gamma^\mu_{\nu\tau,\sigma} + \Gamma^\alpha_{\nu\sigma}\Gamma^\mu_{\alpha\tau} - \Gamma^\alpha_{\nu\tau}\Gamma^\mu_{\alpha\sigma}. \tag{3.13}$$

The symmetric Ricci tensor is obtained by summing up the curvature tensor components as

$$
\begin{aligned}
R_{\nu\tau} &= R^\mu{}_{\nu\mu\tau} \tag{3.14}\\
&= R^0{}_{\nu0\tau} + R^1{}_{\nu1\tau} + R^2{}_{\nu2\tau} + R^3{}_{\nu3\tau}. \tag{3.15}
\end{aligned}
$$

We will see later that $R_{\mu\nu}$ has the following form.
   Let $g = \det g_{\mu\nu}$ be the determinant of the metric tensor. The determinant g is negative in general relativity just as it is in special relativity where $\det(\eta_{\mu\nu}) = -1$. Then,

$$-R_{\mu\nu} = (\ln\sqrt{-g})_{,\mu\nu} - \Gamma^\alpha_{\mu\nu,\alpha} + \Gamma^\beta_{\mu\alpha}\Gamma^\alpha_{\nu\beta} - (\ln\sqrt{-g})_{,\alpha}\Gamma^\alpha_{\mu\nu}. \tag{3.16}$$

3.3   Solving the Einstein Equation

The Einstein equations are solved for the metric tensor components. Usually, one can say something about the $g_{\mu\nu}$ already before attempting a solution, using, for example, the symmetry of the gravitational field in a particular set of coordinates. Symmetries reduce the number of non-zero independent components from ten to a smaller number. For example, in the static, spherically symmetric field in vacuum discussed in this chapter, there are just two unknown functions to be solved for from the Einstein equation.

3.4 Particle Trajectories

After we have solved the equations $R_{\mu\nu} = 0$ for the unknown $g_{\mu\nu}$, the motion of a body in the gravitational field is governed by the geodesic equation

$$\frac{d^2x^\mu}{d\tau^2} + \Gamma^\mu_{\nu\sigma} \frac{dx^\nu}{d\tau} \frac{dx^\sigma}{d\tau} = 0 \tag{3.17}$$

where τ is the **proper time** along the trajectory measured from some point on it. Let $x^\mu(\lambda)$ be the trajectory of the particle, parametrised by variable λ. Then

$$\tau(\lambda) = \frac{1}{c} \int^\lambda \left(-g_{\mu\nu} \frac{dx^\mu}{d\lambda_1} \frac{dx^\nu}{d\lambda_1} \right)^{1/2} d\lambda_1. \tag{3.18}$$

The negative sign occurs because the velocity $dx^\mu/d\tau$ is a time-like vector with negative norm squared.

3.5 Path of Light Rays

For light-rays the proper time along the path is zero: $g_{\mu\nu} dx^\mu dx^\nu = 0$. In this case the trajectory is written in terms of a parameter λ,

$$\frac{d^2x^\mu}{d\lambda^2} + \Gamma^\mu_{\nu\sigma} \frac{dx^\nu}{d\lambda} \frac{dx^\sigma}{d\lambda} = 0 \tag{3.19}$$

and we use the additional constraint that

$$g_{\mu\nu} \frac{dx^\mu}{d\lambda} \frac{dx^\nu}{d\lambda} = 0. \tag{3.20}$$

3.6 Weak Field and Newtonian Limit

A particle moves in a static and weak gravitational field given by $g_{\mu\nu}$. This means that these $g_{\mu\nu}$ are independent of time t and they are close to their Minkowskian values. We also assume that the velocity of the particle which follows the geodesic is very small compared to light velocity c.

In this limit we expect the equations of motion of the particle to reduce to their non-relativistic, Newtonian form.

Minkowski Background

When the field is weak, Cartesian coordinates can be chosen, $x^\mu = \{x^0 = ct, x^i\}$, and the metric tensor is only slightly different from the Minkowski tensor $\eta_{\mu\nu}$. We can therefore write

$$g_{\mu\nu} = \eta_{\mu\nu} + h_{\mu\nu} \tag{3.21}$$

where $h_{\mu\nu}$ is small compared to $\eta_{\mu\nu}$. Also, $h_{\mu\nu}$ should go to zero in an asymptotic region, far away from gravitating bodies, where the metric is purely Minkowskian.

The inverse matrix $g^{\mu\nu}$ is also close to the Minkowski values if $g^{\mu\nu} = \eta^{\mu\nu} + k^{\mu\nu}$ where $k^{\mu\nu}$ are of the order of $h_{\mu\nu}$ and $\eta^{\mu\nu}$ is the inverse of $\eta_{\mu\nu}$ as a matrix. This is so because by definition $g^{\mu\nu}g_{\nu\sigma} = \delta^{\mu}_{\sigma}$.

In the non-relativistic limit we should obtain the Newtonian equation

$$\frac{d^2\mathbf{x^i}}{dt^2} = -\frac{\partial}{\partial x^i}\Phi(x) \tag{3.22}$$

where $\Phi(x)$ is the Newtonian gravitational potential.

The equation of the trajectory for space coordinates x^i is

$$\frac{d^2x^i}{d\tau^2} + \Gamma^i_{\mu\nu}\frac{dx^\mu}{d\tau}\frac{dx^\nu}{d\tau} = 0. \tag{3.23}$$

In the non-relativistic limit $d\tau = dt\sqrt{1 - \mathbf{v}^2/c^2} \sim dt$. This means that $dx^0/d\tau \sim c$ is much larger than the velocities $v^i \sim dx^i/d\tau$. The dominant term in this equation is therefore

$$\frac{d^2x^i}{dt^2} + \Gamma^i_{00}c^2 = 0. \tag{3.24}$$

Because the gravitational field (given here by the metric tensor) is static, all time derivatives are zero and the connection coefficient

$$\Gamma^i_{00} = \frac{1}{2}\eta^{ij}(2h_{j0,0} - h_{00,j}) \sim -\frac{1}{2}h_{00,i}. \tag{3.25}$$

This when compared with the Newtonian equation gives $h_{00} = -2\Phi/c^2$ and

$$g_{00} = -\left(1 + \frac{2\Phi}{c^2}\right). \tag{3.26}$$

Slowing Down of Clocks

The non-relativistic formula obtained in the last section for the 00-component of the metric gives us a result which was first obtained by Einstein using the Equivalence Principle as explained in Chapter 1.

It is important to remember that the time coordinate t is just a way to label the spacetime events. Events which have the same value of t are simultaneous according to the coordinate system x. On the other hand the time shown by a clock following a trajectory is the proper time $\int d\tau = \int \sqrt{-g_{\mu\nu}dx^\mu dx^\nu}/c$ integrated along the trajectory.

In order to compare times shown by two clocks following two different trajectories we choose the same "coordinate time", say, $t = 0$ and note down their readings at the two places where the clocks are present. Then again, at coordinate time t we record their readings at their new positions. These four observations can then be compared.

For stationary clocks, that is clocks not changing their positions, the trajectory is simply $x^i = $ constant. Therefore

$$\int d\tau = \int \sqrt{-g_{00}(x)}dt.$$

As position coordinates, x^i are constant along the trajectory

$$\int d\tau = \sqrt{-g_{00}(x)} \int dt.$$

Therefore the ratio of time intervals $T_x = \int d\tau$ shown by a clock sitting at a point x to $T_y = \int d\tau$ shown by a stationary clock at y for the same interval of "coordinate time" from 0 to t is

$$\frac{T_x}{T_y} = \frac{\sqrt{-g_{00}(x)}}{\sqrt{-g_{00}(y)}} = \frac{(1 + 2\Phi(x)/c^2)^{\frac{1}{2}}}{(1 + 2\Phi(y)/c^2)^{\frac{1}{2}}}. \tag{3.27}$$

This is the same result we derived following Einstein's 1907 argument. Clocks slow down at places where gravitational potential is smaller.

In particular, for a massive spherical body of mass M the Newtonian potential outside the body is $\Phi(r) = -GM/r$ where r is the distance from the center of the gravitating body. Therefore clocks slow down by a factor $\sqrt{1 - 2GM/rc^2}$ compared to clocks very far away, $r \to \infty$ where $\Phi = 0$.

An immediate consequence of this is the **gravitational red shift**. An atom emitting a spectral line of a given frequency is like a clock. A stationary atom on the surface of the Sun, for example, will seem to emit light of lower frequency compared to an identical atom far away (say near earth if the gravitational potential due to earth's gravity on the surface of the earth can be neglected in comparison to the gravitational potential due to the Sun on the surface of the Sun). We have,

$$\omega_* = \left(1 - \frac{2GM}{rc^2}\right)^{\frac{1}{2}} \omega \approx \omega - \frac{\omega GM}{rc^2}. \tag{3.28}$$

This equation has a simple interpretation in quantum theory. A photon with frequency ω has energy $\hbar\omega$ equivalent to a mass $\hbar\omega/c^2$. It loses energy equal to the gravitational potential energy $(\hbar\omega/c^2)GM/r$ in going from r to infinity, becoming a photon of lower energy $\hbar\omega_*$.

3.7 Tutorial on Indexed Quantities

The full mathematical development of vectors and tensors will be done from Chapters 5 onwards. Presently, we can do a number of exercises, treating these as **indexed quantities** like vectors or matrices.

Exercise 21. A_{ij} and B^{ij} .$i, j = 1, \ldots, n$ are two matrices, with A a symmetric matrix $A_{ij} = A_{ji}$ and B antisymmetric $B^{ij} = -B^{ji}$. What is the number of independent elements in A and B? Show that the sum $A_{ij}B^{ij}$ is zero.

Answer 21. A has $n(n+1)/2$ independent elements and B has $n(n-1)/2$. Interchange indices i and j in the sum and use symmetry and antisymmetry properties.

Exercise 22. Two three-indexed quantities are given: ω_{abc} and f_{abc}, $a, b = 1, \ldots, n$. The quantities ω are antisymmetric in the first two indices $\omega_{abc} = -\omega_{bac}$ and f's are antisymmetric the last two $f_{abc} = -f_{acb}$. How many independent elements are there in each set? Given that f is related to ω by

$$f_{abc} = \omega_{abc} - \omega_{acb},$$

solve ω's in terms of f's.

Answer 22. There are $n^2(n-1)/2$ independent quantities in both. There are as many independent linear equations for ω's. These quantities appear in the discussion of Ricci rotation coefficients in section 9.7.2, where the answer is given.

Exercise 23. The Riemann tensor is a four-indexed quantity with the following properties $(i, j, k, l = 1, \ldots, n)$

$$
\begin{aligned}
R_{ijkl} &= -R_{ijlk} & \text{antisymmetry in last two indices,} \\
R_{ijkl} &= -R_{jikl} & \text{antisymmetry in first two indices,} \\
R_{ijkl} &= R_{klij} & \text{symmetry in the composite index } (ij) \text{ and } (kl), \\
R_{ijkl} &+ R_{iklj} + R_{iljk} = 0 & \text{cyclic sum in last three indices.}
\end{aligned}
$$

If there were no symmetries then there would be n^4 independent elements in R_{ijkl}. Show that the number of independent elements reduces to $n^2(n^2-1)/12$ due to these symmetries. There are 20 components in $n = 4$ dimensions, six in three and just one, R_{1212}, in two dimensions.

Answer 23. Divide the set of four indices i, j, k, l into three classes or subsets.

The first class is of those indices in which there are only two distinct indices. These can involve only elements of type R_{ijij}, and there are $n(n-1)/2$ members in this class because of antisymmetry in i and j.

In the second class there are three distinct indices. These elements can only be of type R_{ikjk}. k can take n values and for each of these i, j can take two distinct values from the remaining $n-1$ possibilities. We limit to $i < j$ because the elements with $i > j$ are related to these by interchange of the pairs ik and jk. Thus there are $n(n-1)(n-2)/2$ such independent elements.

The cyclic property in the last three indices is trivial for these two classes and does not reduce the number of elements.

In the third class are elements for which all four indices are different. We can arrange these as R_{ijkl} with $i < j, k < l, i < k$ because all other choices can be related to these by symmetry properties.

There are $n(n-1)/2$ pairs $i < j$ and $(n-2)(n-3)/2$ pairs $k < l$. Thus there are $n(n-1)(n-2)(n-3)/4$ possibilities but these have to be reduced to half (a factor $1/2$)

to only keep elements with $i < k$. The cyclic identities are all non-trivial for this class and can be written (to bring the indices so that $i < k < l, i < j$)

$$R_{ijkl} - R_{ikjl} + R_{iljk} = 0.$$

This further reduces the independent elements of this class by a factor $2/3$ because each cyclic relation allows one of the three elements to be expressed in terms of the other two. The total number of independent components of the Riemann tensor is

$$\frac{n(n-1)}{2} + \frac{n(n-1)(n-2)}{2} + \frac{2}{3} \cdot \frac{1}{2} \cdot \frac{n(n-1)(n-2)(n-3)}{4} = \frac{n^2(n^2-1)}{12}.$$

Chapter 4

Spherically Symmetric Gravitational Field

4.1 The Schwarzschild Solution

Schwarzschild's solution is an **exact solution** of the Einstein equation in empty spacetime which is static and spherically symmetric. It is also the physically most important of the few known exact solutions.

We know we can solve for the Coulomb field of a point charge, or a spherically symmetric static charge distribution at those points where there is no charge by solving the Poisson equation $\nabla^2\phi = 0$. Similarly, Einstein's equation can be solved in the matter-free outer region while the static matter distribution is spherically symmetric and located around the origin. Actually, one can show that even if there is time dependence in matter distribution, but provided it remains spherically symmetric all the time, the gravitational field represented by the metric $g_{\mu\nu}$ in the matter-free region is nevertheless static and given by the Schwarzschild form. This result is known as **Birkhoff's theorem**.

The assumption of a **static** and **spherically symmetric** solution means there exist coordinates $x^0 = ct, x^1 = r, x^2 = \theta, x^3 = \phi$ so that the metric is of the form

$$ds^2 = -a(r)c^2dt^2 + b(r)dr^2 + r^2d\theta^2 + r^2\sin^2\theta d\phi^2 \tag{4.1}$$

where the angular part of the metric is the same as the gravitation-free form because of the spherical symmetry, and $a(r)$ and $b(r)$ are functions of r only. This is enormous simplification as only four diagonal components of the matrix $g_{\mu\nu}$ are non-zero (in the general case there are ten) and there are essentially only two unknown functions of a single variable. So we expect ordinary rather than partial differential equations. As a consequence of time independence and spherical symmetry, metric coefficients are independent of both $x^0 = ct$ and $x^3 =$

ϕ. Therefore, we expect two conserved quantities, related to energy and angular momentum, just as in Newtonian gravity with spherical symmetry.

Starting with this assumed form of $g_{\mu\nu}$ one can calculate $\Gamma^{\mu}_{\nu\sigma}$, and then $R_{\mu\nu}$. We can then write Einstein's equations

$$R_{\mu\nu} = 0$$

and solve for the unknown functions $a(r)$ and $b(r)$. The solution, first found by K. Schwarzschild in 1916 is

$$a(r) = \frac{1}{b(r)} = \left(1 - \frac{2GM}{rc^2}\right) \equiv 1 - \frac{R_S}{r} \tag{4.2}$$

where M is the total mass of the matter distribution as determined by the Newtonian potential $-GM/r$ for $r \to \infty$. The quantity R_S given by

$$R_S = \frac{2GM}{c^2} \tag{4.3}$$

has dimensions of length and is called the **Schwarzschild radius** of the gravitating body.

Because we will refer to it often let us write out the solution in its full traditional form once again:

$$\begin{aligned} ds^2 &= -\left(1 - \frac{2GM}{rc^2}\right)c^2 dt^2 + \left(1 - \frac{2GM}{rc^2}\right)^{-1} dr^2 \\ &\quad + r^2 d\theta^2 + r^2 \sin^2\theta d\phi^2. \end{aligned} \tag{4.4}$$

There are in general forty independent connection coefficients $\Gamma^{\mu}_{\nu\sigma}$ Of these forty, only nine are non-zero. They are listed below for later reference:

$$\begin{aligned} \Gamma^0_{01} = \Gamma^0_{10} &= \frac{1}{2}\left(1 - \frac{R_S}{r}\right)^{-1}\frac{R_S}{r^2}, \\ \Gamma^1_{00} &= \frac{1}{2}\left(1 - \frac{R_S}{r}\right)\frac{R_S}{r^2}, \\ \Gamma^1_{11} &= -\frac{1}{2}\left(1 - \frac{R_S}{r}\right)^{-1}\frac{R_S}{r^2}, \\ \Gamma^1_{22} &= -r\left(1 - \frac{R_S}{r}\right), \\ \Gamma^1_{33} &= -r\sin^2\theta\left(1 - \frac{R_S}{r}\right), \\ \Gamma^2_{12} = \Gamma^2_{21} &= 1/r, \\ \Gamma^2_{33} &= -\sin\theta\cos\theta, \\ \Gamma^3_{23} = \Gamma^3_{32} &= \cot\theta, \\ \Gamma^3_{13} = \Gamma^3_{31} &= 1/r. \end{aligned} \tag{4.5}$$

The derivation of the Schwarzschild solution is in the tutorial at the end of this chapter.

4.2 Conserved Quantities

Calculations for trajectories are greatly simplified due to existence of conserved quantities just as in Newtonian gravity.

We use the following result. If all the components $g_{\mu\nu}$ are independent of a certain coordinate, say x^0, then $g_{0\mu}dx^\mu/d\tau$ is constant *along the geodesic trajectory*.

$$\frac{d}{d\tau}\left(g_{0\mu}\frac{dx^\mu}{d\tau}\right) = 0. \tag{4.6}$$

Proof:

$$\frac{d}{d\tau}\left[g_{0\mu}\frac{dx^\mu}{d\tau}\right] = \frac{dx^\nu}{d\tau}\frac{\partial g_{0\mu}}{\partial x^\nu}\frac{dx^\mu}{d\tau} + g_{0\mu}\frac{d^2x^\mu}{d\tau^2}.$$

Substituting the second derivative from the equation of the geodesic,

$$\frac{d}{d\tau}\left[g_{0\mu}\frac{dx^\mu}{d\tau}\right] = g_{0\sigma,\nu}\frac{dx^\nu}{d\tau}\frac{dx^\sigma}{d\tau} - g_{0\mu}\Gamma^\mu_{\nu\sigma}\frac{dx^\nu}{d\tau}\frac{dx^\sigma}{d\tau}.$$

The coefficient of $g_{0\sigma,\nu}$ is symmetric in ν and σ with summation over both these indices and we write $g_{0\sigma,\nu}$ as the sum of symmetric and antisymmetric parts, that is,

$$g_{0\sigma,\nu} = \frac{1}{2}(g_{0\sigma,\nu} + g_{0\nu,\sigma}) + \frac{1}{2}(g_{0\sigma,\nu} - g_{0\nu,\sigma}).$$

The antisymmetric part when multiplied by $(dx^\nu/d\tau)(dx^\sigma/d\tau)$ and summed gives zero. So using the definition of Γ,

$$\begin{aligned}
\frac{d}{d\tau}\left[g_{0\mu}\frac{dx^\mu}{d\tau}\right] &= \frac{1}{2}(g_{0\sigma,\nu} + g_{0\nu,\sigma})\frac{dx^\nu}{d\tau}\frac{dx^\sigma}{d\tau} - g_{0\mu}\Gamma^\mu_{\nu\sigma}\frac{dx^\nu}{d\tau}\frac{dx^\sigma}{d\tau} \\
&= \left[\frac{1}{2}(g_{0\sigma,\nu} + g_{0\nu,\sigma}) - \frac{1}{2}\{g_{0\sigma,\nu} + g_{0\nu,\sigma} - g_{\sigma\nu,0}\}\right]\frac{dx^\nu}{d\tau}\frac{dx^\sigma}{d\tau} \\
&= \frac{1}{2}[g_{\nu\sigma,0}]\frac{dx^\nu}{d\tau}\frac{dx^\sigma}{d\tau} \\
&= 0
\end{aligned}$$

because all the metric components are independent of x^0.

If there is another coordinate, say x^3, which also does not show up in the expressions for the metric tensor components, then a similar equation

$$\frac{d}{d\tau}\left(g_{3\mu}\frac{dx^\mu}{d\tau}\right) = 0 \tag{4.7}$$

holds as well.

4.3 Planetary Motion

We now work out the consequences of the Schwarzschild metric. The g_{00} in this exact case is identical to the Newtonian limit $-(1 + 2\Phi/c^2)$ which is a remarkable coincidence.

Trajectories of bodies under the influence of a gravitational field produced by the spherically symmetric mass distribution is determined from the geodesic equation

$$\frac{d^2 x^\mu}{d\tau^2} + \Gamma^\mu_{\nu\sigma} \frac{dx^\nu}{d\tau} \frac{dx^\sigma}{d\tau} = 0$$

where Γ's are as given above.

4.3.1 Two Conserved Quantities

In the present case we know that the metric $g_{\mu\nu}$ does not depend on coordinates $x^0 = ct$ and $x^3 = \phi$. Accordingly there are two conserved quantities. The two conserved quantities are

$$C_0 = g_{00} \frac{dx^0}{d\tau} = -\left(1 - \frac{2GM}{rc^2}\right) c \frac{dt}{d\tau} \tag{4.8}$$

and

$$C_3 = g_{03} \frac{dx^3}{d\tau} = r^2 \sin^2\theta \frac{d\phi}{d\tau}. \tag{4.9}$$

The quantities C_0 and C_3 are related to energy and angular momentum. To see this take the non-relativistic limit for $\theta = \pi/2 = \text{const}$,

$$d\tau = dt \left[\left(1 - \frac{2GM}{rc^2}\right) - \frac{1}{c^2}\left(1 - \frac{2GM}{rc^2}\right)^{-1}\left(\frac{dr}{dt}\right)^2 - \frac{r^2}{c^2}\left(\frac{d\phi}{dt}\right)^2\right]^{1/2}.$$

Therefore,

$$\frac{dt}{d\tau} = 1 + \frac{GM}{rc^2} + \frac{1}{2c^2}\left(\frac{dr}{dt}\right)^2 + \frac{r^2}{2c^2}\left(\frac{d\phi}{dt}\right)^2 + \cdots$$

and so,

$$-cC_0 \approx c^2 + \frac{1}{2}\left(\frac{dr}{dt}\right)^2 + \frac{r^2}{2}\left(\frac{d\phi}{dt}\right)^2 - \frac{GM}{r}. \tag{4.10}$$

This expression, if multiplied by the mass m of a particle moving along the geodesic, gives the Newtonian expression of energy in radial coordinates added to the rest-mass energy mc^2. Similarly, multiplication by m of the non-relativistic limit of C_3 gives the expression for angular momentum as we see in the next section.

4.3.2 Motion in the Equatorial Plane

We show that geodesic motion, as in the Newtonian orbit, is confined to the plane $\theta = \pi/2$ if it is initially so confined.

The geodesic equation for $x^2 = \theta$ is

$$\frac{d^2\theta}{d\tau^2} + \frac{2}{r}\frac{d\theta}{d\tau}\frac{dr}{d\tau} - \sin\theta\cos\theta\left(\frac{d\phi}{d\tau}\right)^2 = 0.$$

Therefore, if $\theta = \pi/2$ and $d\theta/d\tau = 0$ at some value of τ, then $d^2\theta/d\tau^2 = 0$ and $\theta = \pi/2$ for all values of τ.

From now on we can restrict ourselves to the "equatorial plane" $\theta = \pi/2$. The conserved quantity

$$h = r^2\frac{d\phi}{d\tau}$$

is just the "**equal area in equal time**" law of Kepler, except that the time is proper time now.

We are looking at a two-dimensional motion with coordinates r, ϕ in the plane. The area law gives the dependence of ϕ on τ and the constancy of C_0 gives that of t on τ. We just need one more equation to completely determine the trajectory giving dependence of r on τ.

It is, of course, possible to write the geodesic equation for $x^1 = r$, but in practice it is much simpler to use the fact that the velocity vector to a trajectory parametrised by the proper time τ has norm $-c^2$,

$$\langle U, U\rangle \equiv g_{\mu\nu}\frac{dx^\mu}{d\tau}\frac{dx^\nu}{d\tau} = -c^2, \tag{4.11}$$

which gives us, substituting for $d\phi/d\tau$ and $dt/d\tau$ from the expressions for h and C_0,

$$\frac{1}{2}\left(\frac{dr}{d\tau}\right)^2 + V(r) = E \tag{4.12}$$

where

$$V(r) = -\frac{GM}{r} + \frac{h^2}{2r^2} - \frac{GM}{rc^2}\frac{h^2}{r^2} \tag{4.13}$$

and E is the constant

$$E = \frac{C_0^2 - c^2}{2}. \tag{4.14}$$

Note that this equation for $dr/d\tau$, written deliberately in the Newtonian form **is an exact result** in the general theory of relativity ! The contribution of general relativity (apart from replacing the time t by proper time τ) is the extra term GMh^2/r^3c^2 in $V(r)$. The equation (as far as dependence of r on proper time τ is concerned) is identical to that of a particle of unit mass in one dimension with total energy E in an "effective potential" $V(r)$.

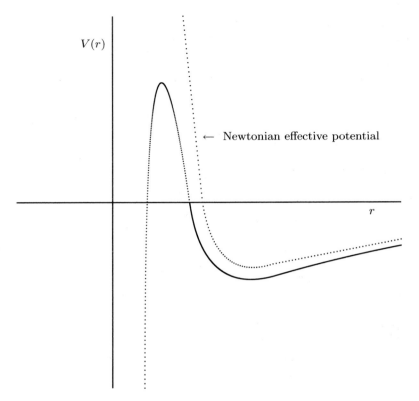

Fig. 4.1: Effective potential $V(r)$ for $h \neq 0$.
The Newtonian effective potential is shown by dotted lines.

4.3.3   Qualitative Features of Orbits

Let us analyse the effective potential $V(r)$ of the previous section for various values of E and h. We can immediately see the following qualitative features of the orbits.

1. For $h = 0$ the radial plunge (or escape) is just as it is for the Newtonian case because V is the same.

2. The general relativistic contribution proportional to $-1/r^3$ eventually dominates over the repulsive core $h^2/2r^2$ as $r \to 0$ and takes the potential back to $-\infty$ instead of $+\infty$ as in the Newtonian case. In general therefore there is both a minimum and a maximum in the effective potential.

3. If the value of E is such that r is at one of the extrema of $V(r)$ and $E = V(r)$ then $dr/d\tau = 0, r =$ constant, and there is a **circular** orbit. There is an **unstable** orbit corresponding to the maximum of V and a **stable** orbit at the minimum.

4. There is now a possibility of a radial plunge if $E > 0$ is greater than the maximum of V even for $h \neq 0$. This is unlike the Newtonian case.

5. For $h \neq 0$ and $E > 0$ but smaller than the maximum, the orbit comes from infinity, goes round the center and goes back to infinity.

6. For $E < 0$ the motion is restricted to within a maximum and a minimum value of r, much like a general central force. The orbits are rotating or *precessing* ellipses, as we shall see in the next section.

4.3.4 Precession of the Perihelion

The equation of the orbit can be obtained from the one-dimensional effective problem by differentiating with respect to τ and factoring out the $dr/d\tau$ factor

$$\frac{d}{d\tau}\left[\frac{1}{2}\left(\frac{dr}{d\tau}\right)^2 + V(r)\right] = 0 \tag{4.15}$$

giving

$$\frac{d^2 r}{d\tau^2} + V'(r) = 0. \tag{4.16}$$

We can convert the derivatives with respect to τ into derivatives with respect to ϕ by using

$$\frac{d}{d\tau} = \frac{d\phi}{d\tau}\frac{d}{d\phi} = \frac{h}{r^2}\frac{d}{d\phi}. \tag{4.17}$$

Finally we can change to the variable $u = 1/r$. The result is

$$\frac{d^2 u}{d\phi^2} + u = \frac{GM}{h^2} + \frac{3GM}{c^2}u^2. \tag{4.18}$$

Compare this equation with the equation for the Keplerian elliptical orbit.

Our aim is to solve this equation approximately in order to estimate the contribution by the second term on the right-hand side. We follow the treatment of P. G. Bergmann.

Recall (from the Tutorial in Chapter 1) that if this term was absent we will get the fixed ellipse

$$u = \frac{GM}{h^2}(1 + \epsilon \cos \phi). \tag{4.19}$$

If we plot u as a function of ϕ the graph will look like a ripple with a period exactly equal to 2π.

With the perturbing general relativistic term $3GMu^2/c^2$ present we still expect the solution to be a *periodic function* although the periodicity may not be 2π. The excess (or deficiency) from 2π of the interval between values of ϕ at which $du/d\phi = 0$ in two successive occurences is the amount by which the perihelion shifts per revolution.

Let us write the equation as

$$\frac{d^2u}{d\phi^2} + u = \frac{GM}{h^2} + \lambda u^2 \qquad (4.20)$$

treating $\lambda = 3GM/c^2$ as a small parameter. The equation is symmetric with respect to $\phi \leftrightarrow -\phi$ so that $u(\phi)$ is an even periodic function. Let the frequency be ω so that we can write the function as

$$u = a_0 + a_1 \cos(\omega\phi) + a_2 \cos(2\omega\phi) + \cdots . \qquad (4.21)$$

Here a_n's are all functions of the parameter λ. As $\lambda \to 0$ we must recover the ellipse $u = GM(1 + \epsilon \cos\phi)/h^2$, therefore

$$\omega \to 1, \qquad a_0(\lambda) \to \frac{GM}{h^2}, \qquad a_1 \to \frac{GM}{h^2}\epsilon, \quad a_n \to 0, n = 2, \ldots . \qquad (4.22)$$

The important point here is that all higher coefficients $a_n, n = 2, \ldots$ are of order λ or higher.

Substituting this ansatz in the equation we get, keeping to first order terms,

$$\begin{aligned} a_0 + a_1(1 &- \omega^2)\cos(\omega\phi) + a_2(1 - 4\omega^2)\cos(2\omega\phi) + \cdots \\ &= GM/h^2 + \lambda(a_0 + a_1\cos(\omega\phi) + O(\lambda))^2 \\ &= GM/h^2 + \lambda(a_0^2 + a_1^2\cos^2(\omega\phi) + 2a_0a_1\cos(\omega\phi)) + \cdots . \end{aligned}$$

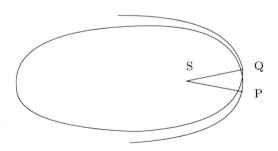

Fig. 4.2: Precession of the perihelion.

The elliptical path of the planet in Newtonian gravity is replaced by a *precessing* ellipse. The perihelion position is shifted from P to Q.

Comparing the coefficients of $\cos(\omega\phi)$ terms on the two sides, (the presence of $\cos^2\omega\phi$ is irrelevant as it is equal to $(1 + \cos 2\omega\phi)/2$),

$$1 - \omega^2 = 2\lambda a_0 \approx 2\frac{3GM}{c^2} \cdot \frac{GM}{h^2}. \tag{4.23}$$

This determines the periodicity to be

$$\frac{2\pi}{\omega} = 2\pi \left(1 - \frac{6(GM)^2}{h^2 c^2}\right)^{-\frac{1}{2}} \approx 2\pi + \frac{6\pi(GM)^2}{h^2 c^2}. \tag{4.24}$$

The additional angle $6\pi(GM)^2/h^2 c^2$ has a positive sign. When a planet returns towards the original perihelion, it has to move this much angle beyond 2π to make the next perihelion. Here h^2/GM is the latus-rectum equal to $L = A(1 - \epsilon^2)$, where A is the semi-major axis and ϵ the eccentricity. So we can write the formula for precession as

$$\Delta\phi = \frac{6\pi GM}{Lc^2} \qquad \text{per revolution,} \tag{4.25}$$

which comes out to be $43''$ of arc per century for the planet Mercury with the smallest value of latus-rectum.

4.4 Deflection of Light in a Gravitational Field

The light-like geodesics in the Schwarzschild field can be discussed just like the time-like geodesics. The geodesics lie in the plane $\theta = \pi/2$ if initially they do so for the same reason.

We use an affine parameter λ to parametrize the trajectory $x^\mu(\lambda)$.

The conserved quantities corresponding to energy and angular momentum can again be written down,

$$C_0 = -\left(1 - \frac{2GM}{rc^2}\right)c\frac{dt}{d\lambda}, \tag{4.26}$$

$$h = r^2\frac{d\phi}{d\lambda}. \tag{4.27}$$

For the third equation for r we use the light-like nature of the tangent vector $\langle U, U \rangle = 0$,

$$-\left(1 - \frac{2GM}{rc^2}\right)c^2\left(\frac{dt}{d\lambda}\right)^2 + \left(1 - \frac{2GM}{rc^2}\right)^{-1}\left(\frac{dr}{d\lambda}\right)^2 + r^2\left(\frac{d\phi}{d\lambda}\right)^2 = 0.$$

Substituting for $dt/d\lambda$ and $d\phi/d\lambda$ from the first two equations we get, after changing to variable $u = 1/r$,

$$\left(\frac{du}{d\phi}\right)^2 = \frac{C_0^2}{h^2} - u^2 + \frac{2GM}{c^2}u^3. \tag{4.28}$$

To convert it into the (traditional) second-order equation of the path, differentiate once with respect to ϕ and cancel the $du/d\phi$ factor,

$$\frac{d^2u}{d\phi^2} + u = \frac{3GM}{c^2} u^2. \tag{4.29}$$

Compare this equation with that for the motion of a particle in the previous section. There is no constant GM/h^2 term on the right-hand side. In the absence of the general relativity term, the equation reduces to the Newtonian case

$$\frac{d^2u}{d\phi^2} + u = 0. \tag{4.30}$$

These 'zeroth-order' solutions are $u = (1/b)\cos\phi$ representing a straight line propagation of light passing at a distance b from the origin. In the first-order correction, we substitute the zeroth-order solution on the right-hand side,

$$\frac{d^2u}{d\phi^2} + u = \frac{3GM}{c^2b^2} \cos^2\phi. \tag{4.31}$$

This equation is linear in u. Writing $\cos^2\phi$ as $(1 + \cos(2\phi))/2$ on the right-hand side and noticing that if $w = \cos(2\phi)$ then $w'' = -4\cos(2\phi)$. So $u_1 = 3 - \cos(2\phi)$ satisfies $u_1'' + u_1 = 3(1 + \cos(2\phi))$. This gives (up to the appropriate constant) a particular solution to our equation. The complete solution is therefore

$$u = \frac{1}{b}\cos\phi + \frac{GM}{2c^2b^2}\left(3 - \cos(2\phi)\right). \tag{4.32}$$

If we convert the trajectory equation into cartesian coordinates $(x = r\cos\phi, y = r\sin\phi)$ it is

$$x = b - \frac{R_S}{2b}\frac{x^2 + 2y^2}{\sqrt{x^2 + y^2}}. \tag{4.33}$$

The path of a light ray looks like a straight line far away from a gravitating body both above and below the x-axis $(|y| >> x)$,

$$x = b - \frac{R_S}{b}|y|. \tag{4.34}$$

The angle between the asymptotes is the angle by which the deflection takes place:

$$\delta = \frac{2R_S}{b}. \tag{4.35}$$

The distance of closest approach of light ray BP to the massive body at the origin is obtained by setting $y = 0$ in the approximation to the path of the light ray. It is

$$\frac{b}{1 + R_s/2b} \approx b - \frac{R_S}{2}.$$

The impact parameter, that is the distance betweeen the straight line representing the incoming asymptotic path of the light ray and the line parallel to it passing from the center of the mass is

$$b \cos(\delta/2) = b \left(1 - \frac{R_S^2}{2b^2} \right) \approx b.$$

The deflection δ is approximately equal to $1.75''$ for rays from a distant star grazing the surface of the Sun so that b is the radius of the Sun (about 7×10^5 Km) and the Schwarzschild radius of the Sun is about 3 Km.

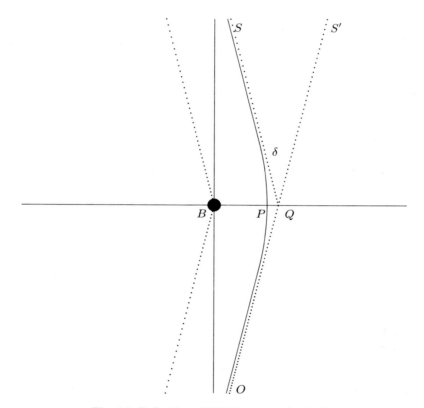

Fig. 4.3: Deflection of light by a massive body.

Light from a far-off source comes in the direction SQ and after deflection is seen by a distant observer O to come from the virtual image S'.

Our treatment of light deflection in this section follows Eddington's exposition in *The Mathematical Theory of Relativity*. We shall rederive this deflection formula in Chapter 13 for weak gravitational fields by a different method.

4.5   Gravitational Lensing

To a good approximation we can assume the bending of light to take place only in the neighbourhood of a massive body. The impact parameter and the closest distance are nearly the same. We can then approximate the light ray trajectory as made up of two straight lines corresponding to incoming and outgoing rays bending sharply at the point of intersection, much like the bending of a light beam in a thin lens.

Let there be a distant source of light (a galaxy for instance) at S in a straight line with the massive body B (which can be another galaxy) and the observer at O.

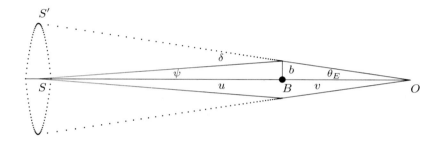

Fig. 4.4: Gravitational Lensing.

Light source(S), massive body(B) and observer(O) in a straight line. The virtual image of the source is a circle S' called the **Einstein Ring** with an angular diameter $2\theta_E$.

Light bends as shown in the figure. The observer O sees the virtual image of the source S at a point S' due to the deflection by an angle δ. Due to axial symmetry, light from all sides is deflected in the same manner and the virtual image of the source is a ring called the **Einstein ring** of angular diameter $2\theta_E$ as shown. The angle θ_E can be calculated as follows. Let u and v be the distances of the source and the observer respectively from the gravitating body B. Let b be the impact parameter, very nearly equal to the distance of the point from where incoming and outgoing rays intersect. From the geometry of the figure we see that $b = \theta_E v = \psi u$ and

$$\delta = \frac{2R_S}{b} = \psi + \theta_E = \frac{b}{u} + \frac{b}{v}$$

or,

$$b = \left(2R_S \frac{uv}{u+v}\right)^{1/2}.$$

The angular diameter of the Einstein ring is therefore $2\theta_E$ with

$$\theta_E = \left(2R_S \frac{u}{v(u+v)} \right)^{1/2}. \tag{4.36}$$

For a source slightly off-axis by an angle β, the circular symmetry is broken and there are only two images in the plane SBO as shown in Figure 4.5. We can easily calculate the angles $\theta_\pm$ of the two images

$$\theta_\pm = \pm\beta + \frac{\theta_E^2}{\theta_\pm}. \tag{4.37}$$

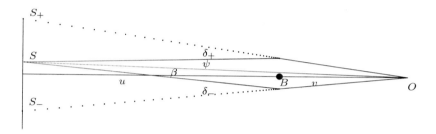

Fig. 4.5: Gravitational Lensing.

Source off the axis by an agle β. The virtual images are at angles $\theta_\pm = \angle BOS_\pm$.

4.6   Tutorial

The Connection Coefficients

Exercise 24. Calculate $g^{\mu\nu}$, $\ln\sqrt{-g}$ and the forty Γ's for the Schwarzschild form of metric.

Answer 24. With only the four diagonal components of the metric tensor non-zero,

$$g_{00} = -a(r), \qquad g_{11} = b(r), \qquad g_{22} = r^2, \qquad g_{33} = r^2\sin^2\theta,$$

$$g^{00} = -a(r)^{-1}, \quad g^{11} = b(r)^{-1}, \quad g^{22} = r^{-2}, \quad g^{33} = r^{-2}(\sin\theta)^{-2},$$

"log-root-minus g" is

$$\ln\sqrt{-g} = \frac{1}{2}(\ln a + \ln b) + 2\ln r + \ln\sin\theta. \tag{4.38}$$

With these the nine non-zero independent connection coefficients can be calculated. We use the convenient abbreviation of a prime to denote differentiation with respect to r, for example $f' = df/dr$ when f is a function of r.

$$\Gamma^0_{01} = \Gamma^0_{10} = (\ln a)'/2,$$

$$\Gamma^1_{00} = a'/2b, \qquad \Gamma^1_{11} = (\ln b)'/2, \qquad \Gamma^1_{22} = -r/b, \quad \Gamma^1_{33} = -r\sin^2\theta/b,$$

$$\Gamma^2_{12} = \Gamma^2_{21} = 1/r, \qquad \Gamma^2_{33} = -\sin\theta\cos\theta,$$

$$\Gamma^3_{23} = \Gamma^3_{32} = \cot\theta, \qquad \Gamma^3_{13} = \Gamma^3_{31} = 1/r.$$

Exercise 25. Calculate the Ricci tensor for the Schwarzschild form of metric.

Answer 25. Frequent occurence of $\ln a$ and $\ln b$ prompts us to define

$$a = e^A, \qquad b = e^B. \tag{4.39}$$

The Ricci tensor $R_{\mu\nu}$ is

$$-R_{\mu\nu} = (\ln\sqrt{-g})_{,\mu\nu} - \Gamma^\alpha_{\mu\nu,\alpha} + \Gamma^\beta_{\mu\alpha}\Gamma^\alpha_{\nu\beta} - (\ln\sqrt{-g})_{,\alpha}\Gamma^\alpha_{\mu\nu}. \tag{4.40}$$

Different parts of the Ricci tensor, after a very short calculation, can be written as a matrix (*only half-filled because it is symmetric*)

$$(\ln\sqrt{-g})_{,\mu\nu} = \begin{bmatrix} 0 & 0 & 0 & 0 \\ & (A''+B'')/2 - 2/r^2 & 0 & 0 \\ & & -1/\sin^2\theta & 0 \\ & & & 0 \end{bmatrix}.$$

Similarly,

$$\Gamma^\alpha_{\mu\nu,\alpha} = \begin{bmatrix} \Gamma^1_{00,1} & 0 & 0 & 0 \\ & \Gamma^1_{11,1} & 0 & 0 \\ & & \Gamma^1_{22,1} & 0 \\ & & & \Gamma^1_{33,1} + \Gamma^2_{33,2} \end{bmatrix}$$

and,

$$\Gamma^\beta_{\mu\alpha}\Gamma^\alpha_{\nu\beta} = \begin{bmatrix} 2\Gamma^0_{01}\Gamma^1_{00} & 0 & 0 & 0 \\ & \sum_{\mu=0}^3 (\Gamma^\mu_{1\mu})^2 & \Gamma^3_{13}\Gamma^3_{23} & 0 \\ & & 2\Gamma^2_{21}\Gamma^1_{22} + (\Gamma^3_{23})^2 & 0 \\ & & & 2(\Gamma^3_{31}\Gamma^1_{33} + \Gamma^3_{32}\Gamma^2_{33}) \end{bmatrix}$$

as well as

$$(\ln\sqrt{-g})_{,\alpha}\Gamma^\alpha_{\nu\beta} = \begin{bmatrix} \gamma^1_{00} & 0 & 0 & 0 \\ & \gamma^1_{11} & \gamma^2_{12} & 0 \\ & & \gamma^1_{22} & 0 \\ & & & \gamma^1_{33} + \gamma^2_{33} \end{bmatrix}$$

where

$$\gamma^1_{\mu\nu} = (\ln\sqrt{-g})_{,1}\Gamma^1_{\mu\nu} = \left(\frac{1}{2}(A'+B') + \frac{2}{r}\right)\Gamma^1_{\mu\nu} \tag{4.41}$$

$$\gamma^2_{\mu\nu} = (\ln\sqrt{-g})_{,2}\Gamma^2_{\mu\nu} = \cot\theta\,\Gamma^2_{\mu\nu}. \tag{4.42}$$

Collecting the terms we calculate

$$-R_{00} = -\frac{a}{2b}\left[A'' + \frac{1}{2}A'(A' - B') + \frac{2A'}{r}\right], \tag{4.43}$$

$$-R_{11} = \frac{1}{2}\left[A'' + \frac{1}{2}A'(A' - B') - \frac{2B'}{r}\right], \tag{4.44}$$

$$-R_{22} = \frac{1}{b}\left[\frac{r}{2}(A' - B') + 1 - b\right], \tag{4.45}$$

$$-R_{33} = \frac{\sin^2\theta}{b}\left[\frac{r}{2}(A' - B') + 1 - b\right]. \tag{4.46}$$

The remaining $R_{\mu\nu}$ are zero.

Exercise 26. Solve the Einstein equation for vacuum for the Schwarzschild case.

Answer 26. As is obvious from the above expressions, the ten equations $R_{\mu\nu} = 0$ are actually four for the diagonal $\mu\nu = 00, 11, 22, 33$.

$$-\frac{a}{2b}\left[A'' + \frac{1}{2}A'(A' - B') + \frac{2A'}{r}\right] = 0, \tag{4.47}$$

$$\frac{1}{2}\left[A'' + \frac{1}{2}A'(A' - B') - \frac{2B'}{r}\right] = 0, \tag{4.48}$$

$$\frac{1}{b}\left[\frac{r}{2}(A' - B') + 1 - b\right] = 0, \tag{4.49}$$

$$\frac{\sin^2\theta}{b}\left[\frac{r}{2}(A' - B') + 1 - b\right] = 0. \tag{4.50}$$

The first two equations give $A' = -B'$ which means $ab =$const. As we expect a and b to have their Minkowski values $a = 1, b = 1$ far away from gravitating matter, as $r \to \infty$ the constant should be equal to 1. Putting $B' = -A'$ and $b = 1/a$ in the third equation (which is the same as the fourth) we get

$$r\frac{da}{dr} = 1 - a$$

which is readily integrated to

$$a(r) = 1 + \frac{\text{const}}{r}.$$

Again, for large values of r the Newtonian limit should prevail, for which we already know

$$g_{00} = -\left(1 + \frac{2\Phi}{c^2}\right) = -\left(1 - \frac{2GM}{rc^2}\right).$$

Therefore we get the values a and b as given.

Exercise 27. Plot the effective potential

$$V(r) = -\frac{GM}{r} + \frac{h^2}{2r^2} - \frac{GMh^2}{r^3c^2}$$

for different values of h. Show that a stable circular orbit (for which $dr/d\tau = 0$) occurs only if

$$h \geq \sqrt{12}\frac{GM}{c}$$

and then its radius is

$$R_{\text{circ.}} = \frac{h^2}{2GM}\left[1 + \left(1 - \frac{12G^2M^2}{h^2c^2}\right)^{1/2}\right].$$

Answer 27. Hint: The circular orbit (as in the Newtonian case) corresponds to the minimum of the effective potential.

Exercise 28. Show that for the stable circular orbit in a Schwarzschild field, the angular velocity $\Omega \equiv d\phi/dt$ (rate of change of ϕ with respect to the Schwarzschild coordinate time) is related to the radius R of the orbit by

$$\Omega^2 = \frac{GM}{R^3}.$$

Answer 28. Let us define the proper angular velocity (rate of change with respect to the *proper time*) to be ω,

$$\omega = \frac{d\phi}{d\tau}$$

so that $h = R^2\omega$. The radius of the stable circular orbit is given by the previous exercise as

$$R = R_{\text{circ.}} = \frac{h^2}{2GM}\left[1 + \left(1 - \frac{12G^2M^2}{h^2c^2}\right)^{1/2}\right].$$

We can solve for h^2 (or equivalently, ω^2 from here in terms of R),

$$\frac{GM}{\omega^2} = R^3\left(1 - \frac{3GM}{Rc^2}\right).$$

Now ω and Ω are related by $dt/d\tau$ which can be calculated from the expression for proper time element

$$-c^2(d\tau)^2 = -(1 - 2GM/Rc^2)c^2(dt)^2 + R^2(d\phi)^2$$

$(dr = 0$ and $\theta = \pi/2)$ we get

$$\frac{\omega^2}{\Omega^2} = \left(\frac{dt}{d\tau}\right)^2 = \frac{1 + R^2\omega^2/c^2}{1 - 2GM/Rc^2}.$$

Substituting the previous expression for ω^2 in terms of R here and simplifying we get $\Omega^2 = GM/R^3$. It is curious that this is just the formula for the frequency of the Newtonian circular orbit.

Part II

Geometry

Part II

Geometry

Chapter 5

Vectors and Tensors

We review the familiar notions of a vector space, the dual vector space, tensor and exterior products.

5.1 Vector Spaces

5.1.1 Vectors

A **real vector space** V is a set whose members, called **vectors**, have two operations defined on them. Two vectors can be added to give a vector called their sum, and, a vector can be multiplied by a real number to give a vector. A vector space has a special vector $\mathbf{0}$, called the **zero vector**, which has the property that when added to any vector of the space the sum is this latter vector.

Two (non-zero) vectors are called **linearly independent** if they are not proportional to each other, that is, one can not be written as a number times the other. More generally, a set of r non-zero vectors $\mathbf{v}_1, \ldots, \mathbf{v}_r$ is called linearly independent if none of them can be written as a **linear combination** of any, some or all the others. A linear combination of vectors is just the sum of these vectors after they have been multiplied by numbers.

If vectors $\mathbf{v}_1, \ldots, \mathbf{v}_r$ are linearly independent then $a_1\mathbf{v}_1 + \cdots + a_r\mathbf{v}_r = \mathbf{0}$ implies that $a_1 = \cdots = a_r = 0$ because if any of the a's fail to be zero, let us say, $a_k \neq 0$, then, by dividing by a_k in the equation we can express $\mathbf{v}_k$ in terms of the others.

Let there be a set of linearly independent vectors. Let us adjoin to this set a new non-zero vector, then the amended set can be linearly independent, or it may fail to be linearly independent. We can try to make a set which is the largest possible linearly independent set by adjoining only those vectors which make the larger set linearly independent.

We will deal only with those spaces, called **finite dimensional** vector spaces, in which this process of finding the largest set of linearly independent set comes to an end and we have a finite set $\mathbf{v}_1, \ldots, \mathbf{v}_n$ of linearly independent vectors. Any other vector of the space can then be written as a linear combination of these vectors. Whichever way we choose the set of linearly independent vectors, it always contains the same number n of vectors.

A set of vectors like these is called a **basis**. The number n of independent vectors is characterstic of the space and is called the **dimension** of the space.

5.1.2   Dual Vector Space

A **functional** α defined on a vector space V is a mapping which assigns to each vector $\mathbf{v} \in V$ a real number $\alpha(\mathbf{v})$. If this mapping satisfies the linearity property, that is for every $\mathbf{v}, \mathbf{u}$ in V and any real number a if

$$\alpha(\mathbf{v} + \mathbf{u}) = \alpha(\mathbf{v}) + \alpha(\mathbf{u}), \tag{5.1}$$

$$\alpha(a\mathbf{v}) = a\alpha(\mathbf{v}), \tag{5.2}$$

then we call α a **linear functional**.

Now consider the set V^* of all linear functionals on the space V. We can define addition and multiplication on linear functionals by

$$(\alpha + \beta)(\mathbf{v}) = \alpha(\mathbf{v}) + \beta(\mathbf{v}),$$

$$(a\alpha)(\mathbf{v}) = a\alpha(\mathbf{v}).$$

With these definitions V^* becomes a vector space whose zero vector is the linear functional which assigns number zero to each vector of V. The space V^* is called the vector space **dual** to V.

5.1.3   Change of Basis

Let V be a vector space of dimension n and let $E = \{\mathbf{e}_i\}_{i=1}^n$ be a basis in V. We write a vector $\mathbf{v} \in V$ in terms of the basis as $\mathbf{v} = x^i \mathbf{e}_i$ where there is a sum over i on the right-hand side from 1 to n. We shall use this tacit assumption of a sum over a repeated index without showing the sign of summation. This is called the **Einstein summation convention**. Any exception to the convention will be explicitly pointed out.

The numbers $x^i, i = 1, \ldots, n$ are called **components** of vector $\mathbf{v}$ with respect to basis E. We are using another convention by writing the components with a superscript.

Let $F = \{\mathbf{f}_i\}_{i=1}^n$ be another basis of V whose vectors can be expanded in terms of the old basis vectors as

$$\mathbf{f}_i = T_i{}^j \mathbf{e}_j. \tag{5.3}$$

The numbers $T_i{}^j$ can be treated as elements of a matrix, with row index i and column index j. It is an invertible matrix with the inverse matrix giving coefficients involved in expanding members of basis E in terms of those of F.

Let the components of a vector $\mathbf{v}$ with respect to basis F be y^i then

$$\mathbf{v} = y^i \mathbf{f}_i = y^i T_i{}^j \mathbf{e}_j = x^j \mathbf{e}_j.$$

Therefore,

$$y^k = x^j (T^{-1})_j{}^k = (T^{-1})^{Tk}{}_j x^j. \tag{5.4}$$

We state the above result as follows: when the basis changes from E to F by matrix T then the components of a vector change by the matrix $(T^{-1})^T$.

5.1.4 Dual Bases

Because of linearity a functional $\alpha \in V^*$ is defined completely if we give its values on a basis of V. As

$$\alpha(\mathbf{v}) = x^i \alpha(\mathbf{e}_i),$$

α is defined if we know the numbers $a_i = \alpha(\mathbf{e}_i)$. This gives us an idea how to define a basis in V^*. Let $\alpha^i \in V^*, i = 1, \ldots, n$ be defined as

$$\alpha^i(\mathbf{e}_j) = \delta^i_j \tag{5.5}$$

where the Kronecker delta δ^i_j has the value 1 if $i = j$ and 0 if $i \neq j$. Any vector $\alpha \in V^*$ can be written $\alpha = a_i \alpha^i$ where $a_i = \alpha(e_i)$, as can be checked by operating α on an arbitrary vector $\mathbf{v} = v^i \mathbf{e}_i$:

$$a_i \alpha^i(v^j \mathbf{e}_j) = a_i v^j \delta^i_j = a_i v^i = v^i \alpha(\mathbf{e}_i) = \alpha(\mathbf{v}).$$

This shows that $\{\alpha^i\}_{i=1}^n$ form a basis in V^*. The basis $A = \{\alpha^i\}_{i=1}^n$ is called the **basis dual** to the basis E. For every basis of V there is a dual basis in V^*.

5.1.5 Change of the Dual Basis

Let $B = \{\beta^i\}_{i=1}^n$ be dual to the basis F discussed above in section 5.1.3,

$$\beta^i(\mathbf{f}_j) = \delta^i_j, \qquad (\mathbf{f}_i = T_i{}^j \mathbf{e}_j),$$

then B is related to A by $(T^{-1})^T$,

$$\beta^i = (T^{-1})^{Ti}{}_k \alpha^k$$

because

$$\beta^i(\mathbf{f}_j) = (T^{-1})^{Ti}{}_k \alpha^k (T_j{}^l \mathbf{e}_l) = (T^{-1})^{Ti}{}_k T_j{}^k = \delta^i_j.$$

This means that if α is any general vector in V^* with components a_i in basis A: $\alpha = a_i \alpha^i$, then the components b_i with respect to basis B ($\alpha = b_i \beta^i$) will be related to a_i by the inverse transpose of $(T^{-1})^T$, that is by T itself.

5.1.6   Contra- and Co-Variant Vectors

The situation in the previous section is described by the following statement.

If we start with the space V and change basis in V by a matrix T, then components of a vector in V transform **contravariantly**, that is, by $(T^{-1})^T$ whereas the components of a vector in dual space V^* (due to corresponding changes in the dual bases) transform **covariantly**, that is, by T itself.

We emphasize that the vectors themselves do not change, *it is only their components that change when bases do.*

Notice the use of *superscripts* for components of vectors in V and *subscripts* for components of those in V^*. This is the standard convention of classical tensor analysis adopted by physicists.

As can be verified immediately the dual $(V^*)^*$ of V^* is V itself with the linear functional $\mathbf{v} \in V^{**} = V$ on V^* acting as

$$\mathbf{v}(\alpha) = \alpha(\mathbf{v}).$$

This makes the designation of contra- and covariant quantities a matter of convention. We must decide which our starting space V is. Then vectors of V will have components transforming contravariantly, and those of V^* will have components transforming covariantly.

In our application of vectors and tensors to differential geometry, fortunately there is a vector space singled out uniquely. That is the **tangent space** at any point of a differentiable manifold.

5.2   Tensor Product

Just as the set of linear functionals on a vector space V form a vector space V^*, the set of all *bilinear* functionals which map a pair of vectors of V into a real number form an n^2 dimensional space $V^* \otimes V^*$.

Let V be a vector space of dimension n.

Consider the Cartesian product set $V \times V$. This is the set whose members are ordered pairs like $(\mathbf{v}, \mathbf{u})$ of vectors $\mathbf{v}, \mathbf{u}$ of V. This cartesian product is *just a set* and not a vector space.

A **bilinear** functional t on $V \times V$ is a mapping which assigns to each pair $(\mathbf{v}, \mathbf{w}) \in V \times V$ a real number $t(\mathbf{v}, \mathbf{w})$ with the following properties

$$t(\mathbf{u} + \mathbf{v}, \mathbf{w}) = t(\mathbf{u}, \mathbf{w}) + t(\mathbf{v}, \mathbf{w}), \qquad t(a\mathbf{v}, \mathbf{w}) = at(\mathbf{v}, \mathbf{w}),$$

$$t(\mathbf{u}, \mathbf{v} + \mathbf{w}) = t(\mathbf{u}, \mathbf{v}) + t(\mathbf{u}, \mathbf{w}), \qquad t(\mathbf{v}, a\mathbf{w}) = at(\mathbf{v}, \mathbf{w}).$$

The set of all bilinear mappings on $V \times V$ forms a vector space W if we define the sum of two bilinear mappings and multiplication by a number as

$$(t + s)(\mathbf{v}, \mathbf{w}) = t(\mathbf{v}, \mathbf{w}) + s(\mathbf{v}, \mathbf{w}), \qquad (at)(\mathbf{v}, \mathbf{w}) = at(\mathbf{v}, \mathbf{w}).$$

We shall presently identify this space W as the n^2 dimensional space called the tensor product of the vector space V^* with itself.

5.2.1 Tensor Product $T_2^0 = V^* \otimes V^*$

Let α and β be two linear functionals on V, that is, members of V^*. We can form a bilinear functional out of these as follows.

Let $\alpha \otimes \beta$, called the **tensor product** of α and β, be given by

$$(\alpha \otimes \beta)(\mathbf{v}, \mathbf{w}) = \alpha(\mathbf{v})\beta(\mathbf{w}). \tag{5.6}$$

It is trivial to check that this indeed is a bilinear functional. This definition also gives us properties of this tensor product.

$$\alpha \otimes (\beta + \gamma) = \alpha \otimes \beta + \alpha \otimes \gamma, \qquad \alpha \otimes (a\beta) = a(\alpha \otimes \beta), \tag{5.7}$$

$$(\alpha + \beta) \otimes \gamma = \alpha \otimes \gamma + \beta \otimes \gamma, \qquad (a\alpha) \otimes \beta = a(\alpha \otimes \beta). \tag{5.8}$$

The vector $\alpha \otimes \beta$ belongs to the vector space W of all bilinear maps on $V \times V$. A vector of this type is called **decomposable** or **factorizable**. An arbitrary bilinear functional on $V \times V$ is not decomposable but can always be written as a linear combination of such vectors.

A bilinear map t is completely determined by its values $t_{ij} = t(\mathbf{e}_i, \mathbf{e}_j)$ on the members of a basis E because of the linear property. It can be actually written as

$$t = t_{ij}\alpha^i \otimes \alpha^j,$$

using the dual basis because t and $t_{ij}\alpha^i \otimes \alpha^j$ give the same result when acting on an arbitrary pair $(\mathbf{v}, \mathbf{u})$:

$$\begin{aligned} (t_{ij}\alpha^i \otimes \alpha^j)(\mathbf{v}, \mathbf{u}) &= t(\mathbf{e}_i, \mathbf{e}_j)\alpha^i(\mathbf{v})\alpha^j(\mathbf{u}) \\ &= t(\mathbf{v}, \mathbf{u}). \end{aligned}$$

The last line follows because for any vector the identity $\mathbf{v} = \alpha^i(\mathbf{v})\mathbf{e}_i$ holds.

We also note from this result that the space W is n^2 dimensional and that $\{\alpha^i \otimes \alpha^j\}_{i,j=1}^n$ form a basis in it. We denote the space W by $V^* \otimes V^*$ or by T_2^0 and call it the tensor product of the space V^* with itself. Members of $T_2^0 = V^* \otimes V^*$ are called **covariant tensors of rank** 2 or (0,2) tensors. The components t_{ij} of tensor t in the basis $\{\alpha^i \otimes \alpha^j\}_{i,j=1}^n$ change to t'_{ij} in basis $\{\beta^i \otimes \beta^j\}_{i,j=1}^n$ and they are related as

$$t'_{ij} = t(\mathbf{f}_i, \mathbf{f}_j) = t(T_i{}^k \mathbf{e}_k, T_j{}^l \mathbf{e}_l) = T_i{}^k T_j{}^l t_{kl}. \tag{5.9}$$

5.2.2   Tensor Product $T_0^2 = V \otimes V$

We have already noted that just as V^* is dual to V, V is dual to V^*. Thus starting with bilinear functionals on the cartesian product $V^* \times V^*$ we can define the tensor product $\mathbf{v} \otimes \mathbf{w}$ of vectors in V exactly in the same manner as in the last section. The resulting space $V \otimes V$ is called the tensor product of spaces V. Its vectors are called **contravariant tensors of second-rank** or $(2,0)$ tensors.

A basis in the space $V \otimes V$ is given by $\{\mathbf{e}_i \otimes \mathbf{e}_j\}_{i,j=1}^n$. A tensor $t \in V \otimes V$ is completely determined by its values on (α^i, α^j) that is by numbers $t^{ij} = t(\alpha^i, \alpha^j)$. It is obvious that we can write $t = t^{ij} \mathbf{e}_i \otimes \mathbf{e}_j$. Under a change of basis from E to F by a matrix T, the components of a contravariant vector transform as

$$t'^{ij} = t(\beta^i, \beta^j) = t((T^{-1T})^i{}_k \alpha^k, (T^{-1T})^j{}_l \alpha^l) = (T^{-1T})^i{}_k (T^{-1T})^j{}_l t^{kl}. \qquad (5.10)$$

5.2.3   Multilinear Functionals and T_r^0

The formalism of the last sections can be generalised to multilinear functionals. A multilinear functional t on $V \times \cdots \times V$ (r-factors) is a map which assigns real number $t(\mathbf{u}, \ldots, \mathbf{w})$ to an ordered set of r vectors of V, $(\mathbf{u}, \ldots, \mathbf{w})$ in such a way that

$$t(\mathbf{u} + \mathbf{v}, \ldots, \mathbf{w}) = t(\mathbf{u}, \ldots, \mathbf{w}) + t(\mathbf{v}, \ldots, \mathbf{w}),$$

$$t(a\mathbf{v}, \ldots, \mathbf{w}) = at(\mathbf{v}, \ldots, \mathbf{w})$$

with similar equations for each of the arguments.

Exactly as in the bilinear case one can define $\alpha \otimes \cdots \otimes \beta$ (r-factors) as the multilinear functional

$$(\alpha \otimes \cdots \otimes \beta)(\mathbf{u}, \ldots, \mathbf{w}) = \alpha(\mathbf{v}) \cdots \beta(\mathbf{w}). \qquad (5.11)$$

The vector space of all such multilinear functionals is called the space $T_r^0 = V^* \otimes \cdots \otimes V^*$ of **covariant tensors of rank** r or $(0,r)$ tensors.

A typical vector in T_r^0 is a linear combination of **decomposable vectors** of type $\alpha \otimes \cdots \otimes \beta$. Indeed the set $\{\alpha^{i_1} \otimes \cdots \otimes \alpha^{i_r}\}, i_1, \ldots, i_r = 1, \ldots, n$ forms a basis in the n^r dimensional space T_r^0. Such tensors t are fully specified by n^r numbers $t_{i\ldots j} \equiv t(\mathbf{e}_i, \ldots, \mathbf{e}_j)$.

5.2.4   Spaces T_0^s

In a similar manner we can define space $T_0^s = V \otimes \cdots \otimes V$ of **contravariant tensors of rank** s as the set of all multilinear functionals on the cartesian product $V^* \times \cdots \times V^*$ (s-factors) with basis $\mathbf{e}_{i_1} \otimes \cdots \otimes \mathbf{e}_{i_s}$ with $i_1, \ldots, i_s = 1, \ldots, n$.

We can see that T_r^0 is dual to T_0^r.

Let $t = \beta \otimes \cdots \otimes \beta_r \in T^0_r = V^* \otimes \cdots \otimes V^*$. Then t can be defined as a *linear* functional on $V \otimes \cdots \otimes V$ as follows: on decomposable vectors it is

$$\begin{aligned} t(\mathbf{v}_1 \otimes \cdots \otimes \mathbf{v}_r) &= (\beta_1 \otimes \cdots \otimes \beta_r)(\mathbf{v}_1 \otimes \cdots \otimes \mathbf{v}_r) \\ &= \beta_1(\mathbf{v}_1) \cdots \beta_r(\mathbf{v}_r), \end{aligned}$$

while on vectors which are sums of these the linearity of the functional is used. This shows that t is in the space dual to T^r_0. Vectors such as t span T^0_r therefore $T^0_r = (T^r_0)^*$.

5.2.5 Mixed Tensor Space T^1_1

Let T be a linear operator $T : V \to V$ on V. Then we can identify T with a bilinear functional on $V^* \times V$ (denoted by the same symbol T) by defining

$$T(\alpha, \mathbf{v}) = \alpha(T(\mathbf{v})).$$

It is easy to see that T indeed is bilinear. Therefore we can regard T as belonging to the tensor space $T^1_1 = V \otimes V^*$.

Just as we defined $\alpha \otimes \beta$ in $V^* \otimes V^*$, we can define, similarly $\mathbf{u} \otimes \beta \in V \otimes V^*$. It is a bilinear functional on $V^* \times V$ given by

$$(\mathbf{u} \otimes \beta)(\alpha, \mathbf{v}) = \alpha(\mathbf{u})\beta(\mathbf{v}).$$

Of course, we can also interpret $\mathbf{u} \otimes \beta$ as a *linear* operator on V by defining $(\mathbf{u} \otimes \beta)(\mathbf{v}) = \beta(\mathbf{v})\mathbf{u}$

Members of the vector space $T^1_1 = V \otimes V^*$ are called mixed tensors of contravariant rank 1 and covariant rank 1 or, $(1, 1)$ tensors. A basis can be chosen in this n^2-dimensional space by choosing $\{\mathbf{e}_i \otimes \alpha^j\}$ with i and j taking values $1, \ldots, n$.

To summarise, tensors of type T^1_1 can be considered as bilinear functionals on $V^* \times V$ or as linear mappings $V \to V$.

5.2.6 Mixed Tensors

The space T^s_r of **mixed tensors** of contravariant rank s and covariant rank r (briefly called (s,r) tensors) can be defined as the set of multilinear maps on $V^* \times \cdots \times V^* \times V \times \cdots \times V$ (there are s factors of V^* and r of V). A typical multilinear map of this type is $\mathbf{v} \otimes \cdots \otimes \mathbf{w} \otimes \alpha \otimes \cdots \otimes \beta$ which acts on $V^* \times \cdots \times V^* \times V \times \cdots \times V$ as

$$(\mathbf{v} \otimes \cdots \otimes \mathbf{w} \otimes \alpha \otimes \cdots \otimes \beta)(\gamma, \ldots, \zeta, \mathbf{u}, \ldots, \mathbf{x}) = \gamma(\mathbf{v}) \cdots \zeta(\mathbf{w})\alpha(\mathbf{u}) \cdots \beta(\mathbf{x}).$$

Such decomposable tensors form the basis

$$\mathbf{e}_{i_1} \otimes \cdots \otimes \mathbf{e}_{i_s} \otimes \alpha^{j_1} \otimes \cdots \otimes \alpha^{j_r}.$$

A general multilinear map t in T_r^s is determined by its values

$$t^{i_1 \ldots i_s}{}_{j_1 \ldots j_r} \equiv t(\alpha^{i_1}, \ldots, \alpha^{i_s}, \mathbf{e}_{j_1}, \ldots, \mathbf{e}_{j_r})$$

which allows us to write

$$t = t^{i_1 \ldots i_s}{}_{j_1 \ldots j_r} \mathbf{e}_{i_1} \otimes \cdots \otimes \mathbf{e}_{i_s} \otimes \alpha^{j_1} \otimes \cdots \otimes \alpha^{j_r}.$$

5.2.7   Interior Product or Contraction

Given a covariant tensor $t \in T_r^0$ of rank r and a vector $\mathbf{v}$ we define a tensor $i_\mathbf{v} t \in T_{r-1}^0$ of rank $r - 1$ as follows:

$$(i_\mathbf{v} t)(\mathbf{v}_1, \ldots, \mathbf{v}_{r-1}) = t(\mathbf{v}, \mathbf{v}_1, \ldots, \mathbf{v}_{r-1}). \tag{5.12}$$

The linear mapping $i_\mathbf{v}$ satisfies the following properties:

$$\begin{aligned}
i_\mathbf{v}(t + s) &= i_\mathbf{v}(t) + i_\mathbf{v}(s), & (5.13) \\
i_\mathbf{v}(at) &= a i_\mathbf{v}(t), & (5.14) \\
i_{\mathbf{v}+\mathbf{u}}(t) &= i_\mathbf{v}(t) + i_\mathbf{u}(t), & (5.15) \\
i_{a\mathbf{v}}(t) &= a i_\mathbf{v}(t). & (5.16)
\end{aligned}$$

Similarly, if we are given a mixed (r, s)-tensor T with contra- and covariant indices then a **contraction** between the k-th contra- and l-th covariant index is defined as an $(r - 1, s - 1)$-tensor $i(k, l)T$ as

$$\begin{aligned}
&(i(k, l)T)(\beta^1, \beta^2, \ldots, \beta^{r-1}, \mathbf{v}_1, \ldots, \mathbf{v}_{s-1}) \\
&= \sum_i T(\beta^1, \beta^2, \ldots, \alpha^i, \ldots, \beta^{r-1}, \mathbf{v}_1, \ldots, \mathbf{e}_i, \ldots, \mathbf{v}_{s-1})
\end{aligned} \tag{5.17}$$

where the dual basis elements α^i and $\mathbf{e}_i$ appear in the k-th and l-th place. Simply said, in terms of components,

$$\begin{aligned}
&(i(k, l)T)^{i_1 \ldots i_{k-1} i_{k+1} \ldots i_r}{}_{j_1 \ldots j_{l-1} j_{l+1} \ldots j_s} \\
&= T^{i_1 \ldots i_{k-1} j i_{k+1} \ldots i_r}{}_{j_1 \ldots j_{l-1} j j_{l+1} \ldots j_s}.
\end{aligned} \tag{5.18}$$

5.2.8   Summary

To summarise, we start with a vector space V with a basis $\{\mathbf{e}_i\}$ and define the dual space V^* with a dual basis $\{\alpha^i\}$. With these spaces and bases as starting point we can define an infinite sequence of vector spaces of higher and higher dimensions with tensor products. The components of a tensor are characterised by the way they transform when we change the basis $\{\mathbf{e}_i\}$ to a new basis. This leads to a change in the dual basis, and to bases in all the tensor spaces.

As a matter of standard established notation, observe carefully the use of super- and sub-scripts for denoting the members of bases in V^* and V (respectively) as well as in the components of contra- and co-variant vectors and tensors.

We can verify the transformation properties of components of the covariant and contravariant tensors: when the basis $\{\mathbf{e}_i\}$ in V is changed to $\{\mathbf{f}_i\}$ as

$$\mathbf{f}_i = T_i{}^j \mathbf{e}_i,$$

the dual basis changes from $\{\alpha^i\}$ to $\{\beta^i\}$,

$$\beta^i = (T^{-1T})^i{}_k \alpha^k,$$

and components of tensors change as

$$t'_{i\ldots j} = T_i{}^k \ldots T_j{}^l t_{k\ldots l}, \tag{5.19}$$

$$t'^{i\ldots j} = (T^{-1T})^i{}_k \ldots (T^{-1T})^j{}_l t^{k\ldots l}, \tag{5.20}$$

$$t'^{i\ldots j}{}_{k\ldots l} = (T^{-1T})^i{}_p \ldots (T^{-1T})^j{}_q T_k{}^m \ldots T_l{}^n t^{p\ldots q}{}_{m\ldots n} \tag{5.21}$$

where the primes denote components with respect to the bases β^i and $\mathbf{f}_i$.

5.3 Wedge or Exterior Product

Antisymmetric covariant tensors are extremely important because of their connection to differential forms, surface and volume integrals and Gauss-Stokes theorem. This importance is reflected in the fact that the antisymmetric part of tensor products has a different symbol and name to denote it.

For second-rank tensors we write

$$\alpha \wedge \beta = \alpha \otimes \beta - \beta \otimes \alpha$$

and generalise it to tensors of higher rank.

5.3.1 Permutations

Let P be a permutation of r objects. This means that P is a one-to-one mapping of the set $\{1, \ldots, r\}$ of first r natural numbers onto itself. There are $r!$ such mappings. Each of these can be considered as composed of more elementary mappings called transpositions, which just exchange two of the integers and map the rest to themselves. The way in which transpositions make a permutation P is not unique but the number of transpositions involved, though not fixed, is always either an even or an odd integer. The permutation is called even or odd accordingly. Let us define $(-1)^P$ to be equal to $+1$ if P is even and -1 if odd. The set of $r!$ permutations can be made a group under composition of mappings as the group law. The identity mapping is even. It is also clear that $(-1)^{P_1 \circ P_2} = (-1)^{P_1}(-1)^{P_2}$. It follows that P^{-1} is even or odd according to the sign of P.

Let us define a linear operator on T_r^0 corresponding to permutation P, also to be denoted by P, as follows. Let $\beta_1, \ldots, \beta_r$ be vectors in V^*. Form the tensor product $\beta_1 \otimes \cdots \otimes \beta_r$. Now define

$$P(\beta_1 \otimes \cdots \otimes \beta_r) \equiv \beta_{P(1)} \otimes \cdots \otimes \beta_{P(r)}. \tag{5.22}$$

It is sufficient to define P on these vectors because any general vector in T_r^0 is a linear combination of such vectors, and P is linear. For two permutations P_1 and P_2 we have clearly $P_2 P_1 = P_2 \circ P_1$ where in this equation the linear operators are on the left and the mappings on the right.

5.3.2   Exterior or Wedge Product

We are now ready to define the wedge product of any number of vectors of V^*.

$$\begin{aligned}
\beta_1 \wedge \beta_2 \wedge \cdots \wedge \beta_r &\equiv \sum_P (-1)^P P(\beta_1 \otimes \cdots \otimes \beta_r) \\
&= \sum_P (-1)^P \beta_{P(1)} \otimes \cdots \otimes \beta_{P(r)}.
\end{aligned} \tag{5.23}$$

If Q is a permutation operator, then

$$\begin{aligned}
\beta_{Q(1)} \wedge \beta_{Q(2)} \wedge \cdots \wedge \beta_{Q(r)} &= \sum_P (-1)^P PQ(\beta_1 \otimes \cdots \otimes \beta_r) \\
&= (-1)^Q \sum_R (-1)^R R(\beta_1 \otimes \cdots \otimes \beta_r)
\end{aligned}$$

where we have used $R \equiv P \circ Q$ and $(-1)^R = (-1)^P(-1)^Q = (-1)^P/(-1)^Q$. Therefore,

$$\beta_{Q(1)} \wedge \beta_{Q(2)} \wedge \cdots \wedge \beta_{Q(r)} = (-1)^Q (\beta_1 \wedge \beta_2 \wedge \cdots \wedge \beta_r). \tag{5.24}$$

In particular, the wedge product like $\beta_1 \wedge \beta_2 \wedge \cdots \wedge \beta_r$ changes sign whenever any two factors in it are exchanged. Therefore if any factor is repeated, the product is the zero vector.

Tensors with this property are called **antisymmetric**. As a multilinear functional on $V \times \cdots \times V$,

$$(\beta_1 \wedge \beta_2 \wedge \cdots \wedge \beta_r)(\mathbf{v}_1, \ldots, \mathbf{v}_r) = \det \|\beta_i(\mathbf{v}_j)\|. \tag{5.25}$$

Linear combinations of antisymmetric tensors are also antisymmetric, therefore the set of all antisymmetric covariant tensors forms a subspace $\Lambda^r(V^*) \subset T_r^0 = V^* \otimes \cdots \otimes V^*$. These tensors are also called r-**forms**.

5.3.3   Bases for $\Lambda^r(V^*)$

A basis in $\Lambda^r(V^*)$ can be chosen by considering all independent tensors of the form $\alpha^{i_1} \wedge \cdots \wedge \alpha^{i_r}$. Obviously, $i_1, \ldots, i_r$ all have to be different, because the antisymmetric wedge product is zero if any two vectors in the product string are the same. Also, a particular combination $i_1, \ldots, i_r$ need be taken only once because any other product with these same indices (though in some other order) is ± 1 times the same vector. There are as many independent tensors of this type as the number of ways to choose a combination of r different indices $i_1, \ldots, i_r$ out of $1, \ldots, n$. The dimension of space $\Lambda^r(V^*)$ is therefore $n!/r!(n-r)!$.

A basis can be chosen consisting of vectors $\{\alpha^{i_1} \wedge \cdots \wedge \alpha^{i_r}\}$ with $i_1 < \cdots < i_r$. For $r = n$ the space $\Lambda^n(V^*)$ is one-dimensional containing multiples of $\alpha^1 \wedge \cdots \wedge \alpha^n$. For $r > n$ the spaces Λ^r are zero, that is, contain only the zero vector.

5.3.4   Space $\Lambda^r(V)$

In exactly the same manner we define the space $\Lambda^r(V)$ of all antisymmetric contravariant tensors. They are called r-**vectors**. This space is spanned by $\{\mathbf{e}_{i_1} \wedge \cdots \wedge \mathbf{e}_{i_r}\}$ with $i_1 < \cdots < i_r$.

5.3.5   Wedge Product of an r- and an s-form

Given an r-form $t \in \Lambda^r(V^*)$ and an s-form $u \in \Lambda^s(V^*)$ we can define an $r+s$-form $t \wedge u$ called the wedge or exterior product of t and u as follows. First define it on decomposable vectors:

$$(\beta_1 \wedge \cdots \wedge \beta_r) \wedge (\gamma_1 \wedge \cdots \wedge \gamma_s) = \beta_1 \wedge \cdots \wedge \beta_r \wedge \gamma_1 \wedge \cdots \wedge \gamma_s$$

and then extend it on general vectors by linearity. Because

$$\beta_1 \wedge \cdots \wedge \beta_r \wedge \gamma_1 \wedge \cdots \wedge \gamma_s = (-1)^{rs} \gamma_1 \wedge \cdots \wedge \gamma_s \wedge \beta_1 \wedge \cdots \wedge \beta_r$$

follows from the antisymmetry of the wedge product, we must have in general

$$t \wedge u = (-1)^{rs} u \wedge t. \tag{5.26}$$

5.3.6   Bases in T_r^0 and $\Lambda^r(V^*)$

Let $t \in \Lambda^r(V^*)$ be written

$$t = \sum_{i_1 < \cdots < i_r} T_{i_1 \ldots i_r} \alpha^{i_1} \wedge \cdots \wedge \alpha^{i_r}.$$

Note that coefficients $T_{i_1 \ldots i_r}$ are defined only for indices $i_1 < \cdots < i_r$.

As $\Lambda^r(V^*) \subset T_r^0$, the r-form t can also be expanded as a member of T_r^0 in the basis $\{\alpha^{j_1} \otimes \cdots \otimes \alpha^{j_r}\}_{j_1,\dots,j_r=1}^n$, with coefficients $t_{j_1,\dots,j_r} = t(\mathbf{e}_{j_1},\dots,\mathbf{e}_{j_r})$,

$$t = \sum_{j_1,\dots,j_r=1}^n t_{j_1,\dots,j_r}\alpha^{j_1} \otimes \cdots \otimes \alpha^{j_r}.$$

By expanding the wedge product basis vectors in terms of a tensor product basis, and then comparing the coefficients, we find the components t's in terms of T's.

$$t_{j_1\dots j_r} = 0 \text{ if any of the indices coincide,}$$

$$t_{j_1\dots j_r} = T_{j_1\dots j_r} \qquad \text{for } j_1 < \cdots < j_r,$$

$$t_{j_1\dots j_r} = (-1)^P T_{P(j_1)\dots P(j_r)}$$

where permutation P brings indices $j_1 \dots j_r$ to increasing order $P(j_1) < \cdots < P(j_r)$

5.3.7   Components of $t \wedge u$ in $\Lambda^{r+s}(V^*)$ Basis

Given that

$$t = \sum_{i_1 < \cdots < i_r} T_{i_1\dots i_r}\alpha^{i_1} \wedge \cdots \wedge \alpha^{i_r},$$

$$u = \sum_{i_1 < \cdots < i_s} U_{i_1\dots i_s}\alpha^{i_1} \wedge \cdots \wedge \alpha^{i_s},$$

we can work out the components $B_{i_1\dots i_{r+s}}$ of $t \wedge u$ in the basis $\{\alpha^{i_1} \wedge \cdots \wedge \alpha^{i_{r+s}}\}$. They are

$$B_{i_1\dots i_{r+s}} = \sum_{(r,s) \text{ shuffles } Q} T_{Q(i_1)\dots Q(i_r)} U_{Q(i_{r+1})\dots Q(i_{r+s})} \qquad (5.27)$$

where the sum is over all (r, s) shuffles defined below.

An (r, s) **shuffle** is defined to be a permutation Q of $(r + s)$ distinct integers $(i_1 < \cdots < i_{r+s})$ such that

$$[i_1, \cdots, i_{r+s}] \rightarrow [Q(i_1), \cdots, Q(i_r) \, ; \, Q(i_{r+1}), \cdots Q(i_{r+s})]$$

where $Q(i_1) < \cdots < Q(i_r)$ and $Q(i_{r+1}) < \cdots < Q(i_{r+s})$.

The total number of (r,s) shuffles is $(r + s)!/r!s!$.

As an example, $(2,4,5,7)$ have the following six $(2,2)$ shuffles:

$$(2, 4, 5, 7) \rightarrow (2, 4; 5, 7), (2, 5; 4, 7), (2, 7; 4, 5), (4, 5; 2, 7), (4, 7; 2, 5), (5, 7; 2, 4).$$

5.3.8 Components of $t \wedge u$ in T^0_{r+s} Basis

Given an r-form t and an s-form u whose antisymmetric components as members of T^0_r and T^0_s respectively are

$$t = t_{i_1,\ldots,i_r} \alpha^{i_1} \otimes \cdots \otimes \alpha^{i_r}$$

and

$$u = u_{j_1,\ldots,j_s} \alpha^{j_1} \otimes \cdots \otimes \alpha^{j_s},$$

we can show that the components $p_{k_1,\ldots,k_{r+s}}$ of $p = t \wedge u$,

$$p = t \wedge u = p_{k_1,\ldots,k_{r+s}} \alpha^{k_1} \otimes \cdots \otimes \alpha^{k_{r+s}}$$

are given by

$$p_{k_1,\ldots,k_{r+s}} = (1/r!s!) \sum_P (-1)^P t_{P(k_1),\ldots,P(k_r)} u_{P(k_{r+1}),\ldots,P(k_{r+s})}.$$

5.4 Tutorial

Exercise 29. $\{e_1, e_2, e_3\}$ are basis vectors in a three-dimensional space V and $\{\alpha^1, \alpha^2, \alpha^3\}$ is the corresponding dual basis in V^*. Choose a new basis $\{f_1, f_2, f_3\}$ where

$$f_1 = e_1 + e_2, \qquad f_2 = e_1 - e_2, \qquad f_3 = e_1 + e_3.$$

Find the basis $\{\beta^1, \beta^2, \beta^3\}$ dual to this new basis.

Answer 29. $\beta^1 = (\alpha^1 + \alpha^2 - \alpha^3)/2, \beta^2 = (\alpha^1 - \alpha^2 - \alpha^3)/2, \beta^3 = \alpha^3$.

Exercise 30. Show that r vectors $\gamma^1, \ldots, \gamma^r \in V^*, \dim V^* = n$ are linearly dependent if and only if $\gamma^1 \wedge \cdots \wedge \gamma^r = 0$.

Answer 30. If $\gamma^1, \ldots, \gamma^r$ are dependent then one can write one of them, say γ^k, as a linear combination of the others. When γ^k is substituted in $\gamma^1 \wedge \cdots \wedge \gamma^r$ the result is zero because each term will have a repeated factor.

On the other hand if $\gamma^1, \ldots, \gamma^r$ are linearly independent then we can take these as the first r vectors of a basis $\gamma^1, \ldots, \gamma^n$ where $\gamma^{r+1}, \ldots, \gamma^n$ are defined appropriately. Let $g_1, \ldots, g_n$ be the dual basis in V. Then by definition of the wedge product

$$(\gamma^1 \wedge \cdots \wedge \gamma^r)(g_1, \ldots, g_r) = 1,$$

therefore $\gamma^1 \wedge \cdots \wedge \gamma^r \neq 0$.

Exercise 31. KRONECKER DELTA

The (1,1) tensor (see section 5.2.5) $\delta \in T^1_1$ is defined in some basis $\{e_i\}$ and its dual basis $\{\alpha^i\}$ as

$$\delta = \sum e_i \otimes \alpha^i.$$

Show that this definition is independent of the basis used and the tensor has constant components

$$\delta^i_j = 0 \text{ if } i \neq j, \quad \text{and} \quad = 1 \text{ if } i = j$$

Answer 31. Change a basis and verify that the inverse-transpose rule for change of bases makes the definition independent of bases.

Exercise 32. Every (1,1) tensor $t \in T^1_1(V)$ determines a linear mapping $T : V \to V$ by the formula (see section 5.2.5)

$$(T\mathbf{v})(\alpha) = t(\alpha, \mathbf{v}).$$

What is the linear mapping corrsponding to the Kronecker delta? And what is the linear mapping corresponding to $t = \mathbf{u} \otimes \beta$ for fixed $\mathbf{u} \in V$ and fixed $\beta \in V^*$.

Answer 32. Identity for the Kronecker delta. For $t = \mathbf{u} \otimes \beta$ the map T is such that $T(\mathbf{v}) = \beta(\mathbf{v})\mathbf{u}$

Chapter 6

Inner Product

In this chapter we introduce the additional structure of an inner product or metric on a vector space and its associated spaces.

6.1 Definition

A vector space is defined by the operations of sum of its vectors and multiplication by real numbers to its vectors.

An inner product or metric is an additional structure on a vector space.

For any two vectors $\mathbf{v}$ and $\mathbf{w}$ in a vector space V their **inner product** is a real number denoted by $\langle \mathbf{v}, \mathbf{w} \rangle$. The function which defines the inner product should have the following properties:

1. It is **linear**, that is,

$$\langle \mathbf{u}, \mathbf{v} + \mathbf{w} \rangle = \langle \mathbf{u}, \mathbf{v} \rangle + \langle \mathbf{u}, \mathbf{w} \rangle, \tag{6.1}$$

$$\langle \mathbf{u}, a\mathbf{v} \rangle = a\langle \mathbf{u}, \mathbf{v} \rangle \tag{6.2}$$

for any $\mathbf{v}, \mathbf{u}, \mathbf{w} \in V$ and any real number a.

2. It is **symmetric**

$$\langle \mathbf{u}, \mathbf{v} \rangle = \langle \mathbf{v}, \mathbf{u} \rangle \tag{6.3}$$

for any $\mathbf{v}, \mathbf{u} \in V$ With this property we can see that the inner product is linear in the first argument as well:

$$\langle \mathbf{u} + \mathbf{v}, \mathbf{w} \rangle = \langle \mathbf{u}, \mathbf{w} \rangle + \langle \mathbf{v}, \mathbf{w} \rangle,$$

$$\langle a\mathbf{u}, \mathbf{v} \rangle = a\langle \mathbf{u}, \mathbf{v} \rangle.$$

3. It is **non-degenerate**, that is, if $\langle \mathbf{u}, \mathbf{v} \rangle = 0$ for all $\mathbf{v} \in V$ then $\mathbf{u} = \mathbf{0}$.

The inner product is also called the **metric**.

An inner product is often defined with a stronger condition of **positive definiteness** which says that $\langle \mathbf{v}, \mathbf{v} \rangle \geq 0$ and $\langle \mathbf{v}, \mathbf{v} \rangle = 0$ if and only if $\mathbf{v} = \mathbf{0}$.

Obviously, a positive definite inner product is non-degenerate (take $\mathbf{v} = \mathbf{u}$ in the condition of non-degeneracy), but not vice versa. **In relativity theory we need inner-products on spacetime which are not positive definite.** In Minkowski space there are time-like vectors whose inner product with themselves is negative or null vectors for which it is zero. But the inner product is always non-degenerate.

Given an inner product we can define the notion of **orthogonality**. Two vectors $\mathbf{v}$ and $\mathbf{u}$ in V are called **orthogonal** if their inner product is zero, that is $\langle \mathbf{v}, \mathbf{u} \rangle = 0$.

For any vector $\mathbf{v}$ the number $\langle \mathbf{v}, \mathbf{v} \rangle$ is its **norm squared**. When the inner product is positive definite, as in Eulidean space, the positive number $\sqrt{\langle \mathbf{v}, \mathbf{v} \rangle}$ is its **norm** or **length**.

When the inner product is not positive definite, norm squared can be positive, negative or zero. A vector with zero norm squared (that is a vector which is orthogonal to itself) is called a **null** vector. A vector with norm squared equal to ± 1 is called **normalized**.

Two non-null, orthogonal vectors are linearly independent. What is interesting is that two non-orthogonal null vectors are also linearly independent. And in (3+1)-dimensional Minkowski space, if two null vectors are orthogonal, then they are necessarily proportional to each other.

6.2 Orthonormal Bases

An inner product or metric as defined above is a bilinear functional on $V \times V$. Therefore it defines a second-rank, symmetric covariant tensor $\mathbf{g}$ called the **metric tensor** through

$$\mathbf{g}(\mathbf{u}, \mathbf{v}) = \langle \mathbf{u}, \mathbf{v} \rangle.$$

We could equally well use the notation with $\mathbf{g}(\mathbf{u}, \mathbf{v})$ in place of $\langle \mathbf{u}, \mathbf{v} \rangle$. In most cases however, $\mathbf{g}$ is a given, fixed tensor and there is ease of notation in using the bracket notation for the inner product. In three-dimensional vector spaces the notation used is $\mathbf{u} \cdot \mathbf{v}$ instead of $\langle \mathbf{u}, \mathbf{v} \rangle$.

Let $\mathbf{e}_i, i = 1, \ldots, n$ be a basis in V. The components of the metric in this basis are

$$g_{ij} = \langle \mathbf{e}_i, \mathbf{e}_j \rangle. \tag{6.4}$$

This symmetric matrix contains all the information about the inner product because if $\mathbf{v} = v^i \mathbf{e}_i$ and $\mathbf{u} = u^j \mathbf{e}_j$ are two vectors, then bilinearitry of the product in its two arguments implies

$$\langle \mathbf{v}, \mathbf{u} \rangle = g_{ij} v^i u^j. \tag{6.5}$$

The non-degeneracy of the inner product means that $g \equiv \det g_{ij} \neq 0$ or, in other words, the matrix of metric tensor components in any basis is non-singular.

We note the important result that despite the existence of null (zero-norm) vectors in the space one can always choose a basis $\{\mathbf{n}_i\}_{i=1}^n$ such that $\eta_{ij} = \langle \mathbf{n}_i, \mathbf{n}_j \rangle = 0$ if $i \neq j$ and the norm squared $\langle \mathbf{n}_i, \mathbf{n}_i \rangle$ is either $+1$ or -1. Such a basis is called an **orthonormal basis** (or o.n. basis). We construct such a basis in the next section.

The number of vectors with norm squared $+1$ and those with norm squared -1 in such an orthonormal basis is fixed by the definition of the inner product. The number of positive norm squared vectors minus the number of negative norm squared vectors in an orthonormal basis is called the **signature** of the metric and is denoted by $\mathrm{sig}(V)$.

There do exist bases which contain (non-zero) null vectors as basis vectors. But these bases are not orthonormal.

6.2.1 Existence of Orthonormal Bases

We go through the standard proof of the existence of orthonormal bases because of its fundamental importance.

Let $\mathbf{a}$ be a non-zero vector with non-zero norm squared $\langle \mathbf{a}, \mathbf{a} \rangle \neq 0$.

There certainly exists a non-null vector of this kind unless the whole space is trivial consisting of just the zero vector $\mathbf{0}$. This is so because if $\langle \mathbf{a}, \mathbf{a} \rangle$ were zero for all $\mathbf{a} \in V$ then by using $\langle \mathbf{a} + \mathbf{b}, \mathbf{a} + \mathbf{b} \rangle = 0$ for any arbitrary $\mathbf{b}$ it follows that $\langle \mathbf{a}, \mathbf{b} \rangle = 0$ for all $\mathbf{b}$. The condition of non-degeneracy then implies $\mathbf{a} = \mathbf{0}$.

Let $\mathbf{n}_1 = \mathbf{a}/\sqrt{|\langle \mathbf{a}, \mathbf{a} \rangle|}$. Depending on the sign of norm squared of $\mathbf{a}$, $\langle \mathbf{n}_1, \mathbf{n}_1 \rangle \equiv \epsilon_1$ is $+1$ or -1.

Let V_1 be the one-dimensional subspace spanned by $\mathbf{n}_1$, and let V_2 be the set of all vectors in V orthogonal to every vector in V_1. Obviously, V_2 is a vector subspace, and every vector $\mathbf{v} \in V$ can be decomposed as

$$\mathbf{v} = \epsilon_1 \langle \mathbf{v}, \mathbf{n}_1 \rangle \mathbf{n}_1 + (\mathbf{v} - \epsilon_1 \langle \mathbf{v}, \mathbf{n}_1 \rangle \mathbf{n}_1)$$

where the first term is in V_1 and the second term is in V_2. The only vector common to V_1 and V_2 is the zero vector $\mathbf{0}$ and this again follows from non-degeneracy.

The inner product restricted to V_2 is again non-degenerate because a vector in V_2 orthogonal to all other vectors of V_2 is moreover orthogonal to V_1 and hence is the zero vector.

We can now start with V_2 as the starting space and find a non-null vector $\mathbf{b} \in V_2$ such that $\langle \mathbf{b}, \mathbf{b} \rangle \neq 0$, and construct $\mathbf{n}_2 = \mathbf{b}/\sqrt{|\langle \mathbf{b}, \mathbf{b} \rangle|}$, with $\langle \mathbf{n}_2, \mathbf{n}_2 \rangle \equiv \epsilon_2$ equal to $+1$ or -1. We proceed in this manner inductively till the whole basis is constructed.

Thus we have a basis $\{\mathbf{n}_i\}$ with the metric components

$$I_{\epsilon ij} = \langle \mathbf{n}_i, \mathbf{n}_j \rangle = \epsilon_i \delta_{ij} \qquad \text{(no summation on } i\text{)} \tag{6.6}$$

or

$$I_\epsilon = \begin{pmatrix} \epsilon_1 & 0 & \cdots & 0 \\ 0 & \epsilon_2 & \cdots & 0 \\ & & \ddots & \\ 0 & 0 & \cdots & \epsilon_n \end{pmatrix}.$$ (6.7)

6.2.2 Signature of the Metric

Note that whichever route we take to choose orthonormal vectors for a basis the number n_+ of vectors with norm $+1$ and the number n_- of vectors with norm -1 is always the same. As $\dim V = n = n_+ + n_-$ is fixed so is the number $t = n_+ - n_-$. t is called the **signature** of the metric. The Minkowski space has one time-like unit vector with $\epsilon_0 = -1$ and three space-like vectors with $(\epsilon_i = 1, i = 1, 2, 3)$ in any orthonormal basis. Thus it has signature $+2$.

Similarly the number $\pm 1 = \epsilon_1 \epsilon_2 \ldots \epsilon_n = \det I_\epsilon$, which is the determinant of the matrix I_ϵ of metric components (in the orthonormal basis), is a characteristic of the metric. If $\{e_j\}$ is any basis with $g_{ij} = \langle e_i, e_j \rangle$ as the metric components then $g \equiv \det \|g_{ij}\|$ (which is always non-zero) has the same sign as $\det I_\epsilon$. We write this number as $\mathrm{sgn}(g)$ in general.

For spacetime in general relativity the sign of $g = \det g_{ij}$ is always negative because the number of vectors with negative norm is odd.

6.3 Correspondence Between V and V^*

A non-degenerate inner product $\langle\ ,\ \rangle$ defined on a vector space V *sets up a one-to-one correspondence* between vectors in V and those in the dual V^*.

Let $\mathbf{v} \in V$ be given. Then every vector $\mathbf{u} \in V$ can be mapped linearly to real numbers by $\mathbf{u} \to \langle \mathbf{v}, \mathbf{u} \rangle$. This helps us define a linear functional $\mathbf{v}^\flat \in V^*$ (called "$\mathbf{v}$-flat") with the help of the inner product as

$$\mathbf{v}^\flat(\mathbf{u}) = \langle \mathbf{v}, \mathbf{u} \rangle.$$

Properties of the inner product ensure that $\mathbf{v}^\flat$ is a linear functional. Obviously, $\mathbf{v}^\flat$ depends on the vector $\mathbf{v}$.

In fact, for finite dimensional spaces, *all* linear functionals arise in this way. In other words, if we are given a linear functional $\alpha \in V^*$ then there exists a vector $\alpha^\sharp \in V$ ("alpha-sharp") such that the number assigned by α to a vector $\mathbf{u}$ is the same number $\langle \alpha^\sharp, \mathbf{u} \rangle$ obtained in taking the inner product with $\alpha^\sharp$.

We can identify $\alpha^\sharp$ as follows.

First choose an orthonormal basis $\{\mathbf{n}_i\}$ with $\langle \mathbf{n}_i, \mathbf{n}_j \rangle = 0$ for $i \neq j$ and $\langle \mathbf{n}_i, \mathbf{n}_i \rangle = \epsilon_i$.

Now define

$$\alpha^\sharp = \sum_i \epsilon_i \alpha(\mathbf{n}_i)\mathbf{n}_i. \tag{6.8}$$

This is indeed the vector with the required property. Expanding $\mathbf{u} = u^j \mathbf{e}_j$ and using $(\epsilon_i)^2 = 1$,

$$
\begin{aligned}
\langle \alpha^\sharp, \mathbf{u} \rangle &= \sum_j u^j \langle \alpha^\sharp, \mathbf{n}_j \rangle = \sum_{ij} u^j \epsilon_i \alpha(\mathbf{n}_i)\langle \mathbf{n}_i, \mathbf{n}_j \rangle \\
&= \sum_{ij} u^j \epsilon_i \alpha(\mathbf{n}_i)\epsilon_i \delta_{ij} = \sum_j u^j \alpha(\mathbf{n}_j) = \alpha(\mathbf{u}).
\end{aligned}
$$

This one-to-one correspondence between the the dual spaces V and V^* is called **raising and lowering of indices** by the matrix g_{ij}. The reason for this nomenclature is as follows.

If $\mathbf{v} \in V$ has components v^i in basis $\mathbf{e}_i$ (not necessarily orthonormal) then the linear functional $\mathbf{v}^\flat \in V^*$ corresponding to it has components

$$a_i = g_{ij} v^j$$

with respect to the dual basis α^i,

$$\mathbf{v}^\flat(\mathbf{u}) = \langle \mathbf{v}, \mathbf{u} \rangle = g_{ij} v^i u^j = g_{ij} v^i \alpha^j(\mathbf{u}) \equiv (a_j \alpha^j)(\mathbf{u}) \tag{6.9}$$

where we have used the property of dual basis: if $\mathbf{u} = u^j \mathbf{e}_j$ then $\alpha^i(\mathbf{u}) = u^i$.

The inverse of this one-to-one correspondence between components of a linear functional $\alpha = a_i \alpha^i$ to a vector $\alpha^\sharp = v^j \mathbf{e}_j$ is similarly given by

$$v^i = g^{ij} a_j$$

where g^{ij} (with matrix indices written as superscripts) is the *inverse* of the matrix g_{ij}. We see this in the next section.

This also explains the musical notation of sharp and flat. A vector $\mathbf{v} \in V$ (contravariant, upper index) corresponds, via the metric, to a form $\mathbf{v}^\flat$ (covariant, lower index=lower pitch=flat), while a form α becomes a vector (contravariant=upper index=higher pitch=sharp) $\alpha^\sharp \in V$.

Remark on Physicists' Notation

Most general relativity physicists use the same letter-symbol for components when a (contravariant) vector $\mathbf{v}$ is put in correspondence with a (covariant) form $\mathbf{v}^\flat$ or vice versa. For example if

$$\mathbf{v} = v^i \mathbf{e}_i,$$

then $\mathbf{v}^\flat$ is written as

$$\mathbf{v}^\flat = v_i \alpha^i$$

in the dual basis. The correspondence itself looks like

$$v_i = g_{ij} v^j, \qquad v^i = g^{ij} v_j.$$

This notation although convenient can be very confusing in geometrical contexts. For example, in dealing with a hypersurface, the normal vector field on the surface and its corresponding 1-form have very different roles to play, and the choice of using the same symbols for components is more a liability than a convenience.

6.4　Inner Product in V^*

An inner product defined on a vector space V, determines in a natural way an inner product on the dual vector space V^* as well as on spaces T_s^r and $\Lambda^r(V^*)$ etc.

　　We saw in the last section that to every $\alpha \in V^*$ corresponds a vector $\alpha^\sharp$ such that $\alpha(\mathbf{u}) = \langle \alpha^\sharp, \mathbf{u} \rangle$ for every $\mathbf{u} \in V$ and conversely every vector in V determines a member of V^* in this manner.

　　This one-to-one correspondence suggests that if α corresponds to $\alpha^\sharp = \sum_i \epsilon_i \alpha(\mathbf{n}_i) \mathbf{n}_i$ for an orthonormal basis $\{\mathbf{n}_i\}$ and similarly if β corresponds to $\beta^\sharp$, then we can define

$$\langle \alpha, \beta \rangle \equiv \langle \alpha^\sharp, \beta^\sharp \rangle = \sum_i \epsilon_i \alpha(\mathbf{n}_i) \beta(\mathbf{n}_i).$$

All properties of the inner-product are satisfied. Perhaps the only property not obvious to see is non-degeneracy. We follow it up in an exercise.

　　In a basis with vectors $\mathbf{e}_i = T_i{}^k \mathbf{n}_k$ the metric is

$$g_{ij} = \langle \mathbf{e}_i, \mathbf{e}_j \rangle = T_i{}^k \epsilon_k T_j{}^k = (T I_\epsilon T^T)_{ij}$$

where I_ϵ is the matrix of the metric in the orthonormal basis

$$(I_\epsilon)_{ij} = \langle \mathbf{n_i}, \mathbf{n}_j \rangle = \epsilon_i \delta_{ij}, \qquad \text{no summation.} \tag{6.10}$$

The convenience of the matrix in the orthonormal basis is that its square is the identity marix and $(I_\epsilon)^{-1} = (I_\epsilon)$.

　　Let $\{\alpha^i\}$ be the basis dual to $\{\mathbf{e}_i\}$. The dual basis is obtained by matrix T^{-1T} acting on the basis $\{\nu^i\}$ dual to $\{\mathbf{n}_i\}$,

$$\alpha^i = \sum_l (T^{-1})_l{}^i \nu^l.$$

The matrix of the metric in the dual space is then

$$
\begin{aligned}
g^{ij} &\equiv \langle \alpha^i, \alpha^j \rangle &\text{(6.11)} \\
&= \sum_k \epsilon_k \alpha^i(\mathbf{n}_k) \alpha^j(\mathbf{n}_k) \\
&= \sum_{k,l,m} (T^{-1})_l{}^i \nu^l(\mathbf{n}_k) \epsilon_k (T^{-1})_m{}^j \nu^m(\mathbf{n}_k) \\
&= (T^{-1T} I_\epsilon T^{-1})^{ij} \\
&= (g^{-1})_{ij} &\text{(6.12)}
\end{aligned}
$$

which follows from the relation $g_{ij} = (T I_\epsilon T^T)_{ij}$ given above.

Therefore we see that the matrix $g^{ij} = \langle \alpha^i, \alpha^j \rangle$ of the naturally determined metric in the dual basis of V^* is the inverse of the matrix $g_{ij} = \langle \mathbf{e}_i, \mathbf{e}_j \rangle$ in the original basis.

6.4.1   Inner Product in Tensor Spaces

We now define the inner product on $T_0^r = V \otimes \cdots \otimes V$.

For $\mathbf{a}_1 \otimes \mathbf{a}_2 \cdots \otimes \mathbf{a}_r \in T_0^r$ define the inner product on decomposable vectors as

$$
\langle \mathbf{a}_1 \otimes \mathbf{a}_2 \cdots \otimes \mathbf{a}_r, \mathbf{b}_1 \otimes \mathbf{b}_2 \cdots \otimes \mathbf{b}_r \rangle \equiv \langle \mathbf{a}_1, \mathbf{b}_1 \rangle \cdots \langle \mathbf{a}_r, \mathbf{b}_r \rangle
$$

and extend by linearity over arbitrary linear combinations.

To check the non-degeneracy property of this inner product let us choose an orthonormal basis $\{\mathbf{n}_i\}$ in V with $\langle \mathbf{n}_i, \mathbf{n}_j \rangle = \pm\delta_{ij}$. Let $t = t^{i_1 \cdots i_r} \mathbf{n}_{i_1} \otimes \cdots \otimes \mathbf{n}_{i_r}$ be such that $\langle t, \mathbf{n}_{j_1} \otimes \cdots \otimes \mathbf{n}_{j_r} \rangle = 0$ for all sets $\{j_1, \ldots j_r\}$. Then it follows from the definition that $t^{i_1 \cdots i_r} = 0$ for all $i_1 \ldots i_r$, that is $t = \mathbf{0}$.

An exactly similar definition can be given for the inner product in $T_r^0(V) = V^* \otimes \cdots \otimes V^*$.

6.4.2   Inner Product in $\Lambda^r(V^*)$

The inner product on the exterior product of spaces $\Lambda^r(V^*) \subset T_r^0$, or $\Lambda^r(V) \subset T_0^r$ is already defined as these are subsets of larger spaces whose inner product they naturally inherit.

But there is a standard convention to **redefine** the inner product on spaces $\Lambda^r(V^*)$ to avoid factors of $r!$.

For $r = 2$, the definition of the wedge product gives,

$$
\langle \eta_1 \wedge \eta_2, \beta_1 \wedge \beta_2 \rangle = \langle \eta_1 \otimes \eta_2 - \eta_2 \otimes \eta_1, \beta_1 \otimes \beta_2 - \beta_2 \otimes \beta_1 \rangle.
$$

This can be written as a determinant

$$
\langle \eta_1 \wedge \eta_2, \beta_1 \wedge \beta_2 \rangle = 2 \det[\langle \eta_i, \beta_j \rangle].
$$

Similarly,

$$\langle \eta_{i_1} \wedge \cdots \wedge \eta_{i_r}, \beta_{j_1} \wedge \cdots \wedge \beta_{j_r} \rangle \equiv r! \det[\langle \eta_{i_m}, \beta_{j_n} \rangle].$$

Notation

The **redefined** inner product is given *without* the factor $r!$

$$\langle \eta_{i_1} \wedge \cdots \wedge \eta_{i_r} | \beta_{j_1} \wedge \cdots \wedge \beta_{j_r} \rangle \equiv \det[\langle \eta_{i_m}, \beta_{j_n} \rangle]. \tag{6.13}$$

In order to remind ourselves of this new normalization, we use the 'bar' notation $\langle \ | \ \rangle$ in place of the comma $\langle \ , \ \rangle$. For $r = 1$, there is no difference and we use the comma notation.

Let $\{\nu^i\}$ be an orthonormal basis in V^* with $\langle \nu^i, \nu^j \rangle = \epsilon_i \delta_{ij}$. Let us use the abbreviation

$$\nu^{i_1 < \cdots < i_r} \equiv \nu^{i_1} \wedge \cdots \wedge \nu^{i_r} \qquad \text{with } i_1 < \cdots < i_r \tag{6.14}$$

for the orthonormal basis vectors in $\Lambda^r(V^*)$. With the re-normalised inner product these basis vectors have

$$\langle \nu^{i_1 < \cdots < i_r} | \nu^{j_1 < \cdots < j_r} \rangle = \epsilon_{i_1} \ldots \epsilon_{i_r} \delta_{i_1 j_1} \ldots \delta_{i_r j_r}.$$

An inner product $\langle \ | \ \rangle$ can be defined for spaces $\Lambda^r(V)$ similarly.

6.5   Orientation and Cartan Tensor

Remember that the space $\Lambda^n(V^*)$ is one-dimensional. Choose $\{\nu^i\}$ as an orthonormal basis in V^*. Then $\Lambda^n(V^*)$ has just two normalized vectors $\eta = \nu^1 \wedge \cdots \wedge \nu^n$ and $-\eta$, with

$$\langle \eta | \eta \rangle = \epsilon_1 \ldots \epsilon_n = \text{sgn}(g).$$

If we remove the zero vector from the one-dimensional space $\Lambda^n(V^*)$ then the space gets divided into two disjoint subsets.

We choose one of these subsets (say the subset to which η belongs) and say that 'we have chosen an orientation' $\eta = \nu^1 \wedge \cdots \wedge \nu^n$ through the orthonormal basis $\{\nu^i\}$.

The vector $\eta \in \Lambda^n(V^*)$ is called the Cartan tensor after we have chosen this orientation.

A different basis (not necessarily orthogonal) $\{\alpha^j\}$ in V^* is defined to **have the chosen orientation** if the vector $\alpha^1 \wedge \cdots \wedge \alpha^n$ is in the chosen subset to which $\eta = \nu^1 \wedge \cdots \wedge \nu^n$ belongs. It is said to have the opposite orientation otherwise.

Let $\{\alpha^j\}$ and $\{\beta^i\}$ be two bases with the chosen orientation. Then the matrix connecting the two bases must have a positive determinant.

$$\beta^i = \sum_j T_{ij} \alpha^j, \qquad \det T > 0.$$

This can be seen from relating both the bases to η,

$$\alpha^i = \sum_j A_{ij}\nu^j, \qquad \beta^i = \sum_j B_{ij}\nu^j$$

and noting that for example

$$\alpha^1 \wedge \cdots \wedge \alpha^n = (\det A)\eta,$$

therefore $\det A > 0$. Similarly, $\det B > 0$. Now $T = BA^{-1}$ therefore $\det T > 0$ as well.

The components of the Cartan tensor in the orthonormal basis (of T_n^0) are just numbers ± 1 depending on the permutation of the factors:

$$
\begin{aligned}
\eta &= \nu^1 \wedge \cdots \wedge \nu^n \\
&= \varepsilon_{i_1 \ldots i_n} \nu^{i_1} \otimes \cdots \otimes \nu^{i_n}, \\
\varepsilon_{1 \ldots n} &= 1, \\
\varepsilon_{i_1 \ldots i_n} &= (-)^P. \qquad (i_1, \ldots, i_n) \text{ are a permutation } P \text{ of}(1, \ldots, n), \\
\varepsilon_{i_1 \ldots i_n} &= 0 \text{ if any two indices coincide.}
\end{aligned}
$$

The components $\bar{\varepsilon}_{j_1 \ldots j_n}$ of the Cartan tensor in a general basis $\{\alpha^i\}$ where $\alpha^i = \sum_j A_{ij}\nu^j$ are given by

$$\bar{\varepsilon}_{j_1 \ldots j_n} = \sqrt{|g|}\varepsilon_{j_1 \ldots j_n}.$$

This is worked out in a tutorial.

6.6 Hodge *-Operator

6.6.1 Definition

Let us choose an orientation, say, $\eta = \nu^1 \wedge \cdots \wedge \nu^n$ with the orthonormal basis $\{\nu^i\}$.

For any $0 \leq r \leq n$ the **star operator** $*$ is a linear map from $\Lambda^r(V^*)$ into $\Lambda^{n-r}(V^*)$. The map is defined as follows:

Let $t \in \Lambda^r(V^*)$ be a fixed r-form and $s \in \Lambda^{n-r}(V^*)$, then $t \wedge s \in \Lambda^n(V^*)$ which is one-dimensional space and therefore $t \wedge s$ is proportional to η. Let the proportionality constant be written as $T(s)$,

$$t \wedge s = T(s)\eta.$$

The map $T : s \to T(s)$ is a linear functional on $\Lambda^{n-r}(V^*)$.

We know that in any vector space W with an inner product $\langle \, , \, \rangle$, a linear functional $c : \mathbf{u} \to c(\mathbf{u})$ will determine a vector $\mathbf{v} \in W$ whose inner product with a vector of the space is the number assigned by the linear functional , $c(\mathbf{u}) = \langle \mathbf{v}, \mathbf{u} \rangle$.

The mapping $T : \Lambda^{n-r}(V^*) \to R$ given by $s \to T(s)$ (for the fixed t chosen) is a linear functional on $\Lambda^{n-r}(V^*)$ and therefore it determines, through the inner product ($\langle \, , \, \rangle$ or $\langle \, | \, \rangle$ or any other) a unique vector $*t$ in $\Lambda^{n-r}(V^*)$ whose inner product with s is this value $T(s)$,

$$\langle *t, s \rangle = T(s).$$

We have denoted the vector by $*t$ because the functional $T(s)$ is determined by t. We can write more transparently but *with a slight abuse of language,*

$$T(s) = \langle *t, s \rangle = \frac{t \wedge s}{\eta}.$$

6.6.2   Formula for the Star Operator

The definition of $*t \in \Lambda^{n-r}(V^*)$ starting from $t \in \Lambda^r(V^*)$ depends on the inner product chosen. On $\Lambda^r(V^*)$ we already have two definitions for the inner product, $\langle \, , \, \rangle$, (which is the direct inheritor of the natural inner product on T_r^0) and $\langle \, | \, \rangle$ differing by a factor of $r!$. To make matters worse there exist different possibilities for defining the map $s \to T(s)$: should $t \wedge s = T(s)\eta$ or $s \wedge t = T(s)\eta$? And we do not even mention the choice of orientation η.

Therefore the definition of the star operator is obtained by fixing an r-dependent proportionality factor.

In the following we define our convention which agrees with most physics texts (including the one in Misner, Thorne and Wheeler) for students' convenience. It is sufficient to define the operator on the orthonormal basis elements of spaces Λ^r.

Our definition corresponds to choosing

$$\langle *t|s \rangle \eta = (\text{sgn}(g))t \wedge s. \tag{6.15}$$

Remember that Λ^0 is the one-dimensional space which can be identified with the set of real numbers R. An orthonormal basis in Λ^0 has only one element which we can identify with the number 1. We define the *-operated vectors in Table 6.1 given below.

This mapping $* : \Lambda^r(V^*) \to \Lambda^{n-r}(V^*)$ is a linear mapping called the **Hodge *-operator**. Actually there is a *-operator for each r, but they are all denoted by the same common symbol.

It is straightforward to see how formulas follow from the definition. We work out the connection between the definition and the formulas given above in the excercises at the end of the chapter.

The star operator maps an r-form into an $(n-r)$-form and vice versa. Therefore applying it twice is the same as mapping r-forms into themselves. Actually it is proportional to identity

$$** = (-1)^{r(n-r)}\mathrm{sgn}(g). \tag{6.16}$$

6.6.3 Star Operator in a General Basis

The star operator is expressed most simply in an orthonormal basis. But we calculate it in other, general bases too. Since an orientation has been chosen, the general basis $\{\alpha^i\}$ should be such that $\alpha^1 \wedge \cdots \wedge \alpha^n$ is a multiple of $\eta = \nu^1 \wedge \cdots \nu^n$ by a *positive* number. This means that if $\alpha^i = S^i_j \nu^j$ then $\det S > 0$.

We start with the convenient form, already a tensor equation

$$*(\nu^{i_1} \wedge \nu^{i_2} \cdots \wedge \nu^{i_r}) = \frac{1}{(n-r)!}\eta^{i_1 k_1} \cdots \eta^{i_r k_r}\varepsilon_{k_1 \ldots k_r j_1 \ldots j_{n-r}}\nu^{j_1} \wedge \cdots \wedge \nu^{j_{n-r}}$$

and change basis. Let

$$\alpha^i = S^i_j \nu^j, \qquad g^{ij} = \langle \alpha^i, \alpha^j \rangle = S^i_k S^j_l \eta^{kl}.$$

Therefore,

$$
\begin{aligned}
*(\alpha^{i_1} \wedge \ldots \alpha^{i_r}) &= \frac{1}{(n-r)!}S^{i_1}_{j_1} \ldots S^{i_r}_{j_r} *(\nu^{j_1} \wedge \ldots \nu^{j_r}) \\
&= \frac{1}{(n-r)!}S^{i_1}_{j_1} \ldots S^{i_r}_{j_r}\eta^{j_1 k_1} \ldots \eta^{j_r k_r} \\
&\quad \times\ \varepsilon_{k_1 \ldots k_r l_1 \ldots l_{n-r}}\nu^{l_1} \wedge \cdots \wedge \nu^{l_{n-r}}.
\end{aligned}
$$

As $S^i_k S^j_l \eta^{kl} = g^{ij}$, we have $S^i_k \eta^{kl} = (S^{-1})^l_j g^{ij}$ therefore

$$S^{i_1}_{j_1} \ldots S^{i_r}_{j_r}\eta^{j_1 k_1} \ldots \eta^{j_r k_r} = (S^{-1})^{k_1}_{p_1} \ldots (S^{-1})^{k_r}_{p_r}g^{i_1 p_1} \ldots g^{i_r p_r}.$$

Next we change the orthonormal vectors ν^i into α^j's,

$$\nu^{l_1} \wedge \cdots \wedge \nu^{l_{n-r}} = (S^{-1})^{l_1}_{q_1} \ldots (S^{-1})^{l_{n-r}}_{q_{n-r}}\alpha^{q_1} \ldots \alpha^{q_{n-r}}$$

and combine all S^{-1} factors to convert the Cartan tensor components $\varepsilon_{k_1 \ldots k_r l_1 \ldots l_{n-r}}$ from the orthonormal basis to the general basis

$$
\begin{aligned}
&(S^{-1})^{k_1}_{p_1} \ldots (S^{-1})^{k_r}_{p_r}(\varepsilon_{k_1 \ldots k_r l_1 \ldots l_{n-r}})(S^{-1})^{l_1}_{q_1} \ldots (S^{-1})^{l_{n-r}}_{q_{n-r}} \\
&= \overline{\varepsilon}_{p_1 \ldots p_r q_1 \ldots q_{n-r}} \\
&= \sqrt{|g|}\varepsilon_{p_1 \ldots p_r q_1 \ldots q_{n-r}}.
\end{aligned}
$$

Therefore we get the general formula

$$
\begin{aligned}
*(\alpha^{i_1} \wedge \ldots \alpha^{i_r}) &= \frac{1}{(n-r)!}\sqrt{|g|}g^{i_1 p_1} \ldots g^{i_r p_r}(\varepsilon_{p_1 \ldots p_r q_1 \ldots q_{n-r}}) \\
&\quad \times\ \alpha^{q_1} \wedge \cdots \wedge \alpha^{q_{n-r}}.
\end{aligned}
$$

Table 6.1
Definition of the Hodge star-operator
Orthonormal basis

r	t	$*t$
0	1	$\nu^1 \wedge \cdots \wedge \nu^n$
1	ν^1 ν^2 $\ldots$ ν^i	$\epsilon_1 \nu^2 \wedge \cdots \wedge \nu^n$ $-\epsilon_2 \nu^1 \wedge \nu^3 \wedge \cdots \wedge \nu^n$ $\ldots$ $(-1)^{i-1}\epsilon_i \nu^1 \wedge \cdots \nu^{i-1} \wedge \nu^{i+1} \cdots \wedge \nu^n$
	$\ldots$	$\ldots$
r	$\nu^{i_1} \wedge \nu^{i_2} \cdots \wedge \nu^{i_r}$ $(i_1 < \cdots < i_r)$	$(-1)^P \epsilon_{i_1} \ldots \epsilon_{i_r} \nu^{j_1} \wedge \cdots \wedge \nu^{j_{n-r}}$ where $j_i < \cdots < j_{n-r}$ is the complimentary set of indices remaining after taking out $i_1 < \cdots < i_r$ from $1, \ldots, n$ and P is the permutation $P : \{i_1, \ldots, i_r, j_1, \ldots, j_{n-r}\} \to \{1, \ldots, n\}$
r	$\nu^{i_1} \wedge \nu^{i_2} \cdots \wedge \nu^{i_r}$	$\frac{1}{(n-r)!} \eta^{i_1 k_1} \ldots \eta^{i_r k_r} \varepsilon_{k_1 \ldots k_r j_1 \ldots j_{n-r}} \nu^{j_1} \wedge \cdots \wedge \nu^{j_{n-r}}$ Another useful form: no restrictions on order of indices
	$\ldots$	$\ldots$
n	$\nu^1 \wedge \cdots \wedge \nu^n$	$\epsilon_1 \ldots \epsilon_n = \mathrm{sgn}(g)$

$$(6.17)$$

Notice that $*(\alpha^1 \wedge \cdots \wedge \alpha^n) = \mathrm{sgn}(g)/\sqrt{|g|}$. In terms of components, a general form $t = t_{i_1 \ldots i_r} \alpha^{i_1} \wedge \cdots \wedge \alpha^{i_r}$ becomes $*t = (*t)_{i_1 \ldots i_{n-r}} \alpha^{i_1} \wedge \cdots \wedge \alpha^{i_{n-r}}$ with

$$(*t)_{q_1 \ldots q_{n-r}} = \sqrt{|g|} t_{i_1 \ldots i_r} g^{i_1 p_1} \ldots g^{i_r p_r} \varepsilon_{p_1 \ldots p_r q_1 \ldots q_{n-r}}.$$

Table 6.2: Hodge star-operator
(General basis)

t	$*t$
1	$\sqrt{|g|} \alpha^1 \wedge \cdots \wedge \alpha^n$
$\alpha^{i_1} \wedge \cdots \wedge \alpha^{i_r}$	$[(n-r)!]^{-1} \sqrt{|g|} g^{i_1 k_1} \ldots g^{i_r k_r} \times$ $\varepsilon_{k_1 \ldots k_r j_1 \ldots j_{n-r}} \alpha^{j_1} \wedge \cdots \wedge \alpha^{j_{n-r}}$
$\alpha^1 \wedge \cdots \wedge \alpha^n$	$\mathrm{sgn}(g)/\sqrt{|g|}$

(6.18)

6.7 Minkowski Space

Minkowski space M is the spacetime of special relativity. It is a four-dimensional space with a real non-degenerate metric $\langle, \rangle$. If $\{\mathbf{e}_\mu\}, \mu = 0, 1, 2, 3$ is a basis then $g_{\mu\nu} = \langle \mathbf{e}_\mu, \mathbf{e}_\nu \rangle$ is a real non-singular symmetric matrix. We say two vectors $\mathbf{u}, \mathbf{v} \in M$ are orthogonal if $\langle \mathbf{u}, \mathbf{v} \rangle = 0$. Because the metric is not positive definite, and merely non-degenerate, our intuition based on orthogonality in the ordinary sense does not work. There are non-zero vectors which are orthogonal to themselves.

A vector $\mathbf{v}$ with negative norm squared $\langle \mathbf{v}, \mathbf{v} \rangle < 0$ is called **time-like**. It is called **space-like** if $\langle \mathbf{v}, \mathbf{v} \rangle > 0$ and **null** or **light-like** if $\langle \mathbf{v}, \mathbf{v} \rangle = 0$. All vectors in M except the zero vector fall into one of these categories.

The basic fact about Minkowski space is:

A non-zero vector orthogonal to a time-like vector must be space-like.

That is, $\mathbf{v} \neq \mathbf{0}$ and $\mathbf{T}$ such that $\langle \mathbf{T}, \mathbf{T} \rangle < 0, \langle \mathbf{T}, \mathbf{v} \rangle = 0$ implies $\langle \mathbf{v}, \mathbf{v} \rangle > 0$. We can rephrase it by saying that there can be no time-like or null vector orthogonal to a time-like vector.

Since the inner product of two time-like vectors $\mathbf{T}$ and $\mathbf{T}'$ can only be non-zero, we define a relation $\mathbf{T} \sim \mathbf{T}'$ on time-like vectors by saying they are related if $\langle \mathbf{T}, \mathbf{T}' \rangle < 0$. A vector is related to itself. If $\mathbf{T}$ is related to $\mathbf{T}'$ then $\mathbf{T}'$ is related to

$\mathbf{T}$. Moreover if $\mathbf{T} \sim \mathbf{T}'$ and $\mathbf{T}' \sim \mathbf{T}''$ then we can prove the transitivity property of this relation (that $\mathbf{T} \sim \mathbf{T}''$) as follows.

First we see that $\mathbf{T} - a\mathbf{T}''$ can be made orthogonal to $\mathbf{T}'$ by choosing $a = \langle \mathbf{T}', \mathbf{T} \rangle / \langle \mathbf{T}', \mathbf{T}'' \rangle > 0$. Since $\mathbf{T}'$ is time-like, a vector orthogonal to it can only be space-like. So,

$$\langle \mathbf{T} - a\mathbf{T}'', \mathbf{T} - a\mathbf{T}'' \rangle > 0.$$

This gives,

$$\langle \mathbf{T}, \mathbf{T}'' \rangle < \frac{1}{2a} \left[\langle \mathbf{T}, \mathbf{T} \rangle + a^2 \langle \mathbf{T}'', \mathbf{T}'' \rangle \right] < 0.$$

This equivalence relation '$\sim$' splits all time-like vectors into two classes called the 'future' and 'past' pointing time-like vectors. There are exactly two equivalence classes because if $\mathbf{T}, \mathbf{T}', \mathbf{T}''$ are three time-like vectors with $\mathbf{T}$ and $\mathbf{T}''$ *not* in the same class as $\mathbf{T}'$, then $\mathbf{T}$ and $\mathbf{T}''$ are in the same equivalence class: $\mathbf{T} \sim \mathbf{T}''$. (The proof is exactly similar to the proof of transitive property given above. We can make $\mathbf{T} - a\mathbf{T}''$ orthogonal to $\mathbf{T}'$ by choosing a positive $a > 0$, then $\mathbf{T} - a\mathbf{T}''$ has to be space-like: $\langle \mathbf{T} - a\mathbf{T}'', \mathbf{T} - a\mathbf{T}'' \rangle > 0$ this gives $\langle \mathbf{T}, \mathbf{T}'' \rangle < 0$.)

A linear combination with *positive* coefficients of two vectors in the same class is again in the class.

Consider the set of all null or light-like vectors in the Minkowski space. As in the case of time-like vectors we can define two null vectors $\mathbf{k}$ and $\mathbf{k}'$ to be related if $\langle \mathbf{k}, \mathbf{k}' \rangle \leq 0$. (This time we have to include zero as well in the inequality otherwise a vector will not be equivalent to itself!). Exactly as in the case of time-like vectors there are two equivalence classes called the 'future' and 'past' **light cones**.

The inner product of two space-like vectors can be positive, negative or zero and there is no further division of space-like vectors.

In Minkowski space there exists an orthonormal basis with one time-like vector $\mathbf{n}_0$ with norm square -1 and three space-like vectors $\mathbf{n}_1, \mathbf{n}_2, \mathbf{n}_3$, with $+1$:

$$\eta_{\mu\nu} \equiv \langle \mathbf{n}_\mu, \mathbf{n}_\nu \rangle = \begin{pmatrix} -1 & & & \\ & 1 & & \\ & & 1 & \\ & & & 1 \end{pmatrix}$$

where μ, ν take values $0, 1, 2, 3$.

As an illustration we construct the star operator for Minkowski space.

The orthonormal basis in M^* dual to the basis $\{\mathbf{n}_\mu\}$ is $\nu^0, \nu^1, \nu^2, \nu^3$ with

$$\langle \nu^0, \nu^0 \rangle = -1, \qquad \langle \nu^1, \nu^1 \rangle = \langle \nu^2, \nu^2 \rangle = \langle \nu^3, \nu^3 \rangle = 1.$$

Let us choose the orientation $\eta = \nu^0 \wedge \nu^1 \wedge \nu^2 \wedge \nu^3$. As $\Lambda^0 = R$ the normal basis vector in this one-dimensional space is the number 1.

We write the *-operated vectors in Table 6.3.

Table 6.3: Star-operator in Minkowski space

t	$*t$
1	$\nu^0 \wedge \nu^1 \wedge \nu^2 \wedge \nu^3$
ν^0	$-\nu^1 \wedge \nu^2 \wedge \nu^3$
ν^1	$-\nu^0 \wedge \nu^2 \wedge \nu^3$
ν^2	$\nu^0 \wedge \nu^1 \wedge \nu^3$
ν^3	$-\nu^0 \wedge \nu^1 \wedge \nu^2$
$\nu^0 \wedge \nu^1$	$-\nu^2 \wedge \nu^3$
$\nu^0 \wedge \nu^2$	$\nu^1 \wedge \nu^3$
$\nu^0 \wedge \nu^3$	$-\nu^1 \wedge \nu^2$
$\nu^1 \wedge \nu^2$	$\nu^0 \wedge \nu^3$
$\nu^1 \wedge \nu^3$	$-\nu^0 \wedge \nu^2$
$\nu^2 \wedge \nu^3$	$\nu^0 \wedge \nu^1$
$\nu^1 \wedge \nu^2 \wedge \nu^3$	$-\nu^0$
$\nu^0 \wedge \nu^2 \wedge \nu^3$	$-\nu^1$
$\nu^0 \wedge \nu^1 \wedge \nu^3$	ν^2
$\nu^0 \wedge \nu^1 \wedge \nu^2$	$-\nu^3$
$\nu^0 \wedge \nu^1 \wedge \nu^2 \wedge \nu^3$	-1

$$(6.19)$$

6.8 Tutorial

Exercise 33. Show that a metric is non-degenerate if and only if the determinant $g = \det g_{ij}$ is non-zero.

Answer 33. If the g were zero then the symmetric matrix g_{ij} would have an eigenvector with eigenvalue zero. That is, there would be numbers $\lambda_j, j = 1, \ldots, n$ *not all zero* such that for all $i = 1, \ldots, n$, $\sum_j g_{ij}\lambda_j = 0 = \langle \mathbf{e}_i, \sum_j \lambda_j \mathbf{e}_j \rangle$. As $\{\mathbf{e}_i\}$ is a basis it follows that $\langle \mathbf{v}, \sum_j \lambda_j \mathbf{e}_j \rangle = 0$ for every $\mathbf{v}$. Because the inner product is non-degenerate, it follows that $\sum_j \lambda_j \mathbf{e}_j = \mathbf{0}$, and because ,again, $\mathbf{e}_i$ form a basis $\lambda_j = 0$ for all $j = 1, \ldots, n$. This contradicts that not all λ's are zero.

Exercise 34. Show that the inner product for the dual space V^* as defined in section 6.4 is non-degenerate.

Answer 34. If $\langle \alpha, \beta \rangle$ were to be zero for every choice of β, then by choosing β to be equal to members of the basis $\{\nu_j\}$ dual to $\{\mathbf{n}_i\}$, one-by-one, we can prove that $\alpha(\mathbf{n}_j) = 0$ for all $j = 1, \ldots, n$. This means that $\alpha(\mathbf{v}) = 0$ for any $\mathbf{v} \in V$. Therefore the inner product as defined is non-degenerate.

Exercise 35. Find the components of the Cartan tensor in a general basis.

Answer 35. Choose an orientation in an o.n. basis $\eta = \nu^1 \wedge \cdots \wedge \nu^n$. Let a general basis $\{\alpha^j\}$ in Λ^1 be related to an o.n. basis $\{\nu^j\}$ as $\alpha^i = \sum_j A_{ij}\nu^j$ and having the same orientation. (So that $\det A > 0$.) We have $g^{ij} = \langle \alpha^i, \alpha^j \rangle = A_{ik}A_{jl}\eta^{kl}$. The components of a vector f in the two bases $f = f_i\alpha^i = k_i\nu^i$ are related by $f_i = k_j B_{ji}$ where $B = A^{-1}$. Thus the components $\bar{\varepsilon}_{j_1\ldots j_n}$ of η in the α basis are obtained by comparison of

$$\eta = \varepsilon_{i_1\ldots i_n}\nu^{i_1} \otimes \cdots \otimes \nu^{i_n} = \bar{\varepsilon}_{j_1\ldots j_n}\alpha^{j_1} \otimes \cdots \otimes \alpha^{j_n}$$

to give

$$\bar{\varepsilon}_{j_1\ldots j_n} = \varepsilon_{i_1\ldots i_n} B_{i_1 j_1} \cdots B_{i_n j_n}.$$

It is obvious from this formula that, like the $\varepsilon_{i_1\ldots i_n}$, the components $\bar{\varepsilon}_{i_1\ldots i_n}$ are also completely antisymmetric; for example

$$
\begin{aligned}
\bar{\varepsilon}_{j_2 j_1 \ldots j_n} &= \varepsilon_{i_1\ldots i_n} B_{i_1 j_2} B_{i_2 j_1} \cdots B_{i_n j_n} \\
&= -\varepsilon_{i_2 i_1 \ldots i_n} B_{i_2 j_1} B_{i_1 j_2} \cdots B_{i_n j_n} \\
&= -\bar{\varepsilon}_{j_2 j_1 \ldots j_n}.
\end{aligned}
$$

All the non-zero components of η in this basis are ± 1 times $\bar{\varepsilon}_{12\ldots n}$. In fact

$$
\begin{aligned}
\bar{\varepsilon}_{12\ldots n} &= \varepsilon_{i_1\ldots i_n} B_{1 i_1} \ldots B_{n i_n} \\
&= \det B = (\det A)^{-1}.
\end{aligned}
$$

From $g^{ij} = A_{ik}A_{jl}\eta^{kl}$ and $g_{ij} = B_{ki}B_{lj}\eta_{kl}$ we know that $(\det B)^2 = g\det(\eta)^{-1}$. As $(\det B)^2 > 0$, we conclude that $g = \det(g_{ij})$ has the same sign as $\det \eta = \epsilon_1 \ldots \epsilon_n$ which is 1 or -1. Therefore

$$\bar{\varepsilon}_{j_1\ldots j_n} = \sqrt{|g|}\varepsilon_{j_1\ldots j_n}.$$

Exercise 36. Starting from the definition of the star operator $\langle *t|s\rangle\eta = (\mathrm{sgn}(g))t \wedge s$ derive the formulas

$$
\begin{aligned}
*(\nu^{i_1} \wedge \cdots \wedge \nu^{i_r}) &= (-1)^P \epsilon_{i_1} \ldots \epsilon_{i_r}\nu^{j_1} \wedge \cdots \wedge \nu^{j_{n-r}} \\
&= \eta^{i_1 k_1} \ldots \eta^{i_r k_r}\varepsilon_{k_1\ldots k_r j_1\ldots j_{n-r}}\nu^{j_1} \wedge \cdots \wedge \nu^{j_{n-r}}.
\end{aligned}
$$

Answer 36. Let $t = \nu^{i_1} \wedge \cdots \wedge \nu^{i_r}$, then the linear functional $T(s)$ on Λ^{n-r} is zero on all vectors s except multiples of $\nu^{j_1} \wedge \cdots \wedge \nu^{j_{n-r}}$ where $j_1 < \cdots < j_{n-r}$ is the complementary set to $i_1 < \cdots < i_r$ in $1, \ldots, n$. Since $\nu^{j_1} \wedge \cdots \wedge \nu^{j_{n-r}}$ belong to an orthonormal basis a vector $*t$ which is orthogonal to all vectors except this one has to be proportional to it. So

$$*t = (\text{constant})\nu^{j_1} \wedge \cdots \wedge \nu^{j_{n-r}}.$$

Applying the definition $\langle *t|s\rangle\eta = (\mathrm{sgn}(g))t \wedge s$ for $s = \nu^{j_1} \wedge \cdots \wedge \nu^{j_{n-r}}$,

$$(\text{constant})\langle \nu^{j_1} \wedge \cdots \wedge \nu^{j_{n-r}}|\nu^{j_1} \wedge \cdots \wedge \nu^{j_{n-r}}\rangle\eta = (\mathrm{sgn}(g))(-)^P\eta$$

where $t \wedge s = \nu^{i_1} \wedge \cdots \wedge \nu^{i_r} \wedge \nu^{j_1} \wedge \cdots \wedge \nu^{j_{n-r}} = (-)^P \eta$, and P is the permutation $\{i_1, \ldots, i_r, j_1, \ldots, j_{n-r}\} \to \{1, \ldots, n\}$. The inner product

$$\langle \nu^{j_1} \wedge \cdots \wedge \nu^{j_{n-r}} | \nu^{j_1} \wedge \cdots \wedge \nu^{j_{n-r}} \rangle = \epsilon_{j_1} \ldots \epsilon_{j_{n-r}},$$

therefore it cancels the factors in $(\mathrm{sgn}(g)) = \epsilon_1 \ldots \epsilon_n$ to give the required formula.

The other formula is straightforward when we realize that $(-)^P = \varepsilon_{i_1 \ldots i_r j_1 \ldots j_{n-r}}$ and $\epsilon_i \delta^{ij} = \eta^{ij}$.

Exercise 37. Show that $** = (-1)^{r(n-r)} \mathrm{sgn}(g)$.

Answer 37.

$$
\begin{aligned}
(\nu^{i_1} \wedge \cdots \nu^{i_r}) &= (-1)^P \epsilon_{i_1} \ldots \epsilon_{i_r} * (\nu^{j_1} \wedge \cdots \nu^{j_{n-r}}) \\
&= (-1)^P (-1)^Q (\epsilon_1 \ldots \epsilon_n) \nu^{i_1} \wedge \cdots \nu^{i_r} \\
&= (-1)^P (-1)^Q \mathrm{sgn}(g) \nu^{i_1} \wedge \cdots \nu^{i_r}
\end{aligned}
$$

where Q is the permutation

$$
\begin{aligned}
Q &= \{j_1 < \cdots < j_{n-r}; i_1 < \cdots < i_r\} \to \{1, \ldots, n\}, \\
P &= \{i_1, \ldots, i_r, j_1, \ldots, j_{n-r}\} \to \{1, \ldots, n\}.
\end{aligned}
$$

As Q and Q^{-1} have the same parity, $(-1)^P (-1)^Q$ is the same as the parity of $Q^{-1} P$: $\{i_1 < \cdots < i_r; j_1 < \cdots < j_{n-r}\} \to \{j_1 < \cdots < j_{n-r}; i_1 < \cdots < i_r\}$, which is $(-1)^{r(n-r)}$.

Exercise 38. Show that in the Minkowski space two orthogonal null vectors must be proportional to each other.

Answer 38. Choose an orthonormal basis. Without loss of generality we can assume that one of the null vectors has components $(1, 1, 0, 0)$ and the other (t, a, b, c) with $a^2 + b^2 + c^2 = t^2$. Orthogonality of the two vectors implies that $-t + a = 0$. Therefore the second vector is $(t, t, 0, 0) = t(1, 1, 0, 0)$.

Chapter 7

Elementary Differential Geometry

The model of spacetime used by the general theory of relativity is that of a differential manifold with a Riemannian geometry. In this and the next few chapters we develop the necessary background with emphasis on tools necessary for a physicist. This introduction is not rigorous from a mathematician's point of view. We would assume that these manifolds have all the nice mathematical properties (Hausdorff nature, paracompactness etc.) which are needed for the existence of various geometrical quantities and procedures.

7.1 Coordinates and Functions

In physics we regularly encounter sets whose points can be labelled by a fixed number (say n) of coordinates. Let M be such a set. By coordinates of a point $p \in M$ we mean numbers $x^1, x^2, \ldots, x^n$ which we associate with the point. In order to be useful this rule of association or mapping (call it ϕ) should have some obviously desirable properties.

1. First of all, coordinates should be independent. It should not happen that one of the coordinates can be calculated by some formula in terms of the remaining coordinates. The minimum number of coordinates needed to specify the points of M is called the **dimension** of M.

2. Moreover, different points should have different coordinates. The mapping ϕ which associates coordinates to the points of M has to be one-to-one.

3. Further, the mapping should be continuous both ways so that neighbouring points have coordinates which are close to each other and vice versa.

4. Lastly, in order to discuss the way quantities change from one point to another, every point should have coordinates defined for all other points in its neighbourhood.

This is expressed by saying that coordinates should be defined on **open sets** of M. The result of this condition is that points of our set M are mapped onto open sets of the n-dimensional Euclidean space R^n and we know how to do the calculus there.

If we can define such a single mapping ϕ for the whole set M we are lucky because then the set M is mapped onto an open set of n-dimensional Euclidean space R^n and all mathematical analysis on M can be transferred to that open subset of R^n. But often it is not possible to cover the entire set M with one such mapping. In such cases we consider coordinates in patches or "charts".

A **coordinate chart** (U, ϕ) is specified by an open subset U of M on which a one-to-one, bothways continuous mapping ϕ is defined which takes points $p \in U \subset M$ to points $\phi(p) = x = (x^1, \ldots, x^n) \in R^n$. We write $x^i = \phi^i(p)$.

We cover the whole set M by several such charts (U, ϕ), $(V, \psi) \ldots$ etc such that every point of M falls in one or the other of these sets $U, V, \ldots$. Such a collection of charts is called an **atlas**.

If (U, ϕ) and (V, ψ) are two charts in an atlas then (unless there is no point common between U and V) a point p in $U \cap V$ can be specified by coordinates $x = \phi(p)$ as well as by $y = \psi(p)$. As p moves within $U \cap V$ the pair of numbers $(x(p), y(p))$ determine a functional relationship between the two coordinate variables:

$$y = \psi(p) = (\psi \circ \phi^{-1})(x), \qquad x = \phi(p) = (\phi \circ \psi^{-1})(y). \tag{7.1}$$

All these functional relationships between all coordinates with overlapping domains in an atlas are continuous by definition.

If, moreover, they happen to be infinitely differentiable as well, then the set M is called a **differentiable manifold** and the atlas is said to determine a **differentiable structure**.

On the same set M we can define charts and atlases in a variety of ways. Two atlases whose charts are mutually compatible with each other, i.e., determine infinitely differentiable coordinates relaionships, have the same differential stucture.

It is quite possible for a set to have two mutually incompatible differential stuctures. For physical applications the choice of the differential structure is usually determined by physics and one does not seem to need the exotic differential structures that can possibly also be defined.

A **smooth function** f defined on a manifold M is a real-valued function which when expressed in all the local coordinates is infinitely differentiable.

Let (U, ϕ) be a chart containing point p. Then $f(p) = f(\phi^{-1}(x)) = f \circ \phi^{-1}(x)$ is infinitely differentiable as a function of coordinates x.

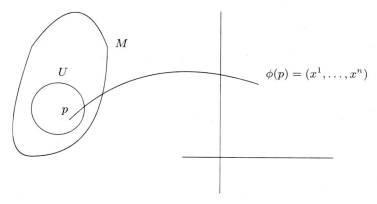

Fig. 7.1: A coordinate chart (U, ϕ) on a manifold mapping a point p into numbers $(x^1, \ldots, x^n)$.

The definition is independent of which coordinate system is chosen to define it. If there are several charts defined around the point, the differentiability of f can be checked in any one of them. The differentiability in other charts follows because all coordinates are smooth functions of each other.

We denote by $\mathcal{F}(M)$ the class of all smooth funtions defined on M. $\mathcal{F}(M)$ is obviously a vector space because the sum of smooth functions is smooth, as is the function obtained by multiplying by a real number.

Actually the set has moreover a multiplication defined on it because members of $\mathcal{F}(M)$ can be multiplied by each other to give another smooth function: if $f, g \in M$ are smooth then $(fg)(p) \equiv f(p)g(p)$ is also smooth.

7.2 Curves and Tangent Vectors

A **smooth curve** c is a mapping from an interval I of the real line into a differentiable manifold M such that the coordinates of the mapped point $c(t), t \in I$ are smooth, i.e., infinitely differentiable functions of t.

Let us assume that I contains a point $t = t_0$ which is mapped to the point $p_0 = c(t_0)$. Let (U, ϕ) be a coordinate chart containing the point p_0. Then the smoothness of the curve is determined by the infinite differentiability of the functions $x^i(t) = \phi^i(c(t))$ with respect to t.

The rate of change of these coordinates with respect to parameter t,

$$v^i \equiv \left. \frac{dx^i}{dt} \right|_{t_0} \tag{7.2}$$

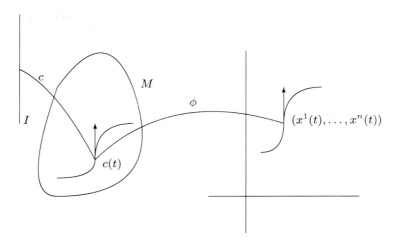

Fig. 7.2: A curve $c : I \to M$ is represented by coordinates as n functions $t \to (x^1(t), \ldots, x^n(t))$. The arrows represent the tangent vectors to the curve in M and its image in R^n given by the components $(dx^1/dt, \ldots, dx^n/dt)$.

are called the components of a **tangent vector** to the curve c at p_0 in terms of coordinates (U, ϕ). If p_0 belongs to the domain of some other chart (V, ψ) then the components in that chart will be

$$w^i = \left.\frac{dy^i}{dt}\right|_{t_0} = \left.\frac{\partial y^i}{\partial x^j}\right|_{x(p_0)} \left.\frac{dx^j}{dt}\right|_{t_0}.$$

In other words, the same tangent vector has components in the two coordinates related by

$$w^i = \frac{\partial y^i}{\partial x^j} v^j. \tag{7.3}$$

The student might have seen this formula before. It is known as the **law of transformation of a contravariant vector field**.

It is important to picture the tangent vector to a curve c as a geometric entity independent of the coordinate systems that may be defined in the neighbourhood of the point in question. Actually, the tangent vector determines the rate at which the point $c(t)$ changes with respect to t. As we cannot measure the extent by which the abstract point $c(t)$ moves because there is no distance function to tell how far the point has moved, we measure this rate indirectly through the coordinates.

Coordinates are a convenient set of n real-valued functions on M. If we take function $f \in \mathcal{F}(M)$ we can measure how $f(c(t))$ changes with t. By knowing how all functions $f \in \mathcal{F}(M)$ change with respect to t, we know what the tangent vector is.

For an arbitrary smooth function f,

$$\frac{df}{dt}\bigg|_{t_0} = v^i \frac{\partial}{\partial x^i}\bigg|_{\phi(p_0)} f(\phi^{-1}(x)).$$

For all possible functions f what remains common is the linear differential operator

$$\mathbf{v} = v^i \frac{\partial}{\partial x^i}\bigg|_{\phi(p_0)}. \tag{7.4}$$

In this form it is independent of the coordinate chart because

$$\mathbf{v} = v^i \frac{\partial}{\partial x^i}\bigg|_{\phi(p_0)} = v^i \frac{\partial y^j}{\partial x^i}\frac{\partial}{\partial y^j}\bigg|_{\psi(p_0)} = w^j \frac{\partial}{\partial y^j}\bigg|_{\psi(p_0)}.$$

We can also look at the tangent vector $\mathbf{v}$ at p_0 as a rule which assigns to every smooth function f a real number, namely, its rate of change $\mathbf{v}(f) = df/dt$ at t_0 along the curve to which it is tangent. We discuss this in the next section.

7.3 Tangent Space

Let $\mathcal{F}(\mathcal{M})$ be the class of smooth functions defined above. Then a tangent vector $\mathbf{v}$ at p as discussed in the last section gives for any smooth function f (defined in a neighbourhood of the point p) its rate of change. The function $f \to \mathbf{v}(f)$ has the following properties:

$$\mathbf{v}(f + g) = \mathbf{v}(f) + \mathbf{v}(g), \tag{7.5}$$

$$\mathbf{v}(af) = a\mathbf{v}(f), \tag{7.6}$$

$$\mathbf{v}(fg) = \mathbf{v}(f)g(p) + f(p)\mathbf{v}(g), \tag{7.7}$$

where f, g are smooth functions and a is any real number. These properties can be verified immediately from the definition of $\mathbf{v}$ given as a differentiable operator in the last section.

Conversely, we can use these properties to *define* a tangent vector. Since coordinate functions x^i are (trivially) infinitely differentiable functions in a neighbourhood of point p, we can recover the components of the vector by acting with $\mathbf{v}$ on x^i:

$$\mathbf{v} = v^i \frac{\partial}{\partial x^i}, \qquad v^i \equiv \mathbf{v}(x^i). \tag{7.8}$$

We would use whichever definition is more convenient for the purpose.

The sum $\mathbf{v_1} + \mathbf{v_2}$ of two tangent vectors at the same point $p \in M$ can be defined in an obvious fashion as

$$(\mathbf{v_1} + \mathbf{v_2})(f) = \mathbf{v_1}(f) + \mathbf{v_2}(f).$$

Similarly we define multiplication by a real number a,

$$(a\mathbf{v})(f) = a\mathbf{v}(f).$$

These definitions make the set $T_p(M)$ of all tangent vectors associated with the point $p \in M$ a vector space, called the **tangent space** at p.

As every vector $\mathbf{v}$ can be written

$$\mathbf{v} = v^i \frac{\partial}{\partial x^i}\bigg|_{\phi(p_0)}, \tag{7.9}$$

the n operators

$$\left\{ \frac{\partial}{\partial x^1}, \frac{\partial}{\partial x^2}, \ldots, \frac{\partial}{\partial x^n} \right\}$$

form a basis in $T_p(M)$. This shows that the tangent space at any point of the manifold has the same dimension n as the manifold itself. It is the basis naturally associated with the coordinate chart (U, ϕ) and is called the **natural** or **coordinate basis** associated with the coordinates.

One must not think that these operators have a meaning only when they have something to act on. They should, instead, be treated as symbols for the basis vectors which also happen to represent the role of a tangent vector as a differential operation.

7.4   Vector Fields on a Manifold

Smooth functions and vector fields are the fundamental building blocks of a differentiable manifold. All other geometric quantities: tensor fields, differential forms, Lie derivatives, connections, curvature, metric etc. are defined using these.

We have seen that there is a tangent space associated with every point of the manifold. A vector field is a smooth way of choosing one tangent vector at each of these spaces.

A **tangent vector field** or simply a **vector field** X on M is an assignment of a tangent vector $X(p) \in T_p(M)$ at each point $p \in M$ in a smooth way. This 'smooth way' means that in terms of local coordinates, the components of $X(p)$ change smoothly with coordinates of p. For a chart (U, ϕ) the vector field can be written

$$X(p) = v^i(x)\frac{\partial}{\partial x^i}$$

where v^i are now smooth functions of coordinates x. The components of the vector field are smooth in all coordinate systems if they are smooth in one. This is so because of the law of transformation which relates the components of the vector at the same point in the two coordinate systems.

A tangent vector field is defined on the whole manifold, whereas the coordinate systems are defined in patches. But this does not create problems because one can pass from one coordinate system to the other on the overlapping part.

We again emphasize: notice the way components change when coordinate systems are changed:

$$X = v'^j(x')\frac{\partial}{\partial x'^j} = v^i(x)\frac{\partial}{\partial x^i} = v^i(x)\frac{\partial x'^j}{\partial x^i}\frac{\partial}{\partial x'^j}$$

or

$$v'^j(x') = \frac{\partial x'^j}{\partial x^i}v^i(x). \tag{7.10}$$

For a smooth vector field X on M the action of $X(p)$ on a smooth function $f \in \mathcal{F}(M)$ gives another smooth function $X(f) \in \mathcal{F}$:

$$(X(f))(p) = (X(p))(f) = v^i(x)\frac{\partial f}{\partial x^i}. \tag{7.11}$$

We can use this fact to give a definition of vector field X on M as a mapping of smooth functions $\mathcal{F}(M)$ on M into themselves with the following properties. $X : \mathcal{F}(M) \to \mathcal{F}(M)$ is such that

$$X(f + g) = X(f) + X(g), \tag{7.12}$$

$$X(fg) = X(f)g + fX(g) \tag{7.13}$$

for every $f, g \in \mathcal{F}$.

It is easy for physicists to imagine vector fields, familiar as they are with physical fields of force like magnetic, electrical or gravitational fields. All these fields assign a vector quantity at each spacetime point. We shall see in a later chapter that physicists' idea of lines of force: that is, a family of curves whose tangent vectors give the force field is also very naturally defined and useful.

The class $\mathcal{X}(M)$ of all smooth vector fields on a manifold M is a vector space because we can define the sum of two vector fields X and Y as the vector field obtained by adding the vectors $X(p)$ and $Y(p)$ assigned to a point $p \in M$ and denoting this as the value of $X + Y$ at p. Similarly we can say $(aX)(p) = aX(p)$.

The student should appreciate that here we are dealing with an infinite number of vector spaces; there is a tangent space $T_p(M)$ at each point $p \in M$. A vector field is an infinite collection of vectors, one in each tangent space. It is a function. The class or collection of vector fields is itself a vector space $\mathcal{X}(M)$ because we can define addition and multiplication by a real number for these objects. But this vector space, like most function spaces, is certainly an infinite dimensional vector space. We can multiply a vector field X by a smooth function $f \in \mathcal{F}(M)$ instead of multiplying by a constant number a and get another smooth field. This is expressed by saying that $\mathcal{X}(M)$ is not just a vector space but an $\mathcal{F}(M)$-module where the job of multiplication by numbers is taken over by the ring of functions $\mathcal{F}(M)$.

7.5   Local Basis Fields

Let there be a coordinate system around point $p \in M$ with coordinate functions x^i. We have seen that any vector field can be written in these coordinates as

$$X(p) = v^i(x)\frac{\partial}{\partial x^i}.$$

This shows that vector fields $E_i, i = 1, \ldots, n$ given by

$$E_i(p) = \left.\frac{\partial}{\partial x^i}\right|_{\phi(p)}$$

which are defined in the domain U of the coordinate system form a basis for vector fields. We call these **local basis fields**.

In general if there are n vector fields $X_1, \ldots, X_n$ such that at each point $p \in U \subset M$ in a domain U, the vectors $X_1(p), \ldots, X_n(p)$ are linearly independent then we call the vectors a local basis field.

One must understand the symbolic nature of the notation used for the natural basis vectors in the coordinate chart (U, ϕ). If p and q are two points, then we must write

$$\left.\frac{\partial}{\partial x^i}\right|_{\phi(p)}, \qquad \left.\frac{\partial}{\partial x^i}\right|_{\phi(q)}$$

to distinguish the tangent vectors at the two points. They belong to entirely different tangent spaces T_p and T_q respectively. In practice however one writes this local basis as

$$\left\{\frac{\partial}{\partial x^1}, \ldots, \frac{\partial}{\partial x^n}\right\}$$

to simplify the notation.

A simple way to picture the natural basis vectors is to realise that these vectors are tangent vectors to the coordinate 'mesh' of the curvilinear coordinates. For example, $\partial/\partial x^i$ is the tangent vector passing through the given point to the curve all of whose coordinates except x^i are fixed.

In the case of radial r, θ coordinates in the plane, the $\partial/\partial r$ are along the radial direction and $\partial/\partial \theta$ along circles of constant r in the direction of increasing θ. Of course, there is no concept of the "length" or norm of these vectors. So one can picture them as little arrows pointing in the direction determined by the coordinate mesh.

7.6   Lie Bracket

Given two vector fields X and Y we can define a new vector field $[X,Y]$ called their **Lie bracket** as follows:

$$[X,Y](f) = X(Y(f)) - Y(X(f)) \qquad \forall f \in \mathcal{F}(M). \tag{7.14}$$

We check easily that $[X,Y]$ too satisfies the defining properties given in the previous section: the second ('Leibnitz') property need only be verified.

$$
\begin{aligned}
X(Y(fg)) &= X(Y(f)g + fY(g)) \\
&= X(Y(f))g + Y(f)X(g) + X(f)Y(g) + fX(Y(g)).
\end{aligned}
$$

The second and third terms in the final expression cancel when $Y(X(fg))$, written similarly, is subtracted from $X(Y(fg))$.

The Lie bracket being a 'commutator' of two vector fields acting on a smooth function is antisymmetric,

$$[X,Y] = -[Y,X] \tag{7.15}$$

and satisfies the **Jacobi identity**

$$[X,[Y,Z]] + [Y,[Z,X]] + [Z,[X,Y]] = 0 \tag{7.16}$$

for any vector fields X, Y, Z.

In some local coordinates (U, ϕ) if the fields X and Y are expressed as

$$X = v^i(x) \frac{\partial}{\partial x^i}, \qquad Y = w^i(x) \frac{\partial}{\partial x^i},$$

then $[X,Y]$ is given by

$$[X,Y] = \left(v^i(x) \frac{\partial w^j}{\partial x^i} - w^i(x) \frac{\partial v^j}{\partial x^i} \right) \frac{\partial}{\partial x^j} \tag{7.17}$$

as follows from direct computation. Note that for coordinate basis fields $[\partial/\partial x^i, \partial/\partial x^j] = 0$ for any i, j.

7.7   Cotangent Space

The space T_p^* dual to the tangent space T_p is called the **cotangent space**. We can associate it in an abstract way as also a space attached to the point p. Actually there is a virtual crowding at the point p because we will associate the whole infinite set of tensor spaces $(T_p)_s^r$ and differential forms $\Lambda(T_p^*)$ to the point.

The basis dual to the standard basis

$$\left\{ \frac{\partial}{\partial x^1}, \frac{\partial}{\partial x^2}, \ldots, \frac{\partial}{\partial x^n} \right\}$$

is denoted by

$$\{dx^1, \ldots, dx^n\}.$$

It consists of linear functionals dx^i which act on the basis vectors in T_p as

$$dx^i \left(\frac{\partial}{\partial x^j} \right) = \delta^i_j. \tag{7.18}$$

Just as the symbols $\partial/\partial x^i$ are chosen to give correct transformation laws under change of coordinates,

$$\frac{\partial}{\partial x'^i} = \frac{\partial x^j}{\partial x'^i} \frac{\partial}{\partial x^j} \equiv A_i{}^j \frac{\partial}{\partial x^j},$$

the dual basis changes as naturally,

$$dx'^i = \frac{\partial x'^i}{\partial x^j} dx^j \equiv (A^{-1})_j{}^i dx^j = (A^{-1T})^i{}_j dx^j \tag{7.19}$$

according to the 'inverse-transposed' matrix rule.

Analogous to a vector field we can define a **cotangent vector field**, also called a **one-form** or the **covariant vector field**, as a smooth assignment of a vector in T_p^*. Such a field can be written as

$$\alpha = a_i(x)dx^i.$$

The components of a cotangent vector field transform as expected,

$$\alpha = a'_j(x')dx'^j = a_i(x)dx^i = a_i(x)\frac{\partial x^i}{\partial x'^j}dx'^j$$

therefore

$$a'_j(x') = \frac{\partial x^i}{\partial x'^j}a_i(x). \tag{7.20}$$

If α is a cotangent vector field and X a vector field, we can take the pairing of dual vectors at each point, getting a smooth function

$$\alpha(X) = a_i(x)v^j(x)dx^i \left(\frac{\partial}{\partial x^j} \right) = a_i(x)v^i(x). \tag{7.21}$$

7.8 Tensor Fields

Starting with the tangent vector space T_p at a point of a manifold we can construct the infinite set of tensor spaces

$$T_s^r(p) = \underbrace{T_p \otimes \cdots \otimes T_p}_{r \text{ factors}} \otimes \underbrace{T_p^* \otimes \cdots \otimes T_p^*}_{s \text{ factors}}. \tag{7.22}$$

In this notation the tangent space T_p is $T_0^1(p)$ and the cotangent space T_p^* is $T_1^0(p)$.

A smooth choice of a tensor at each point gives tensor fields

$$t = t^{i_1 \ldots i_r}{}_{j_1 \ldots j_s}(x) \frac{\partial}{\partial x^{i_1}} \otimes \cdots \otimes \frac{\partial}{\partial x^{i_r}} \otimes dx^{j_1} \otimes \cdots \otimes dx^{j_s}. \tag{7.23}$$

We will denote tensor fields on a manifold by $\mathcal{X}_s^r(M)$. The set $\mathcal{X}(M)$ of vector fields is just $\mathcal{X}_0^1(M)$ in this notation. The set of cotangent vector fields, also called **differential 1-forms**, is $\mathcal{X}_1^0(M)$ which will be denoted also by $\Lambda^1(M)$. The smooth functions in $\mathcal{F}(M)$ are tensors of rank zero as well as differential 0-forms, therefore $\mathcal{F}(M) = \mathcal{X}_0^0(M) = \Lambda^0(M)$

7.9 Defining Tensors Fields

We have defined tensor fields through their components in the natural basis. If the components are smooth functions of coordinates, the fields are smooth or differentiable.

We have seen that a vector field maps a smooth function to a smooth function.

Let α be a covariant vector field and X a smooth vector field. Then α (which assigns a vector $\alpha(p) \in T_p^*$ for each p in the dual space) determines a smooth function $\alpha(X)$ as follows:

$$\big(\alpha(X)\big)(p) = \alpha(p)(X(p)).$$

This mapping of a vector field into smooth functions is such that

$$\alpha(X + Y) = \alpha(X) + \alpha(Y), \tag{7.24}$$
$$\alpha(fX) = f\alpha(X), \tag{7.25}$$

for $X, Y \in \mathcal{X}$ and $f \in \mathcal{F}(M)$. Notice that here the linearity property in the second equation is not just on multiplication with real numbers, but *with respect to smooth functions* in $\mathcal{F}(M)$.

One can, in fact, *define* a smooth cotangent vector field as a mapping of this kind.

It is crucial to demand linearity with respect to smooth functions and not just real numbers. As a counterexample, let $a_i(x), \lambda^j(x), i, j = 1, \ldots, n$ be smooth functions in some coordinate system. Define for any $X = v^i(x)\partial/\partial x^i$ a mapping

$$\alpha(X)(x) = \left(v^i(x) + \frac{\partial v^i}{\partial x^j} \lambda^j(x) \right) a_i(x)$$

then this mapping is linear when X is multiplied by a number but not when multiplied by a function. Linearity with respect to functions ensures the proper transformation property for these fields.

The basic idea is that although the mappings are defined on spaces of *fields* defined in a region, we must make sure that the mappings determine a quantity in the appropriate tensor space $(T_s^r)_p$ which depends only on vectors in T_p, T_p^* etc. and not spaces at neighbouring points. In the counterexample above the presence of derivatives of components v^i of the vector field X shows dependence of the mapping on points in the neighbourhood, because a derivative involves comparing values at neighbouring points.

A covariant tensor field T of second-rank (i.e., type $\mathcal{X}_2^0(M)$) is a bilinear mapping of $\mathcal{X}(M) \times \mathcal{X}(M)$ into $\mathcal{F}(M)$ such that

$$\begin{align}
T(X+Y,Z) &= T(X,Z)+T(Y,Z), &(7.26)\\
T(fX,Y) &= fT(X,Y), &(7.27)\\
T(X,Y+Z) &= T(X,Y)+T(X,Z), &(7.28)\\
T(X,fY) &= fT(X,Y), &(7.29)
\end{align}$$

for fields X, Y and functions f.

A tensor field S of type $\mathcal{X}_2^1(M)$ can be defined as a bilinear mapping $S : \mathcal{X}(M) \times \mathcal{X}(M) \to \mathcal{X}(M)$,

$$\begin{align}
S(X+Y,Z) &= S(X,Z)+S(Y,Z), &(7.30)\\
S(fX,Y) &= fS(X,Y), &(7.31)\\
S(X,Y+Z) &= S(X,Y)+S(X,Z), &(7.32)\\
S(X,fY) &= fS(X,Y). &(7.33)
\end{align}$$

A case where a mapping $\mathcal{X} \times \mathcal{X} \to \mathcal{X}$ **does not** define a tensor field is the mapping $(X,Y) \to [X,Y]$ of two vector fields giving the third vector field which is their Lie bracket. We have

$$[X,fY] = f[X,Y] + X(f)Y$$

where the extra term spoils the linearity.

Similarly, the mapping $X,Y \to D_X Y$ giving the covariant derivative or connection (to be discussed in the next chapter) is a celebrated example of a mapping not defining a tensor field. But we shall see that a combination $T : (X,Y) \to D_X Y - D_Y X - [X,Y]$ where the last term cancels the non-tensorial terms of the first two, does define a tensor called torsion.

7.10   Differential Forms and Exterior Derivative

The antisymmetric covariant tensor fields, which amount to choosing members of $\Lambda^r(T_p^*)$ smoothly, are specially important and are called **differential r-forms** on the manifold M and denoted by $\Lambda^r(M)$.

Differential r-Forms

We define **differential 0-forms** as the class of smooth functions $\Lambda^0(M) = \mathcal{F}(M)$ and as noted above the differential 1-forms are just the cotangent vectors $\Lambda^1(M) = \chi_1^0(M)$

An r-form is a tensor field of the form

$$\alpha = \sum_{i_1 < \cdots < i_r} A_{i_1 \dots i_r} dx^{i_1} \wedge \cdots \wedge dx^{i_r}$$

with coefficients $A_{i_1 \dots i_r}$ smooth functions of coordinates.

The wedge product of an r-form and an s-form is the wedge product at each point. It naturally satisfies

$$\alpha \wedge \beta = (-1)^{rs} \beta \wedge \alpha$$

where α is an r-form and β an s-form.

Exterior Derivative

The entire machinery of differential forms revolves round the concept of **exterior derivative** which we now define.

For a 0-form (that is a function) $f \in \mathcal{F}$ defines the cotangent field

$$df = \frac{\partial f}{\partial x^i} dx^i. \tag{7.34}$$

For α, an r-form $(r > 0)$ the exterior derivative $d\alpha$ is the (r+1)-form

$$d\alpha = \sum_{j=1}^{n} \sum_{i_1 < \cdots < i_r} \frac{\partial A_{i_1 \dots i_r}}{\partial x^j} dx^j \wedge dx^{i_1} \wedge \cdots \wedge dx^{i_r}. \tag{7.35}$$

We must use the antisymmetry of the wedge product in the above expression to bring it into the standard form of a basis, that is with dx^i factors in order of increasing i.

For example, if $\alpha = A_i dx^i$ is a 1-form, the exterior derivative is

$$d\alpha = \frac{\partial A_i}{\partial x^j} dx^j \wedge dx^i = \sum_{j<i} \left(\frac{\partial A_i}{\partial x^j} - \frac{\partial A_j}{\partial x^i} \right) dx^j \wedge dx^i.$$

A crucial property of the exterior derivative operator is that applying it twice gives zero on any differential form:

$$
\begin{aligned}
d(d\alpha) &= d\left(\sum_{j=1}^{n} \sum_{i_1 < \cdots < i_r} \frac{\partial A_{i_1 \dots i_r}}{\partial x^j} dx^j \wedge dx^{i_1} \wedge \cdots \wedge dx^{i_r} \right) \\
&= \sum_{k} \sum_{j} \sum_{i_1 < \cdots < i_r} \frac{\partial^2 A_{i_1 \dots i_r}}{\partial x^k \partial x^j} dx^k \wedge dx^j \wedge dx^{i_1} \wedge \cdots \wedge dx^{i_r} \\
&= 0. \tag{7.36}
\end{aligned}
$$

The double sum over k, j gives zero because the second partial derivatives of $A_{i_1 \ldots i_r}$ are symmetric in these indices while there is antisymmetry in these indices due to $dx^k \wedge dx^j$.

It is clear that if α is an r-form and β an s-form,

$$d(\alpha \wedge \beta) = (d\alpha) \wedge \beta + (-1)^r \alpha \wedge d\beta \tag{7.37}$$

which follows directly from the formula for the wedge product and the definition of the exterior derivative.

For a one-form α the exterior derivative can be defined as the antisymmetric bilinear functional on a pair of vector fields X_0 and X_1 as

$$d\alpha(X_0, X_1) = X_0(\alpha(X_1)) - X_1(\alpha(X_0)) - \alpha([X_0, X_1]) \tag{7.38}$$

which is easily verified. Let $X_0 = u^i \partial/\partial x^i$ and $X_1 = v^i \partial/\partial x^i$ Then,

$$
\begin{aligned}
d\alpha(X_0, X_1) &= \frac{\partial A_i}{\partial x^j}(dx^j \wedge dx^i)(X_0, X_1) \\
&= \frac{\partial A_i}{\partial x^j}(u^j v^i - u^i v^j) \\
&= u^j \frac{\partial A_i v^i}{\partial x^j} - v^j \frac{\partial A_i u^i}{\partial x^j} - A_i \left(u^j \frac{\partial v^i}{\partial x^j} - v^j \frac{\partial u^i}{\partial x^j} \right) \\
&= X_0(\alpha(X_1)) - X_1(\alpha(X_0)) - \alpha([X_0, X_1]).
\end{aligned}
$$

This formula can be generalised to an r-form β. $d\beta$ is the (r+1)-form defined by

$$
\begin{aligned}
d\beta(X_0, X_1, \ldots, X_r) = {} &X_0(\beta(X_1, X_2, \ldots, X_r)) - X_1(\beta(X_0, X_2, \ldots, X_r)) \\
&+ X_2(\beta(X_0, X_1, \ldots, X_r)) - \cdots + (-1)^r X_r(\beta(X_0, X_1, \ldots, X_{r-1})) \\
&+ \sum_{i<j}(-1)^{i+j}\beta([X_i, X_j], X_0, \ldots, X_{i-1}, X_{i+1}, \ldots, X_{j-1}, X_{j+1}, \ldots, X_r).
\end{aligned}
\tag{7.39}
$$

7.11   Closed and Exact Differential Forms

A differential r-form α is called **closed** if the (r + 1)-form $d\alpha$ is zero. An r-form α is called **exact** if it can be written as $\alpha = d\beta$ where β is an (r − 1)-form. The set of all closed r-forms $C^r(M)$ is a subspace in $\Lambda^r(M)$ and the set of all exact r-forms is still a smaller subspace of $C^r(M)$.

Every exact form is obviously closed because $d(d(\beta)) = 0$. The converse is *only locally* true in the following sense. Let B be an open ball in coordinate space R^n, and let $U = \phi^{-1}(B)$. If α is a closed form $d(\alpha) = 0$ then there exists an $(r − 1)$-form β such that restricted to U, $\alpha = d(\beta)$. This result is called the **Poincare lemma**.

7.12 Tutorial

Manifolds

Exercise 39. CHOICE OF COORDINATES
Choose an appropriate atlas of coordinate charts for the following manifolds and find the change of coordinates maps:

    1. one-dimesional 'sphere', that is the circle S^1,

    2. two-dimensional sphere S^2,

    3. two-dimensional torus T^2,

    4. the two-dimensional surface of a cone with vertex angle α.

Answer 39. Hints and suggestions
(Draw diagrams to understand the hints !)

The circle
S^1 can be considered as the subset of the two-dimensional plane with points whose coordinates satisfy $x^2 + y^2 = 1$. One chart (U_1, ϕ_1) assigns to a point p the usual angle $\phi_1(p) = \theta_1, 0 < \theta_1 < 2\pi$. This is valid for all points except $(x = 1, y = 0)$. So U_1 is the set of all points on the circle except $(1, 0)$. Another chart (U_2, ϕ_2) can be given covering all points except $(x = -1, y = 0)$ mapping the point p to the angle $\theta_2 = \phi_2(p), -\pi < \theta_2 < \pi$ going counterclockwise from just below the point $(x = -1, y = 0)$ to just above it. The two coordinate patches intersect in two disjoint arcs: $U_1 \cap U_2 = U \cup L$, the upper semicircle (U) where the transformation law for coordinates is $\theta_1 = (\phi_1 \circ \phi_2^{-1})(\theta_2) = \theta_2, (0 < \theta_1 < \pi)$, and lower semicircle (L) where $\theta_1 = (\phi_1 \circ \phi_2^{-1})(\theta_2) = \theta_2 + 2\pi, (\pi < \theta_1 < 2\pi)$.

    Another atlas can be constructed for S^1 consisting of four charts corrsponding to points on the upper (U, ϕ_U), lower (W, ϕ_W), right (R, ϕ_R) and left (L, ϕ_L) semicircles. For upper and lower semicircles assign x as one coordinate of the point on the circle. For right and left semicircles define y as such a coordinate. The coordinate transformation for a point lying in both U and L is, for example,

$$y = (\phi_L \circ \phi_U^{-1})(x) = +\sqrt{1 - x^2}, \qquad x = (\phi_U \circ \phi_L^{-1})(y) = -\sqrt{1 - y^2}.$$

Similarly for other overlaps.

The sphere
Stereographic projection is one choice for constructing charts for S^2. Consider S^2 the set of points $\{(x, y, z) \in R^3 | x^2 + y^2 + z^2 = 1\}$ in three dimensions. Let P_S be the tangent plane at the south pole $(0, 0, -1)$. To find coordinates of a given point $p \in S^2$ join the north pole $(0, 0, 1)$ to p and extend the line till it meets the plane P_S at , say $(x^1, x^2, -1)$. The coordinates of the point p are taken to be (x^1, x^2). This coordinate system includes all points of S^2 except the north pole. Another chart with coordinates (y^1, y^2) can be contructed similarly by switching the roles of the north and south poles. The coordinate transformation between the two charts can be found by straightforward algebra.

The torus
As a set it is the cartesian product of two circles $T^2 = S^1 \times S^1$. We can take the atlas
of two charts U_1, U_2 defined above for one circle and similarly U_1', U_2' for the other circle.
Then the torus is specified by four charts $U_{ij} \equiv U_i \times U_j$ where $i, j = 1, 2$:

$$U_{11} \equiv U_1 \times U_1' \quad : \quad p \to (\theta_{11}, \theta_{11}') : 0 < \theta_{11}, \theta_{11}' < 2\pi,$$
$$U_{12} \equiv U_1 \times U_2' \quad : \quad p \to (\theta_{12}, \theta_{12}') : 0 < \theta_{12} < 2\pi, -\pi < \theta_{12}' < \pi,$$
$$U_{21} \equiv U_2 \times U_1' \quad : \quad p \to (\theta_{21}, \theta_{21}') : 0 < \theta_{21}' < 2\pi, -\pi < \theta_{21} < \pi,$$
$$U_{22} \equiv U_2 \times U_2' \quad : \quad p \to (\theta_{22}, \theta_{22}') : -\pi < \theta_{22}, \theta_{22}' < \pi.$$

The coordinate change for overlap between charts U_{11} and U_{12} for example, is as follows:

$$\text{On the overlap} \qquad U_{11} \cap U_{12},$$
$$\theta_{11} = \theta_{12},$$
$$\theta_{11}' = \theta_{12}', \qquad 0 < \theta_{11}' < \pi,$$
$$\theta_{11}' = \theta_{12}' + 2\pi, \qquad \pi < \theta_{11}' < 2\pi.$$

Similarly for other chart overlaps.

The cone
The cone with a vertex angle α can be built up from a plane sheet of paper from which an
angular wedge with angle $\beta = 2\pi(1 - \sin\alpha)$ is cut away and then the two straight edges
are joined together. On the cone we can choose coordinates r, θ with $0 < r < \infty$ with r
measured from the vertex point of the cone and $0 < \theta < 2\pi \sin\alpha$ measured along the circle
$r = \text{const.}$ as one would have done on the original sheet. This covers all points except those
on the edges which have been joined. To include these another chart is required. One can
choose another radial line (for example that belonging to $\theta = \pi$) and define coordinates
r, θ' with $\theta' = \theta - \pi, (\pi < \theta < 2\pi \sin\alpha - \pi)$ and $\theta' = \theta + 2\pi \sin\alpha - \pi, (0 < \theta < \pi)$.

Exercise 40. Find a suitable atlas for the n-dimensional sphere S^n.

Answer 40. Generalise the stereographic projection map to this case.

Exercise 41. Consider the union of the X- and Y-axes as a subset R of points on the
plane

$$R = \{(x, y) \in R^2 | \text{either } x = 0 \text{ or } y = 0\}.$$

Does this set form a differentiable manifold?

Answer 41. Hint: can one define a chart containing the point $(0,0)$?

Vector Fields

Exercise 42. Plot the following vector fields (by showing the direction of the vectors by
an arrow) on the two-dimensional plane with coordinates (x^1, x^2):

$$X = x^1 \frac{\partial}{\partial x^1} + x^2 \frac{\partial}{\partial x^2},$$
$$Y = x^1 \frac{\partial}{\partial x^2} - x^2 \frac{\partial}{\partial x^1}.$$

Find the integral curves of the field: that is curves $t \to (x^1(t), x^2(t))$ such that the tangent vectors to those curves give the vectors of the fields.

Answer 42. For X the curves are $t \to (C_1 \exp(t), C_2 \exp(t))$ where C_1, C_2 are constants. For Y the curves are $t \to (A \cos(t + B), A \sin(t + B))$ with A, B constants.

Interpretation of the Lie Bracket

Two steps forward, two steps backward

Two vector fields $X = a^i(x)\partial/\partial x^i$ and $Y = b^i(x)\partial/\partial x^i$ are given. We take a point P with coordinates x in its neighbourhood and consider an integral curve of vector field X passing through P. Let the curve be given by $t \to x^i(t)$ so that the point P corresponds to $t = 0$. Take $t > 0$ to be infinitesimally small and keep quantities up to second-order in t. The point Q close to P along the curve then has coordinates

$$y^i = x^i(t) = x^i(0) + ta^i(x) + \frac{t^2}{2!}(a^i)_{,j}a^j + \cdots .$$

We shall write this symbolically as $y = x + ta(x) + t^2 a'.a/2$. In this notation quantities of second-order will not be shown with their argument because any change in their argument in the infinitesimal region will be one higher order.

This point Q is close to P in the direction of the vector field X at P. Now starting from Q locate a point R in the direction of Y with an infinitesimal parameter s. The point R has coordinates $z = y + sb(y) + s^2 b'.b/2$. Now from R go in the reverse direction $-X$ (by the same parameter t that was used for bringing P to Q) to point S which has coordinates $w = z - ta(z) + t^2 a'.a/2$. Finally from S come to a point T along $-Y$ with parameter s. The point T has coordinates $u = w - sb(w) + s^2 b'.b/2$.

Are we back to the starting point P after these four steps (two in the forward directions of X and Y respectively, and then two in the backward direction) or is there a shortfall? The Lie bracket measures that.

Exercise 43. Find the difference in the coordinates of T and P keeping quantities up to second-order. Show that T is in the direction of $[X, Y]$ by infinitesimal parameter st.

Answer 43. We have

$$
\begin{aligned}
y &= x + ta(x) + t^2 a'.a/2, \\
z &= y + sb(y) + s^2 b'.b/2, \\
w &= z - ta(z) + t^2 a'.a/2, \\
u &= w - sb(w) + s^2 b'.b/2.
\end{aligned}
$$

Therefore,

$$
\begin{aligned}
u &= w - sb(w) + s^2 b'.b/2 \\
&= z - ta(z) + t^2 a'.a/2 - sb(z - ta(z)) + s^2 b'.b/2 \\
&= z - ta(z) + t^2 a'.a/2 - sb(z) + stb'.a + s^2 b'.b/2 \\
&= y + sb(y) + s^2 b'.b/2 - ta(y + sb) + t^2 a'.a/2 \\
&\quad - sb(y + sb) + stb'.a + s^2 b'.b/2 \\
&= y + sb(y) + s^2 b'.b/2 - ta(y) - tsa'.b + t^2 a'.a/2 \\
&\quad - sb(y) - s^2 b'.b + stb'.a + s^2 b'.b/2 \\
&= y - ta(y) - tsa'.b + t^2 a'.a/2 + stb'.a \\
&= x + ta + t^2 a'.a/2 - ta(x + ta) - tsa'.b + t^2 a'.a/2 + stb'.a \\
&= x + ta + t^2 a'.a/2 - ta - t^2 a'.a - tsa'.b + t^2 a'.a/2 + stb'.a \\
&= x - tsa'.b + stb'.a.
\end{aligned}
$$

In fuller notation $u^i = x^i + st[(b^i)_{,j} a^j - (a^i)_{,j} b^j]$ which shows that the point T with coordinates u is in the direction of the field

$$
\left(a^j \frac{\partial b^i}{\partial x^j} - b^j \frac{\partial a^i}{\partial x^j} \right) \frac{\partial}{\partial x^i} = [X, Y].
$$

Chapter 8

Connection and Curvature

We have introduced Γ_{ij}^k in Chapter 2 as basic quantities in terms of which the Riemann-Christoffel curvature tensor and the equation of a geodesic curve are expressed.

It was found by Levi-Civita that geodesic and curvature can be defined using the notion of **parallel displacement** which governs the geometry of the space. Parallel displacement or parallel transport became the starting point of differential geometry of not just Riemannian spaces where an infinitesimal distance is defined, but also of more general spaces.

In this chapter we define connection in this general setting.

8.1 Directional Derivative

To find the rate of change of a real-valued function f along a curve we use the tangent vector. If the tangent vector to a curve at a given point $p \in M$ is $\mathbf{v}$ then the required rate is $\mathbf{v}(f)$. This is the **directional derivative** $D_\mathbf{v}$ of the function

$$D_\mathbf{v} f = \mathbf{v}(f) = v^i \frac{\partial f}{\partial x^i} = df(\mathbf{v}) \tag{8.1}$$

where v^i are the components of the vector in local coordinates.

Now suppose that $X = a^i(x)\partial/\partial x^i$ is a vector field on the manifold M, and we are interested in finding the rate of change of the vector field in the direction $\mathbf{v} \in T_p$ at the point $p \in M$.

Is it alright if we take the partial derivatives $\mathbf{v}(a^i) = v^j \partial a^i/\partial x^j$ of the components a^i as a measure of the change in the vector field along $\mathbf{v}$?

Components of a vector are determined by the basis used. How can we be sure that the change in the components is not due to the change in basis vectors themselves?

Consider the example of the radial coordinates r, θ in the two-dimensional Euclidean plane. We can judge if two vectors are the same or not even if they are located at different places in a euclidean space. We already have a concept of parallel vectors in this space. A constant vector field on the euclidean plane has vectors at each point parallel and of the same constant length.

The components of such a constant vector field change from point to point in the $\{\partial/\partial r, \partial/\partial \theta\}$ basis because the direction of the basis vectors changes.

The basic problem in defining the derivative of a vector field is that if p and q are two neighbouring points, vectors $X(p)$ and $X(q)$ belong to two different vector spaces T_p and T_q respectively. How can we compare these? It is not like finding the change in a function f. In that case an observer at q can telephone the colleague at p and communicate the value of the function f at q so that the observer at p can find the difference or the change in f.

But it is not possible to communicate on the telephone the direction of a vector. How can one identify a vector in a vector space which is supposedly equal to a given vector in some other vector space so that the two can be compared?

There is no alternative but to **define** *the concept of* **parallel displacement** *by giving a reasonable rule for identifying a vector in T_p when a given vector in T_q is shifted or displaced and brought to T_p 'without change'.*

The hints for defining such a rule are obtained from our experience of two-dimensional surfaces. For example, to define the parallel transport of a tangent vector at a point of the two-dimensional spherical surface to an infinitesimally close point we can use the concept of parallelism in the surrounding three-dimensional Euclidean space and, after having shifted the vector to the new point as if it were a three-dimensional vector, take its projection tangent to the surface and define that as the parallel shifted vector.

8.1.1   Connection Coefficients

We now discuss the reasonable rule mentioned above for identifying a vector in T_p when a vector in T_q is given and brought parallel to itself to point p.

Let the points p and q be in a chart with coordinates x and let $\partial/\partial x^i$ be the natural basis vector fields. Let p and q have coordinates x and $x + \Delta x$ respectively where Δx is infinitesimal change in the coordinates. Let

$$\left.\frac{\partial}{\partial x^i}\right|_x , \quad \text{and} \quad \left.\frac{\partial}{\partial x^i}\right|_{x + \Delta x}$$

be the natural basis vectors (for some fixed i) at point p, q respectively and let

$$\left.\frac{\partial}{\partial x^i}\right|_{\|} \quad \text{be the vector} \quad \left.\frac{\partial}{\partial x^i}\right|_{x + \Delta x}$$

brought from T_q to T_p by parallel displacement.

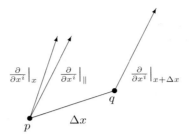

Fig. 8.1

What can we say about this vector?
(1) If $\Delta x \to 0$ then

$$\left.\frac{\partial}{\partial x^i}\right|_{\|} \to \left.\frac{\partial}{\partial x^i}\right|_x .$$

(2) The difference between

$$\left.\frac{\partial}{\partial x^i}\right|_{\|} \qquad \text{and} \qquad \left.\frac{\partial}{\partial x^i}\right|_x$$

is of first order in Δx. Since the difference is a vector in T_p it can be expanded in the natural basis $\partial/\partial x^i$ in T_p.

Accordingly we write

$$\left.\frac{\partial}{\partial x^i}\right|_{\|} = \left.\frac{\partial}{\partial x^i}\right|_x + \Delta x^j \Gamma_{ji}^k \left.\frac{\partial}{\partial x^k}\right|_x \tag{8.2}$$

where Γ_{ij}^k are coefficients of expansion (dependent on x in general). These coefficients are called **connection coefficients** in the coordinate system x. It is important to emphasize here that Γ_{ij}^k is **not assumed to be symmetric** in indices i, j.

Given this rule for displacement of basis vectors, we can transport any general vector

$$\lambda^i \left.\frac{\partial}{\partial x^i}\right|_{x+\Delta x} \quad \text{at } q \text{ to } \quad \lambda^i \left.\frac{\partial}{\partial x^i}\right|_{\|} = \lambda^i \left(\left.\frac{\partial}{\partial x^i}\right|_x + \Delta x^j \Gamma_{ji}^k \left.\frac{\partial}{\partial x^k}\right|_x \right) \tag{8.3}$$

in T_p.

8.1.2 Directional Derivative

Let $t \to x(t)$ be a curve such that $x(0) = x$ and $x(t + \Delta t) = x + \Delta x$. The change in the vector field $X = a^i \partial/\partial x^i$ from point x to $x + \Delta x$ can now be calculated.

We parallel transport the vector at $x + \Delta x$

$$a^i(x + \Delta X) \left. \frac{\partial}{\partial x^i} \right|_{x+\Delta x} \qquad \text{to the point } x \text{ as} \qquad a^i(x + \Delta X) \left. \frac{\partial}{\partial x^i} \right|_{\|}$$

and compare with

$$a^i(x) \left. \frac{\partial}{\partial x^i} \right|_x .$$

The result is

$$
\begin{aligned}
X|_\| - X|_x &= a^i(x + \Delta X) \left. \frac{\partial}{\partial x^i} \right|_\| - a^i(x) \left. \frac{\partial}{\partial x^i} \right|_x \\
&= \Delta x^j \left(\frac{\partial a^k}{\partial x^j} + \Gamma^k_{ji} a^i \right) \left. \frac{\partial}{\partial x^k} \right|_x .
\end{aligned}
$$

Dividing by the parameter change Δt and taking the limit $\Delta t \to 0$ we get the covariant derivative of the vector field X in the direction of the tangent vector $\mathbf{v}$ to the curve

$$
\begin{aligned}
D_\mathbf{v} X = \lim_{\Delta t \to 0} (X|_\| - X|_x)/\Delta t &= \lim_{\Delta t \to 0} \frac{\Delta x^j}{\Delta t} \left(\frac{\partial a^k}{\partial x^j} + \Gamma^k_{ji} a^i \right) \frac{\partial}{\partial x^k} \\
&= v^j \left(\frac{\partial a^k}{\partial x^j} + \Gamma^k_{ji} a^i \right) \frac{\partial}{\partial x^k} .
\end{aligned}
$$

8.1.3   Components of a Parallel Displaced Vector

We can restate the content of equation (8.3) by (reversing the role of points q and p) saying that when a vector with components a^i at p (with coordinates x) is parallel displaced to point q (with coordinates $x + \Delta x$) its components (in the natural basis at $x + \Delta x$) become

$$a^k|_\| = a^k - \Delta x^j \Gamma^k_{ji} a^i. \tag{8.4}$$

8.2   Transformation Formula for Γ^k_{ij}

The definition of parallel transport is specific to a coordinate system, and the connection coefficients have meaning only for this coordinate system. In order to be useful this procedure should be independent of coordinates. We find that insisting on independence of coordinates gives us the transformation formula for the connection coefficients Γ^k_{ij}. Although we will give a much simpler proof of this important formula later on, (twice in fact!) it is worthwhile to go through the following calculation as it teaches certain manipulation skills.

We calculate $D_{\mathbf{v}} X$ in some other coordinate system x'^i. We call the new coordinate system x', and the connection coefficients Γ'^k_{ji}. The tangent vector is $\mathbf{v} = v'^j \partial/\partial x'^j = v^i \partial/\partial x^i$ with $v^i = v'^j \partial x^i/\partial x'^j$ and $a^i(x) = a'^j(x')\partial x^i/\partial x'^j$.

Comparing the two expressions for $D_{\mathbf{v}} X$,

$$
\begin{aligned}
D_{\mathbf{v}} X &= v'^m \left(\frac{\partial a'^l}{\partial x'^m} + \Gamma'^l_{mr} a'^r \right) \frac{\partial}{\partial x'^l} \\
&= v^j \left(\frac{\partial a^k}{\partial x^j} + \Gamma^k_{ji} a^i \right) \frac{\partial}{\partial x^k} \\
&= v'^m \frac{\partial x^j}{\partial x'^m} \left[\frac{\partial}{\partial x^j} \left(a'^l \frac{\partial x^k}{\partial x'^l} \right) + \Gamma^k_{ji} a'^r \frac{\partial x^i}{\partial x'^r} \right] \frac{\partial}{\partial x^k} \\
&= v'^m \left[\frac{\partial}{\partial x'^m} \left(a'^l \frac{\partial x^k}{\partial x'^l} \right) + \frac{\partial x^j}{\partial x'^m} \Gamma^k_{ji} a'^r \frac{\partial x^i}{\partial x'^r} \right] \frac{\partial}{\partial x^k} \\
&= v'^m \left[\frac{\partial a'^l}{\partial x'^m} \frac{\partial x^k}{\partial x'^l} + a'^l \frac{\partial^2 x^k}{\partial x'^m \partial x'^l} + \frac{\partial x^j}{\partial x'^m} \Gamma^k_{ji} a'^r \frac{\partial x^i}{\partial x'^r} \right] \frac{\partial}{\partial x^k}.
\end{aligned}
$$

Take (i) $\partial/\partial x^k$ inside the square bracket and the first term becomes $(\partial a'^l/\partial x'^m)\partial/\partial x'^l$; (ii) replace the dummy l index in the second term by r in anticipation of comparison with the first line; (iii) write $\partial/\partial x^k = (\partial x'^l/\partial x^k)\partial/\partial x'^l$ and take it inside the square bracket to multiply by the second and third terms; and (iv) factor $\partial/\partial x'^l$ from all terms. We get

$$
\begin{aligned}
D_{\mathbf{v}} X &= v'^m \left(\frac{\partial a'^l}{\partial x'^m} + \Gamma'^l_{mr} a'^r \right) \frac{\partial}{\partial x'^l} \\
&= v'^m \left[\frac{\partial a'^l}{\partial x'^m} + \left(\frac{\partial x'^l}{\partial x^k} \frac{\partial^2 x^k}{\partial x'^m \partial x'^r} + \frac{\partial x'^l}{\partial x^k} \frac{\partial x^j}{\partial x'^m} \Gamma^k_{ji} \frac{\partial x^i}{\partial x'^r} \right) a'^r \right] \frac{\partial}{\partial x'^l}.
\end{aligned}
$$

Comparing the two sides of the equation, the transformation formula is obtained:

$$
\Gamma'^l_{mr}(x') = \Gamma^k_{ji}(x) \frac{\partial x^i}{\partial x'^r} \frac{\partial x^j}{\partial x'^m} \frac{\partial x'^l}{\partial x^k} + \frac{\partial x'^l}{\partial x^k} \frac{\partial^2 x^k}{\partial x'^m \partial x'^r}. \tag{8.5}
$$

The three-index notation at first seems to suggest that Γ^k_{ij} is a third-rank tensor. But this is not so because of the extra term in the formula above. In older notation the coefficients Γ^k_{ji} were written $\{^k_{ji}\}$ to emphasize their non-tensor nature. They are called **Christoffel symbols**.

On any reasonable manifold there are an infinite number of independent definitions of connections. Practically, all we need to do is to choose smooth functions Γ^i_{jk} in each coordinate neighbourhood and make sure that they transform in the appropriate way when coordinates are changed in overlapping domains.

Each connection gives its own concept of change of vector under parallel transport and so defines the geometry of the manifold or space.

However, we shall see later that in a Riemannian space where the parallel transport is required to satisfy the condition that the inner product of two vectors does not change when they are both parallel tranported, there is a unique symmetric connection, called the Levi-Civita connection.

Note that if Γ^i_{jk} are components of a connection in some coordinates, then the transpose $\tilde{\Gamma}^i_{jk} \equiv \Gamma^i_{kj}$ defines components of another connection. This follows from the above formula. But, $-\Gamma^i_{jk}$ are not components of any connection if Γ^i_{jk} are. This again follows from the transformation formula, and shows the non-tensor nature.

8.3   Geodesics

A **geodesic** is a curve whose tangent vector at any point when parallel transported to a neighbouring point on the curve *coincides* with the tangent vector at that point. When this happens the parameter to the curve is called an **affine** parameter.

A geodesic is the "straightest possible" curve whose direction, as fixed by the tangent vector, does not change. It remains parallel to itself.

If the curve is represented by $x(t)$ in coordinates, then it is clear from the formula for parallel transport that

$$\left.\frac{dx^k}{dt}\right|_{x+\Delta x} = \left.\frac{dx^k}{dt}\right|_{x} - \Delta x^j \Gamma^k_{ji} \frac{dx^i}{dt}.$$

Dividing by Δt and taking the limit we get the equation that the representative coordinates of a geodesic curve must satisfy:

$$\frac{d^2 x^k}{dt^2} + \Gamma^k_{ji} \frac{dx^j}{dt} \frac{dx^i}{dt} = 0. \tag{8.6}$$

Note the role of the affine parameter t in the equation of the geodesic. If $s = s(t)$ is another parameter for the same curve $x(t) = x(t(s))$, the equation of the geodesic will become

$$\frac{d^2 x^k}{ds^2} + \Gamma^k_{ij} \frac{dx^i}{ds} \frac{dx^j}{ds} = -\frac{dx^k}{ds} \frac{d^2 s}{dt^2} \left(\frac{ds}{dt}\right)^{-2}.$$

The equation for the geodesic retains its form if

$$\frac{d^2 s}{dt^2} = 0, \qquad \text{or } s = at + b$$

where a and b are constants. Therefore the parameters which all give the same geodesic equation are related to each other by the affine transformation $s = at + b$.

8.4 Covariant Derivative

We introduced the connection coefficients and parallel transport to define the concept of directional derivative of a vector field. In fact the transformation formula for the connection coefficients Γ's was obtained from the requirement that the directional derivative should not depend on which coordinate system is used to calculate it.

For this reason, the directional derivative of a vector field $X = a^i \partial/\partial x^i$ in the direction of a vector $\mathbf{v}$,

$$D_{\mathbf{v}} X = v^j \left(\frac{\partial a^k}{\partial x^j} + \Gamma_{ji}^k a^i \right) \frac{\partial}{\partial x^k}$$

is also called the covariant derivative.

8.4.1 Semicolon Notation

There is a standard notation used by physicists to define the components of derivatives in individual directions,

$$a^k_{;j} \equiv a^k_{,j} + \Gamma_{ji}^k a^i. \tag{8.7}$$

Another way to describe the covariant derivative is to say that the two-index quantities $a^k_{,j}$, that is, the ordinary partial derivatives of components of X, are not tensor components because they do not transform like one. But the quantities $a^k_{;j}$ are components of a second-rank (1,1) tensor.

8.4.2 Zero Covariant Derivative Means Parallel Displacement

Given a law for parallel displacement we can define a covariant derivative. Given a rule for calculating covariant derivatives we can define a vector field parallel transported along a curve if its covariant derivative with respect to the tangent vector vanishes.

If we look at components of a vector field, the vanishing of covariant derivative

$$a^k_{;j} = a^k_{,j} + \Gamma_{ji}^k a^i = 0$$

means derivative vanishing in all directions. Writing

$$a^k_{,j} = \lim_{\Delta x \to 0} \frac{a^k(x + \Delta x) - a^k(x)}{\Delta x^j}$$

shows that for infinitesimal Δx^j (in any direction)

$$a^k(x + \Delta x) = a^k(x) - \Delta x^j \Gamma_{ji}^k a^i$$

which is the formula for parallel displacement of a vector from point x to $x + \Delta x$.

This means that if the covariant derivative vanishes in all directions, the vector field values in a neighbourhood can all be obtained by parallel translation starting from any single point.

8.5   Abstract Definition

The notion of covariant or directional derivative $D_{\mathbf{v}}X$ of a vector field X in the direction $\mathbf{v}$ can be generalised to define $D_V X$, the directional derivative of a vector field X at all points of a region where the directions are determined by another vector field V.

A **connection** or **covariant derivative** is a rule or mapping D by which we associate to a vector field X and a vector field V, a third vector field $D_V X$ such that the following conditions are satisfied,

$$
\begin{aligned}
D_{V+W}X &= D_V X + D_W X, & (8.8) \\
D_{fV}X &= f D_V X, & (8.9) \\
D_V(X+Y) &= D_V X + D_V Y, & (8.10) \\
D_V(fX) &= f D_V X + V(f)X, & (8.11)
\end{aligned}
$$

where W, Y are any vector fields and f any smooth function.

Recall the remarks made in section 7.9 of the last chapter. The mapping from a pair of vector fields V, X to field $D_V X : (V, X) \to D_V X$ is **not linear** when functions f are multiplied by X because of the last condition. Therefore it does not define a tensor. **A connection is not a (1,2) tensor.** But for a fixed X, the mapping $\chi \to \chi$ given by $V \to D_V X$ *is* linear because of the first two conditions. So, this mapping does define a (1,1) tensor. Indeed that is what is called the "covariant derivative" DX. We will come to this point again later.

Let us take $X = \partial/\partial x^i$ to be the basis vector field whose components are constant, and let $V = \partial/\partial x^j$. Then expanding $D_V X$ in the basis the coefficients of expansion Γ_{ji}^k are:

$$
D_{\partial/\partial x^j}\left(\frac{\partial}{\partial x^i}\right) = \Gamma_{ji}^k \frac{\partial}{\partial x^k}. \tag{8.12}
$$

The relation of $D_V X$ and Γ_{ji}^k is established by this formula.

The transformation formula for Γ's can be rederived using this definition:

$$
\begin{aligned}
D_{\partial/\partial x'^j}\left(\frac{\partial}{\partial x'^i}\right) &= \Gamma_{ji}'^k \frac{\partial}{\partial x'^k} \\
&= D_{(\partial x^m/\partial x'^j)\partial/\partial x^m}\left(\frac{\partial x^r}{\partial x'^i}\frac{\partial}{\partial x^r}\right) \\
&= \frac{\partial x^m}{\partial x'^j} D_{\partial/\partial x^m}\left(\frac{\partial x^r}{\partial x'^i}\frac{\partial}{\partial x^r}\right) \\
&= \frac{\partial x^m}{\partial x'^j}\left[\frac{\partial x^r}{\partial x'^i} D_{\partial/\partial x^m}\left(\frac{\partial}{\partial x^r}\right) + \left(\frac{\partial}{\partial x^m}\frac{\partial x^r}{\partial x'^i}\right)\frac{\partial}{\partial x^r}\right] \\
&= \left[\frac{\partial x^m}{\partial x'^j}\frac{\partial x^r}{\partial x'^i}\Gamma_{mr}^s + \left(\frac{\partial^2 x^s}{\partial x'^j \partial x'^i}\right)\right]\frac{\partial}{\partial x^s}
\end{aligned}
$$

where we have renamed the dummy index r by s in the second term of the last line. In order to compare the last line with the first, we reconvert the last factor, the partial derivative $\partial/\partial x^k$ with respect to x, into partial derivative with respect to x':

$$\frac{\partial}{\partial x^s} = \frac{\partial x'^k}{\partial x^s} \frac{\partial}{\partial x'^k}.$$

So that

$$\Gamma'^k_{ji} \frac{\partial}{\partial x'^k} = \left[\frac{\partial x^m}{\partial x'^j} \frac{\partial x^r}{\partial x'^i} \Gamma^s_{mr} + \left(\frac{\partial^2 x^s}{\partial x'^j \partial x'^i} \right) \right] \frac{\partial x'^k}{\partial x^s} \frac{\partial}{\partial x'^k}$$

and for comparison

$$\Gamma'^k_{ji} = \frac{\partial x'^k}{\partial x^s} \frac{\partial x^m}{\partial x'^i} \frac{\partial x^r}{\partial x'^j} \Gamma^s_{mr} + \frac{\partial^2 x^s}{\partial x'^i x'^j} \frac{\partial x'^k}{\partial x^s}$$

which is the formula for transformation for connection coefficients we obtained earlier.

8.6 Torsion Tensor

It is easy to see that the **difference of two connections is a tensor**.

Let $D^{(1)}$ and $D^{(2)}$ be two connections. Then the terms $V(f)X$ in the fourth condition, which prevents the connection from becoming a tensor, cancel. From the definitions of $D_V^{(1)}$ and $D_V^{(2)}$ we deduce

$$(D_V^{(1)} - D_V^{(2)})(fX) = f(D_V^{(1)} - D_V^{(2)})X$$

which makes the difference a tensor.

The map $T : \mathcal{X} \times \mathcal{X} \to \mathcal{X}$ (vector fields are denoted by $\mathcal{X}$) is defined by

$$T(X, Y) = D_X Y - D_Y X - [X, Y]. \tag{8.13}$$

It is bilinear in both arguments, as can be checked immediately — the term preventing the linearity is compensated by the Lie bracket. Hence this mapping defines a $(1,2)$ tensor called the **torsion tensor**.

Recall that connection coefficients are not assumed to be symmetric. Therefore Γ^k_{ij} and Γ^k_{ji} define two independent connections. (The transformation property of Γ^k_{ji} is the same as Γ^k_{ij}). The *difference* of these two connections is a tensor with components T^k_{ij},

$$T^k_{ij} \equiv \Gamma^k_{ij} - \Gamma^k_{ji} \tag{8.14}$$

equal to the torsion tensor defined above. Actually,

$$T\left(\frac{\partial}{\partial x^i}, \frac{\partial}{\partial x^j} \right) = T^k_{ij} \frac{\partial}{\partial x^k}. \tag{8.15}$$

8.7   Cartan Equations

It is useful to look at what happens to bases, rather than to components of a vector field by a covariant derivative. This is Cartan's approach.

8.7.1   Connection Matrix

The tangent vector space T_p at any point p has an infinite number of bases. Any two bases, say, $\{e_i\} = \{e_1, \ldots, e_n\}$ and $\{f_i\} = \{f_1, \ldots, f_n\}$ are related by a transformation matrix

$$f_i = e_k A^k{}_i.$$

When we choose bases in all the tangent spaces in a region in a smooth fashion, then we have **basis vector fields**. The best example is the natural or coordinate basis fields which are written (for tangent space T_p where p has coordinates x) as

$$\left. \frac{\partial}{\partial x^i} \right|_x, \qquad i = 1, \ldots, n.$$

Consider two neighbouring points on a curve passing through p with coordinates $x = x(t)$ and $x(t + \Delta t) = x + \Delta x$. We have seen that when

$$\left. \frac{\partial}{\partial x^i} \right|_{x+\Delta x} \qquad \text{is brought to point } x, \text{ it becomes} \qquad \left. \frac{\partial}{\partial x^i} \right|_{\|} \qquad \text{equal to}$$

$$\left. \frac{\partial}{\partial x^i} \right|_{\|} = \left. \frac{\partial}{\partial x^i} \right|_x + \Delta x^j \Gamma^k_{ji} \left. \frac{\partial}{\partial x^k} \right|_x.$$

The parallel transported vectors $(\partial/\partial x^i)_{\|}, i = 1, \ldots, n$ also form a basis in T_p which is related to the basis $(\partial/\partial x^i)_x$ by the transformation matrix

$$\left. \frac{\partial}{\partial x^i} \right|_{\|} = \left(\delta^k{}_i + \Delta x^j \Gamma^k_{ji} \right) \left. \frac{\partial}{\partial x^k} \right|_x.$$

In the limit of $\Delta t \to 0$ the tangent vector to the curve has components $v^i = \lim \Delta x^i / \Delta t$ and the infinitesimal amount by which the matrix relating the parallel transported basis $(\partial/\partial x^i)_{\|}, i = 1, \ldots, n$ and the original basis differs from the identity matrix is

$$\Delta t A^k{}_i = v^j \Delta t \Gamma^k_{ji}.$$

We can look upon these numbers as a rule which assigns to every 'direction', that is every tangent vector $\mathbf{v} \in T_p$, a matrix $A^k{}_i$ of numbers in a linear fashion. For each fixed k and i this is a differential one-form belonging to T_p^*.

We define a **connection matrix** ω of one-forms

$$\omega^k{}_i \equiv \Gamma^k_{ji} dx^j. \tag{8.16}$$

Then all information about the connection is contained in this matrix of one-forms.

In particular, we can say that when the natural basis vectors are brought from a neighbouring point to a given point along a curve which has tangent vector $\mathbf{v}$, then the two bases are related by a transformation matrix which differs from the identity by the matrix $\Delta t A^k{}_i$ where $A^k{}_i = \omega^k{}_i(\mathbf{v})$

The two ways of defining connection are related by

$$D_{\mathbf{v}}\left(\frac{\partial}{\partial x^i}\right) = \omega^k{}_i(\mathbf{v})\frac{\partial}{\partial x^k}. \tag{8.17}$$

8.7.2 General Bases

It is not essential to use the natural basis $\partial/\partial x^i$ for defining the connection matrix. Any set of non-vanishing vector fields $X_a, a = 1, \ldots, n$ in a region can be used as a basis provided the vectors $X_1, \ldots, X_n$ are linearly independent at each point of the region.

The connection matrix is again defined by finding the infinitesimal amount by which the matrix (relating the parallel transported basis vectors from a neighbouring point and the original basis vectors at the point) differs from the identity matrix.

A connection matrix ω_X for the basis fields $X_a, a = 1, \ldots, n$ is defined as the matrix of one-forms obtained by expanding the vector $D_{\mathbf{v}}(X_a)$ for a fixed a as a linear combination of the basis vectors: $D_{\mathbf{v}}(X_a) = X_b C^b{}_a$ and realising that these coefficients $C^b{}_a$ depend linearly on $\mathbf{v}$. The matrix ω_X of 1-forms is defined as $(\omega_X)^b{}_a(\mathbf{v}) = C^b{}_a$. Thus

$$D_{\mathbf{v}}(X_a) = X_b(\omega_X)^b{}_a(\mathbf{v}). \tag{8.18}$$

8.7.3 Change of Bases

Let the basis fields $X_a, a = 1, \ldots, n$ be replaced by another set of basis fields $Y_a, a = 1, \ldots, n$ such that

$$Y_a(x) = X_b(x)U^b{}_a(x) \tag{8.19}$$

where $U^b{}_a(x)$ are the coefficients by which fields $Y_a(x), a = 1, \ldots, n$ are expanded in $X_b(x), b = 1, \ldots, n$.

The connection matrix for the basis fields Y is defined as

$$D_{\mathbf{v}}(Y_a) = Y_b(\omega_Y)^b{}_a(\mathbf{v}). \tag{8.20}$$

The left-hand side can be written

$$
\begin{aligned}
D_{\mathbf{v}}(X_b U^b{}_a) &= D_{\mathbf{v}}(X_b)U^b{}_a + X_b \mathbf{v}(U^b{}_a) \\
&= D_{\mathbf{v}}(X_b)U^b{}_a + X_b d(U^b{}_a)(\mathbf{v}) \\
&= X_c(\omega_X)^c{}_b U^b{}_a + X_c dU^c{}_a(\mathbf{v}) \\
&= Y_d U^{-1}{}^d{}_c(\omega_X)^c{}_b(\mathbf{v})U^b{}_a + Y_d U^{-1}{}^d{}_c d(U^c{}_a)(\mathbf{v}).
\end{aligned}
$$

Using matrix notation, and suppressing indices

$$
\omega_Y = U^{-1}\omega_X U + U^{-1}dU. \tag{8.21}
$$

The student may recognise this formula as the transformation formula of a non-abelian gauge potential. This is no coincidence because gauge potentials too are connection 1-forms, although in a different 'bundle'.

When basis X_a is the natural basis $\partial/\partial x^i$ and Y_a is the basis $\partial/\partial x'^i$, the matrix $U^k{}_i$ is

$$
\frac{\partial}{\partial x'^i} = U^k{}_i \left(\frac{\partial}{\partial x^k} \right) = \frac{\partial x^k}{\partial x'^i} \left(\frac{\partial}{\partial x^k} \right)
$$

and its inverse is

$$
(U^{-1})^m{}_k = \frac{\partial x'^m}{\partial x^k}.
$$

Obviously the formula for the trasformation of connection coefficients is the same as $\omega' = U^{-1}\omega U + U^{-1}dU$ for this U. This is the third time we have derived the transformation formula for the connection components.

8.7.4   Cartan's Structural Equations

The connection matrix of 1-forms $\omega_X{}^b{}_a$ defined for the basis vector fields $X_1, \ldots, X_n$ can be related to its **dual basis** $\alpha^1, \ldots, \alpha^n$ of 1-forms (or covariant vectors) defined as $\alpha^a(X_b) = \delta^a_b$.

We introduced the torsion tensor in the last section. Now we define n **torsion tensors** T^a of rank $(0,2)$ with respect to the basis X_a by expanding the vector field $T(X,Y)$ in this basis. The coefficients are n bilinear functions T^a,

$$
T(X,Y) = D_X Y - D_Y X - [X,Y] \equiv T^a(X,Y)X_a. \tag{8.22}
$$

In this formula write the vector fields X and Y as (using the dual nature of basis $\{\alpha^a\}$)

$$
X = \alpha^a(X)X_a, \qquad Y = \alpha^a(Y)X_a,
$$

then
$$T^a(X,Y)X_a = D_X(\alpha^a(Y)X_a) - D_Y(\alpha^a(X)X_a) - X_a\alpha^a([X,Y]).$$
But
$$D_X(\alpha^c(Y)X_c) = \alpha^c(Y)X_a\omega_X{}^a{}_c(X) + X(\alpha^c(Y))X_c$$
with a similar equation for $D_Y X$. Substituting and using formulas
$$d\alpha(X,Y) = X(\alpha(Y)) - Y(\alpha(X)) - \alpha([X,Y])$$
and $(\alpha \wedge \beta)(X,Y) = \alpha(X)\beta(Y) - \alpha(Y)\beta(X)$ for 1-forms, we get **Cartan's first structural equation**

$$d\alpha^a = -\omega_X{}^a{}_c \wedge \alpha^c + T^a. \tag{8.23}$$

Let ω_X be a connection matrix for a set of basis fields X_a. Define the **curvature two-form matrix**

$$\Omega_X{}^a{}_b \equiv d\omega_X{}^a{}_b + \omega_X{}^a{}_c \wedge \omega_X{}^c{}_b. \tag{8.24}$$

This equation is known as Cartan's **second structural equation**.

8.8   Curvature 2-Form

We defined the **curvature two-form matrix**

$$\Omega_X{}^a{}_b \equiv d\omega_X{}^a{}_b + \omega_X{}^a{}_c \wedge \omega_X{}^c{}_b$$

in the previous section. The transformation property of the curvature two-form is simple;

$$
\begin{aligned}
d\omega_Y &= d(U^{-1}\omega_X U + U^{-1}dU) \\
&= dU^{-1} \wedge \omega_X U + U^{-1}d\omega_X U - U^{-1}\omega_X \wedge dU + dU^{-1} \wedge dU.
\end{aligned}
$$

We have used the properties of the exterior derivative

$$d(\alpha \wedge \beta) = d\alpha \wedge \beta + (-1)^r \alpha \wedge d\beta$$

for any r-form α and the fact that $d \circ d = 0$. Similarly,

$$\omega_Y \wedge \omega_Y = (U^{-1}\omega_X U + U^{-1}dU) \wedge (U^{-1}\omega_X U + U^{-1}dU)$$

can be written

$$U^{-1}\omega_X \wedge \omega_X U + U^{-1}\omega_X \wedge dU + U^{-1}dUU^{-1} \wedge \omega_X U + U^{-1}dUU^{-1} \wedge dU.$$

Using $U^{-1}dU = -dU^{-1}U$ which follows from $d(U^{-1}U) = d(\mathbf{1}) = 0$, we see that the curvature two-form Ω transforms as

$$\Omega_Y = U^{-1}\Omega_X U. \tag{8.25}$$

This, too is an important equation and shows that Ω is a matrix whose indices are tensor indices. Unlike the connection matrix, the curvature 2-form is a tensor. In gauge theories the matix Ω is called the "field strength" tensor.

8.9 Riemann-Christoffel Curvature Tensor

The curvature 2-form has an intimate relation with an antisymmetric $(1,3)$ tensor defined below.

Let us define a tensor which maps three vector fields X, Y, Z multilinearly into a vector field $R(X,Y)Z$ defined by

$$R(X,Y)Z = D_X D_Y Z - D_Y D_X Z - D_{[X,Y]}Z. \tag{8.26}$$

Checking the multilinearity is short and straightforward algebra. Obviously $R(X + X', Y)Z = R(X,Y)Z + R(X',Y)Z$ and similarly in Y and Z. For multiplication by a function we have for example

$$R(fX,Y)Z \quad = \quad fD_X D_Y Z - D_Y(fD_X Z) - D_{[fX,Y]}Z.$$

Now $D_Y(fD_X Z) = fD_Y D_X Z + Y(f)D_X Z$. And $[fX,Y] = f[X,Y] - Y(f)X$ so $D_{[fX,Y]} = fD_{[X,Y]} - Y(f)D_X$. Therefore f factors out in second and third terms as well. Similarly, $R(X, fY)Z = fR(X,Y)Z$ and $R(X,Y)(fZ) = fR(X,Y)Z$.

From the definition it is obvious that the tensor is antisymmetric;

$$R(X,Y)Z = -R(Y,X)Z. \tag{8.27}$$

8.9.1 Relation of R and Ω

Let us explore the relation of tensor R and 2-form Ω. We fix a set of basis fields X_a and drop the basis label X from $\omega_X{}^c{}_a$ and write simply $\omega^c{}_a$. Calculate $R(X,Y)X_a$ step by step:

$$D_Y X_a = \omega^c{}_a(Y)X_c,$$

next

$$
\begin{aligned}
D_X D_Y X_a \quad &= \quad X(\omega^c{}_a(Y))X_c + \omega^c{}_a(Y)D_X(X_c) \\
&= \quad X(\omega^c{}_a(Y))X_c + \omega^c{}_a(Y)\omega^d{}_c(X)X_d \\
&= \quad X(\omega^c{}_a(Y))X_c + \omega^d{}_a(Y)\omega^c{}_d(X)X_c
\end{aligned}
$$

where we have interchanged the dummy indices c and d in the second term so that both terms have factor X_c. The second term in the definition of R is just the negative of this term with the role of X and Y interchanged. The third term is

$$-D_{[X,Y]}X_a = -\omega^c{}_a([X,Y])X_c.$$

We can combine all the terms now and use the formula for exterior derivative of a 1-form,

$$d\alpha(X,Y) = X(\alpha(Y)) - Y(\alpha(X)) - \alpha([X,Y])$$

to get

$$R(X,Y)X_a = \left[d\omega^c{}_a(X,Y) + (\omega^c{}_d \wedge \omega^d{}_a)(X,Y)\right]X_c.$$

Thus the 2-form Ω defined by

$$\Omega^c{}_a = d\omega^c{}_a + \omega^c{}_d \wedge \omega^d{}_a$$

is indeed related to the tensor R by

$$R(X,Y)X_a = X_c\Omega^c{}_a(X,Y). \tag{8.28}$$

8.9.2 The (Second) Bianchi Identity

The following identity is trivial to see;

$$
\begin{aligned}
d\Omega &= d(d\omega + \omega \wedge \omega) \\
&= d\omega \wedge \omega - \omega \wedge d\omega \\
&= (\Omega - \omega \wedge \omega) \wedge \omega - \omega \wedge (\Omega - \omega \wedge \omega) \\
&= \Omega \wedge \omega - \omega \wedge \Omega.
\end{aligned}
$$

The relation

$$d\Omega + \omega \wedge \Omega - \Omega \wedge \omega = 0 \tag{8.29}$$

is called the **Bianchi identity**. For historical reasons it is named the **second** Bianchi identity. This general identity holds for *all* connections and their associated curvatures, whereas the first Bianchi identity, given in the next section, holds only for curvatures whose connection has torsion tensor equal to zero.

8.9.3 The (First) Bianchi Identity

Every torsion-free (or symmetric) connection, (that is one for which the torsion tensor is zero) satisfies the following identity, called the **first Bianchi identity**.

$$R(X,Y)Z + R(Y,Z)X + R(Z,X)Y \equiv 0 \qquad (T = 0). \tag{8.30}$$

Recall that for torsion-free connection $T = 0$, we must have

$$D_X Y - D_Y X = [X,Y],$$

therefore writing fully,

$$R(X,Y)Z + \text{cyclic} = D_X D_Y Z - D_Y D_X Z - D_{[X,Y]}Z + \text{cyclic}.$$

Of the six double-D terms on the right-hand side we can combine $D_X D_Y Z - D_X D_Z Y = D_X(D_Y Z - D_Z Y) = D_X([Y,Z])$. This can then combine with the single-D term $-D_{[Y.Z]}X$ to give $[X,[Y.Z]]$. Therefore the entire right-hand side vanishes as a consequence of the Jacobi identity.

8.10   Components of the Curvature Tensor

We now calculate the components of the curvature 2-form Ω or, equivalently, the Riemann curvature tensor R in the *natural coordinate basis* $\partial/\partial x^i$.

Recall that the connection matrix ω has components Γ_{ij}^k in the natural basis for coordinates x given by

$$\omega^k{}_i = \Gamma_{ji}^k dx^j.$$

The curvature matrix $\Omega = d\omega + \omega \wedge \omega$ is therefore,

$$\Omega^k{}_i = d(\Gamma_{ji}^k dx^j) + \Gamma_{ml}^k dx^m \wedge \Gamma_{ni}^l dx^n = \left(\frac{\partial \Gamma_{ni}^k}{\partial x^m} + \Gamma_{ml}^k \Gamma_{ni}^l \right) dx^m \wedge dx^n$$

where we have changed the dummy index j to n in the first term on the right-hand side. As $dx^m \wedge dx^n = dx^m \otimes dx^n - dx^n \otimes dx^m$, we can write

$$\Omega^k{}_i = R^k{}_{imn} dx^m \otimes dx^n \tag{8.31}$$

with R given by

$$-R^k{}_{imn} = \Gamma_{mi,n}^k - \Gamma_{ni,m}^k + \Gamma_{mi}^l \Gamma_{nl}^k - \Gamma_{ni}^l \Gamma_{ml}^k. \tag{8.32}$$

Here we have used, as before, a comma followed by an index i as an abbreviation for partial derivative with respect to x^i.

The components of the curvature tensor are related to the tensor R by

$$R\left(\frac{\partial}{\partial x^i}, \frac{\partial}{\partial x^j} \right) \frac{\partial}{\partial x^k} = R^m{}_{kij} \frac{\partial}{\partial x^m}. \tag{8.33}$$

This is seen most easily from the formula connecting R to Ω.

8.11   Covariant Derivative of Tensor Fields

8.11.1   Scalar Field

Recall the definition of the directional derivative D_V given at the beginning of this chapter. On a function it merely gives the ordinary derivative

$$D_V(f) = df(V).$$

For a fixed f it is linear on vector field V and as it maps vector fields into scalars, it defines a differential 1-form. We can write this as

$$D(f) = df. \tag{8.34}$$

8.11.2 Vector Field

For a fixed X, $D_V X$ is linear in V. It maps vector field V into vector field $D_V X$. Thus it defines a tensor of type $(1,1)$ whose components in the natural basis are $a^i_{;j}$. We write

$$
\begin{aligned}
D(X) &= a^i_{;j} \frac{\partial}{\partial x^i} \otimes dx^j \\
&= (a^i_{,j} + \Gamma^i_{jk} a^k) \frac{\partial}{\partial x^i} \otimes dx^j \\
&= a^k \frac{\partial}{\partial x^i} \otimes \omega^i{}_k + \frac{\partial}{\partial x^i} \otimes da^i
\end{aligned}
$$

where the components $a^i(x)$ are the scalar functions whose exterior derivative is $da^i = a^i_{,j} dx^j$.

We can summarise the definition simply as

$$
\begin{aligned}
D(X + Y) &= D(X) + D(Y), & (8.35) \\
D(fX) &= f D(X) + X \otimes df, & (8.36)
\end{aligned}
$$

which defines the connection matrix as

$$
D\left(\frac{\partial}{\partial x^k}\right) = \frac{\partial}{\partial x^i} \otimes \omega^i{}_k. \tag{8.37}
$$

Note that our directional covariant derivative $D_X Y$ can be written in terms of this derivative as

$$
D_X Y = i_X DY \tag{8.38}
$$

where the operator i_X is the interior multiplication or contraction by a vector field X (see section 5.2.7 of Chapter 5).

8.11.3 Cotangent Vector Field

In order to generalise the concept of covariant differentiation to all tensor fields we must first define it on cotangent vector fields through the components method.

We have defined parallel displacement of a vector with components a^i at x to point $x + \Delta x$ as the vector with components

$$
a^k_\| = a^k - \Delta x^j \Gamma^k_{ji} a^i.
$$

Let b_k be components of a cotangent vector at x. As a vector belonging to the dual space it maps the tangent vector with components a^k to the number $b_k a^k$.

The **parallelly displaced covariant vector** is defined with components $b_{\|k}$ such that at the point $x + \Delta x$ it maps the parallel displaced tangent vector $a_{\|}^k$ to the same number as the original vector does to the tangent vector:

$$b_k a^k = b_{\|k} a_{\|}^k = b_{\|k} (a^k - \Gamma_{ji}^k a^i \Delta x^j).$$

Now we expect the parallel displaced components to be given by a formula similar to that for tangent vectors. That is, a formula of the form

$$b_{\|k} = b_k + \gamma_{jk}^m b_m \Delta x^j;$$

then the above requirement implies $\gamma_{ji}^k = \Gamma_{ji}^k$ giving

$$b_{\|k} = b_k + \Gamma_{jk}^m b_m \Delta x^j. \tag{8.39}$$

The parallel displacement of a covariant vector defines the **covariant derivative of a covariant vector field** $\beta = b_i dx^i$.

Let $x(t)$ be a curve, with $x^j(t + \Delta t) = x^j + \Delta x^j$. The covariant derivative (in direction j) is obtained as the limit of the difference of b_i at $x + \Delta x$ with b_i at x parallel displaced from x to $x + \Delta x$ divided by Δx^j. As

$$b_i(x + \Delta x) - b_{\|i} = b_i(x + \Delta x) - b_i(x) - \Gamma_{ji}^m b_m \Delta x^j$$

the derivative is

$$b_{i;j} \equiv b_{i,j} - \Gamma_{ji}^k b_k \tag{8.40}$$

for this parallel displacement. One can check that $b_{i;j}$ do indeed transform as components of a second-rank covariant tensor.

We can write the operator D of covariant differentiation on covariant vector fields as

$$D\beta = b_{i;j} dx^i \otimes dx^j \tag{8.41}$$

which is equivalent to

$$Ddx^k = -dx^i \otimes \omega^k{}_i. \tag{8.42}$$

Like the operator D on vector fields this operator too has the following properties:

$$D(\alpha + \beta) = D\alpha + D\beta, \tag{8.43}$$
$$D(f\alpha) = f D\alpha + \alpha \otimes df. \tag{8.44}$$

8.11.4 Tensor Fields

The process of covariant differentiation that has been defined on scalar, contra- and co-variant vector fields, can be extended to all tensor fields.

We define a mapping D which maps tensor fields T, S etc. to tensor fields of one higher covariant rank such that

$$
\begin{align}
D(T+S) &= D(T) + D(S), & (8.45)\\
D(fT) &= fD(T) + T \otimes df, & (8.46)\\
D(T \otimes Q) &= D(T) \otimes Q + T \otimes D(Q). & (8.47)
\end{align}
$$

For practical calculations we just need to remember

$$
D\left(\frac{\partial}{\partial x^k}\right) = \frac{\partial}{\partial x^i} \otimes \omega^i{}_k, \tag{8.48}
$$

$$
D dx^k = -dx^i \otimes \omega^k{}_i. \tag{8.49}
$$

For example

$$
\begin{aligned}
&D\left(T^{i\dots j}{}_{k\dots l}(x)\frac{\partial}{\partial x^i} \otimes \cdots \otimes \frac{\partial}{\partial x^j} \otimes dx^k \otimes \cdots \otimes dx^l\right)\\
=\quad & T^{i\dots j}{}_{k\dots l}(x)\frac{\partial}{\partial x^m} \otimes \cdots \otimes \frac{\partial}{\partial x^j} \otimes dx^k \otimes \cdots \otimes dx^l \otimes \omega^m{}_i + \cdots\\
+\quad & T^{i\dots j}{}_{k\dots l}(x)\frac{\partial}{\partial x^i} \otimes \cdots \otimes \frac{\partial}{\partial x^m} \otimes dx^k \otimes \cdots \otimes dx^l \otimes \omega^m{}_j\\
-\quad & T^{i\dots j}{}_{k\dots l}(x)\frac{\partial}{\partial x^i} \otimes \cdots \otimes \frac{\partial}{\partial x^j} \otimes dx^m \otimes \cdots \otimes dx^l \otimes \omega^k{}_m - \cdots\\
-\quad & T^{i\dots j}{}_{k\dots l}(x)\frac{\partial}{\partial x^i} \otimes \cdots \otimes \frac{\partial}{\partial x^j} \otimes dx^k \otimes \cdots \otimes dx^m \otimes \omega^l{}_m\\
+\quad & \frac{\partial}{\partial x^i} \otimes \cdots \otimes \frac{\partial}{\partial x^j} \otimes dx^k \otimes \cdots \otimes dx^l \otimes d(T^{i\dots j}{}_{k\dots l}(x)). \quad (8.50)
\end{aligned}
$$

In each of these terms except the last interchange the two dummy indices occur in ω. In the last term write

$$
d(T^{i\dots j}{}_{k\dots l}(x)) = (T^{i\dots j}{}_{k\dots l}(x))_{,m} dx^m.
$$

This gives the general formula for the covariant derivative components for any tensor field

$$
\begin{aligned}
T^{i\dots j}{}_{k\dots l;m} =\ & T^{i\dots j}{}_{k\dots l,m}\\
& + \Gamma^i_{mn} T^{n\dots j}{}_{k\dots l} + \cdots + \Gamma^j_{mn} T^{i\dots n}{}_{k\dots l}\\
& - \Gamma^n_{mk} T^{i\dots j}{}_{n\dots l} - \cdots - \Gamma^n_{ml} T^{i\dots j}{}_{k\dots n}.
\end{aligned} \tag{8.51}
$$

8.12   Transport Round a Closed Curve

Given a closed curve we can start at a point on the curve with some tangent vector and ask what will happen when the vector is parallel transported round the curve, brought back and compared with the original vector.

We do this for an infinitesimal "rectangle" $ABCD$ made up from the point A with coordinates x^i, point B with coordinates $x^i + \Delta_1^i$, D with $x^i + \Delta_2^i$ and C with $x^i + \Delta_1^i + \Delta_2^i$. Let v^i be the components of a tangent vector which is to be taken round the closed curve $ABCD$. Parallel transport from point D to C is obtained by reversing the sign of Δ_1 in the formula for transport from C to D and similarly A to D by reversing the sign of Δ_2 in the formula for D to A. Therefore to find out if the vector remains the same after going round $ABCDA$ it is sufficient to check the difference of the vector transported from A to C via B and the same vector transported from A to C via D.

The components v^i at A become at B,

$$v_B^i = v^i - \Gamma_{kj}^i(x)v^j\Delta_1^k.$$

This vector v_B^i becomes, on further transporting to C,

$$
\begin{aligned}
v_C^i &= v_B^i - \Gamma_{kj}^i(x+\Delta_1)v_B^j\Delta_2^k \\
&= v^i - \Gamma_{kj}^i(x)v^j\Delta_1^k - \Gamma_{kj}^i(x)v^j\Delta_2^k - \Gamma_{kj,l}^i(x)v^j\Delta_1^l\Delta_2^k \\
&\quad + \Gamma_{lj}^n(x)\Gamma_{kn}^i v^j\Delta_1^l\Delta_2^k + \cdots.
\end{aligned}
$$

On the other hand, the same vector taken to C from A via D becomes, by a similar reasoning,

$$
\begin{aligned}
v_C'^i &= v_D^i - \Gamma_{kj}^i(x+\Delta_2)v_D^j\Delta_1^k \\
&= v^i - \Gamma_{kj}^i(x)v^j\Delta_1^k - \Gamma_{kj}^i(x)v^j\Delta_2^k - \Gamma_{kj,l}^i(x)v^j\Delta_1^k\Delta_2^l \\
&\quad + \Gamma_{lj}^n(x)\Gamma_{kn}^i v^j\Delta_1^k\Delta_2^l + \cdots.
\end{aligned}
$$

Therefore the difference is, to lowest order,

$$v'^i - v^i = R^i{}_{jkl}v^j\Delta_1^k\Delta_2^l. \tag{8.52}$$

This equation shows that if $R^i{}_{jkl}$ is zero in a region, then we can set up parallel vector fields which can then define the covariant derivative totally in terms of ordinary derivatives of components. When the Riemann tensor is not zero, transport of a vector around a curve does not match with the initial vector at the starting point, the difference being a measure of curvature.

8.13   Tutorial

Exercise 44. RICCI FORMULAS Show that if v^i are the components of a contravariant vector field and a_i those of a covariant vector field in some coordinate basis, then the

Riemann tensor is related to the failure of repeated covariant derivatives to commute as follows:

$$v^i{}_{;j;k} - v^i{}_{;k;j} = -R^i{}_{ljk}v^l, \tag{8.53}$$

$$a_{i;j;k} - a_{i;k;j} = R^l{}_{ijk}a_l. \tag{8.54}$$

Answer 44. Straightforward calculation. Hint: first write the covariant derivative of $w^i_j \equiv v^i{}_{;j}$ as a second-rank mixed tensor and only then substitute its value.

Exercise 45. The Ricci formula is fundamental. In fact the defintion of curvature tensor $(D_X D_Y - D_Y D_X - D_{[X,Y]})Z = R(X,Y)Z$ is based on it. Show that the second Ricci formula above can be written in the abstract notation as (with $D_X \alpha \equiv i_X D\alpha$)

$$[(D_X D_Y - D_Y D_X - D_{[X,Y]})\alpha](Z) = -\alpha(R(X,Y)Z). \tag{8.55}$$

Answer 45. Hint: use the fact that because the derivative D_X respects the tensor product and contractions and satisfies Leibnitz's rule,

$$(d(\alpha(Z))(X) = X(\alpha(Z)) = D_X(\alpha(Z)) = (D_X\alpha)(Z) + \alpha(D_X Z).$$

Exercise 46. Show that for a covariant vector field with components a_i the antisymmetric tensor $a_{i;j} - a_{j;i}$ constructed with covariant derivatives is the same as with ordinary derivatives if torsion is zero.

Answer 46. Obviously if $\Gamma^i_{jk} = \Gamma^i_{kj}$ then

$$a_{i;j} - a_{j;i} = a_{i,j} - a_{j,i}.$$

Exercise 47. Show that for any covariant vector field α the following equality holds:

$$(D\alpha)(X,Y) - (D\alpha)(Y,X) = d\alpha + \alpha(T(X,Y)). \tag{8.56}$$

Answer 47. Follows directly from the definition of D and the torsion tensor T.

Chapter 9

Riemannian Geometry

We have discussed connection, covariant derivative, and curvature in the previous chapter without even mentioning the metric tensor g_{ij}. This has been done to emphasize that these concepts are independent of the existence of a metric. We first discuss Riemannian geometry in its classical (component-index) form in the coordinate basis. Only later do we put in the abstract and elegant form.

The fundamental result of this chapter is that in a differentiable manifold on which a metric is defined, a symmetric connection which 'respects' the metric can only be the standard Levi-Civita connection.

We now explain these terms.

9.1 Riemannian Space

A **Riemannian space** is a manifold in which every tangent space has an inner product $\langle \ , \ \rangle$ defined on it. The inner product is such that for two smooth vector fields X and Y the number $\langle X(x), Y(x) \rangle$ at a point with coordinates x changes smoothly with x. The geometry of Riemannian space is called the Riemannian geometry.

By the linearity property of the inner product it follows that it is enough to specify the metric for a basis, that is, to specify what the smooth functions

$$g_{ij}(x) \equiv \left\langle \frac{\partial}{\partial x^i}, \frac{\partial}{\partial x^j} \right\rangle \tag{9.1}$$

are for the natural coordinate basis vectors $\partial/\partial x^i$.

If $X(x)$ has components $a^i(x)$ and $Y(x)$ has $b^i(x)$ in the natural basis, then

$$\langle X(x), Y(x) \rangle = g_{ij}(x) a^i b^j. \tag{9.2}$$

The symmetry of the inner product $\langle X(x), Y(x) \rangle = \langle Y(x), X(x) \rangle$ implies $g_{ij} = g_{ji}$. Similarly the non-degeneracy of the inner product implies that g_{ij}, as a matrix is non-singular, that is, $g \equiv \det \|g_{ij}\| \neq 0$.

Although the discussion in this chapter is quite general, we eventually would apply it to the physical spacetime. Therefore we assume the metric g_{ij} has the signature of Minkowski space. This means that if we choose an orthonormal basis in any tangent space the form of the metric will become diagonal η_{ij} with one -1 and the rest $+1$. We use the Minkowski terminology and call a tangent vector **time-like**, **space-like** or **null** according as its inner product with itself is negative, positive or zero.

9.1.1 Metric Tensor

The metric is a bilinear form on tangent spaces. So it defines a second-rank covariant tensor.

We have studied in Chapter 6 that the inner product in one space V defines an inner product in all the associated spaces, V^*, T^r_s etc. In particular the inner product in the space V^* with basis $\{\alpha^i\}$ dual to the basis $\{\mathbf{e}_i\}$ of V has the matrix $g^{ij} = \langle \alpha^i, \alpha^j \rangle$. The matrix g^{ij} is equal to the inverse of the matrix $g_{ij} = \langle \mathbf{e}_i, \mathbf{e}_j \rangle$.

In coordinate basis we have

$$g^{ij} = \langle dx^i, dx^j \rangle. \tag{9.3}$$

The covariant nature of the metric tensor components can be seen from their transformation under change of coordinates:

$$g'_{ij}(x') = \left\langle \frac{\partial}{\partial x'^i}, \frac{\partial}{\partial x'^j} \right\rangle = \frac{\partial x^k}{\partial x'^i} \frac{\partial x^l}{\partial x'^j} \left\langle \frac{\partial}{\partial x^k}, \frac{\partial}{\partial x^l} \right\rangle = \frac{\partial x^k}{\partial x'^i} \frac{\partial x^l}{\partial x'^j} g_{kl}(x).$$

Using the matrix notation

$$\|U\| = \|U\|^k{}_i = \left\| \frac{\partial x^k}{\partial x'^i} \right\|,$$

we can write the above equation as

$$\|g'\| = U^T \|g\| U.$$

Taking the matrix inverse of this equation, we get

$$\|g'\|^{-1} = U^{-1} \|g\|^{-1} U^{-1^T}.$$

The inverse matrix U^{-1} is actually

$$\|U^{-1}\|^k{}_i = \left\| \frac{\partial x'^k}{\partial x^i} \right\|$$

because

$$\frac{\partial x'^k}{\partial x^i} \frac{\partial x^i}{\partial x'^j} = \frac{\partial x'^k}{\partial x'^j} = \delta^k_j.$$

Recall from Chapter 5 that if covariant vectors or tensors transform with a matrix U, then contravariant vectors or tensors transform with $U^{-1}{}^{T}$ and vice-versa. This means that if we write the elements of the matrix

$$\|g^{-1}\| = \|g^{kl}\|,$$

then g^{kl} transform as components of a second-rank contravariant tensor

$$g'^{ij}(x') = \frac{\partial x'^{i}}{\partial x^{k}} \frac{\partial x'^{j}}{\partial x^{l}} g^{kl}(x).$$

Moreover, as the inverse of a symmetric matrix is symmetric, therefore

$$g^{kl} = g^{lk}.$$

The symmetric covariant second-rank tensor with components g_{ij} determines everything in Riemannian geometry and is called the **fundamental** or **metric** tensor. We can write this tensor as

$$\mathbf{g} = g_{ij} dx^{i} \otimes dx^{j}. \tag{9.4}$$

There is a reason for the name metric tensor, at least for the case when the inner product is positive definite. If x and $x + \Delta x$ are two infinitesimally close points, then

$$(\Delta s)^{2} \equiv g_{ij} \Delta x^{i} \Delta x^{j} \tag{9.5}$$

represents the square of infinitesimal length or distance of the diplacement vector with components Δx^{i}. As is clear from the transformation formula for g_{ij} the length Δs of the displacement vector is not dependent on which coordinates we use (up to second-order in Δx), and depends only on the two points. This formula is the generalisation of the Pythogorean theorem

$$(\Delta s)^{2} = \Delta x \Delta x + \Delta y \Delta y + \Delta z \Delta z \tag{9.6}$$

of Euclidean geometry. Specification of g_{ij} either through the infinitesimal distance square or the inner product is the starting point of Riemannian geometry.

Another way to look at it is to note that if $x(t)$ are the coordinates of points on a curve, then the length of the curve between points $x(t_{1})$ and $x(t_{2})$ can be written as

$$s = \int_{t_{1}}^{t_{2}} \left[g_{ij} \frac{dx^{i}}{dt} \frac{dx^{j}}{dt} \right]^{\frac{1}{2}} dt. \tag{9.7}$$

9.1.2   Induced Metric

Riemannian spaces commonly occur as subspaces of Euclidean or Minkowskian spaces as indeed we have seen in Chapter 2 and will see still more in Chapter 16.

The Euclidean space of dimension n is the Riemannian space R^n whose points are labelled by coordinates $x^i, i = 1, \ldots, n$, each going from $-\infty$ to $+\infty$, and the metric is given by $\delta_{ij} dx^i \otimes dx^j$, where δ_{ij} is the Kronecker delta. The Minkowskian space is defined by using coordinates x^μ (in physics we number them with $\mu = 0, 1, 2, 3$ and call x^0 the *time* coordinate). The metric for Minkowski space is $\eta_{\mu\nu} dx^\mu \otimes dx^\nu$, where $\eta_{00} = -1, \eta_{0i} = 0, \eta_{ij} = \delta_{ij}$ for $i, j = 1, 2, 3$.

Let a subspace S (an m-dimensional surface) of an n-dimensional Riemannian space M with metric g_{ij} be given. We can choose coordinates $u^r = u^1, \ldots, u^m$ on the surface. Let those points on the surface treated as points of M have coordinates $x^i, i = 1, \ldots, n$. Then x^i are given as functions $x^i(u), i = 1, \ldots n$ of coordinates u^r of the surface. Then the surface has metric

$$
\begin{aligned}
\overline{g}_{rs} &= \left\langle \frac{\partial}{\partial u^r}, \frac{\partial}{\partial u^s} \right\rangle \\
&= \left\langle \frac{\partial x^i}{\partial u^r} \frac{\partial}{\partial x^i}, \frac{\partial x^j}{\partial u^s} \frac{\partial}{\partial x^j} \right\rangle \\
&= \frac{\partial x^i}{\partial u^r} \frac{\partial x^j}{\partial u^s} g_{ij}.
\end{aligned}
$$

We can write this relation also as

$$
g_{ij} dx^i \otimes dx^j = g_{ij} \frac{\partial x^i}{\partial u^r} \frac{\partial x^j}{\partial u^s} du^r \otimes du^s = \overline{g}_{rs} du^r \otimes du^s.
$$

We have taken the example of Euclidean space for illustration. Any subspace of a Riemannian space will get an induced metric as well as the connection in this way. We discuss these issues in the last chapter.

9.2   Levi-Civita Connection

9.2.1   The Fundamental Theorem

The **fundamental theorem of Riemannian geometry** says that there is a unique, symmetric (i.e., torsion-free) connection on a Riemannian manifold which "preserves the inner product" under parallel transport. This means if two tangent vectors are taken in the tangent space of a point, and both are parallel displaced by the connection to a neighbouring point, then the inner product of the vectors at the first point is the same as the inner product at the displaced point (as determined by the value of the metric tensor at that neighbouring point). This unique connection, known as **Levi-Civita** or **Riemann connection** is given by components in the natural basis by

$$\Gamma^k_{ij} = \frac{1}{2}g^{kl}(g_{il,j} + g_{jl,i} - g_{ij,l}). \tag{9.8}$$

This is the same familiar formula we have been working with from Chapter 2 onwards.

The proof of the fundamental theorem is very simple.

Since torsion is zero, we will assume $\Gamma^k_{ij} = \Gamma^k_{ji}$. Let v^i and w^i be components of two tangent vectors. Then

$$v^i_{\parallel}(x + \Delta x) = v^i(x) - \Gamma^i_{kj}\Delta x^k v^j.$$

Similarly,

$$w^i_{\parallel}(x + \Delta x) = w^i(x) - \Gamma^i_{kj}\Delta x^k w^j.$$

The condition on the connection is

$$g_{ij}(x + \Delta x)v^i_{\parallel}w^j_{\parallel} = g_{ij}(x)v^i w^j. \tag{9.9}$$

Keeping to lowest order we get

$$\Delta x^k v^i w^j (g_{ij,k} - g_{lj}\Gamma^l_{ki} - g_{il}\Gamma^l_{kj}) = 0,$$

so

$$g_{ij,k} = g_{lj}\Gamma^l_{ki} + g_{il}\Gamma^l_{kj}. \tag{9.10}$$

Let us define for convenience $[ki,j] \equiv g_{jl}\Gamma^l_{ki}$ and write the above equation as

$$g_{ij,k} = [ki,j] + [kj,i]. \tag{9.11}$$

Interchanging j and k and again i and k we rewrite the above equations;

$$g_{ik,j} = [ji,k] + [jk,i], \tag{9.12}$$

$$g_{kj,i} = [ik,j] + [ij,k]. \tag{9.13}$$

We now add the last two equations, subtract the first equation from that sum and get, using the symmetry $[ij,k] = [ji,k]$,

$$2[ij,k] = 2g_{kl}\Gamma^l_{ij} = g_{ik,j} + g_{jk,i} - g_{ij,k}. \tag{9.14}$$

The expression for Γ is obtained on multiplying by the inverse g^{mk} and summing over k.

9.2.2   Covariant Derivative of the Metric

As the parellel transport given by the Levi-Civita connection preserves the inner product, it is not surprising that the covariant derivative of the metric tensor $g_{ij;k}$ is zero,

$$g_{ij;k} = g_{ij,k} - g_{rj}\Gamma^r_{ik} - g_{ir}\Gamma^r_{jk} = 0. \tag{9.15}$$

The same happens for the contravariant counterpart g^{ij} of the metric tensor

$$g^{ij}{}_{;k} = g^{ij}{}_{,k} + g^{rj}\Gamma^i_{rk} + g^{ir}\Gamma^j_{rk} = 0. \tag{9.16}$$

Therefore we can freely take g_{ij} or g^{ij} in or outside covariant differentiations as if it were a constant,

$$g^{ij}(K^{l\dots m}{}_{r\dots s})_{;k} = (g^{ij}K^{l\dots m}{}_{r\dots s})_{;k}. \tag{9.17}$$

9.2.3   A Useful Formula

An expression in which the upper and one of the lower indices of the connection coefficient are contracted occurs often. The useful formula for it is

$$\Gamma^i_{ij} = (\ln\sqrt{-g})_{,j}. \tag{9.18}$$

The proof is easy:

$$
\begin{aligned}
2\Gamma^i_{ij} &= g^{ik}[g_{ki,j} + g_{kj,i} - g_{ij,k}] \\
&= g^{ik}g_{ik,j} \\
&= \mathrm{Tr}\left(\|g\|^{-1}\frac{\partial}{\partial x^j}\|g\|\right).
\end{aligned}
$$

The last two terms in the first line above cancel if dummy indices i, k are switched in one of the terms. Now we use the formula for the derivative of the determinant of a symmetric non-singular matrix M whose elements depend on a parameter x:

$$\frac{d}{dx}\det M = \det M\,\mathrm{Tr}\left(M^{-1}\frac{d}{dx}M\right).$$

This formula follows by simply diagonalising the matrix by a similarity transformation: $D = SMS^{-1}$ where D is the diagonal matrix. The formula is obviously true for a diagonal non-singular matrix. If D is a matrix with non-zero diagonal elements $\lambda_1, \dots, \lambda_n$,

$$
\begin{aligned}
\frac{d}{dx}\det D &= \sum \lambda_1 \dots \frac{d\lambda_i}{dx} \dots \lambda_n \\
&= \det D \sum_i \frac{1}{\lambda_i}\frac{d\lambda_i}{dx} \\
&= \det D\,\mathrm{Tr}\left(D^{-1}\frac{d}{dx}D\right).
\end{aligned}
$$

Now we know that the determinant is unchanged, $\det D = \det M$. And in the trace,

$$\mathrm{Tr}\left(D^{-1}\frac{d}{dx}D\right) = \mathrm{Tr}\left(SM^{-1}S^{-1}\frac{d}{dx}(SMS^{-1})\right)$$

$$= \mathrm{Tr}\left(SM^{-1}S^{-1}\frac{dS}{dx}(MS^{-1})\right)$$

$$+ \mathrm{Tr}\left(SM^{-1}\frac{dM}{dx}S^{-1}\right)$$

$$+ \mathrm{Tr}\left(S\frac{dS^{-1}}{dx}\right).$$

In the first term on the right-hand side we can bring the matrix MS^{-1} as the first factor inside the trace using the property of the trace $\mathrm{Tr}(AB) = \mathrm{Tr}(BA)$. Then the first term becomes

$$\mathrm{Tr}\left(S^{-1}\frac{dS}{dx}\right) = \mathrm{Tr}\left(\frac{dS}{dx}S^{-1}\right)$$

which can be combined with the third term to give the trace of

$$\frac{dS}{dx}S^{-1} + S\frac{dS^{-1}}{dx} = \frac{dSS^{-1}}{dx} = 0.$$

The middle term becomes the required expression in the formula when the last factor is brought to the left as first.

Our expression for the Christoffel symbol Γ^i_{ij} is thus proved because g^{ik} is the matrix inverse to g_{ik}. The sign inside the square root is taken to be negative because the spacetime signature makes the determinant of the metric tensor negative.

9.3 Bianchi Identity in Components

The (second) Bianchi identity $d\Omega + \omega \wedge \Omega - \Omega \wedge \omega = 0$ for a symmetric, that is, torsion-free, connection such as in the present case of Riemannian geometry, can be written in component form in a special way which is used in Einstein's theory of gravitation.

Recall that (section 8.10)

$$\omega^i{}_j = \Gamma^i_{kj}dx^k,$$

$$\Omega^i{}_j = \frac{1}{2}R^i{}_{jkl}dx^k \wedge dx^l = R^i{}_{jkl}dx^k \otimes dx^l.$$

Therefore,

$$d\Omega^i{}_j = \frac{1}{2}R^i{}_{jkl,m}dx^m \wedge dx^k \wedge dx^l$$

while

$$\omega^i{}_r \wedge \Omega^r{}_j = \frac{1}{2}\Gamma^i_{mr}R^r{}_{jkl}dx^m \wedge dx^k \wedge dx^l$$

and

$$\begin{aligned}
\Omega^i{}_r \wedge \omega^r{}_j &= \frac{1}{2}\Gamma^r_{kj}R^i{}_{rml}dx^m \wedge dx^l \wedge dx^k \\
&= -\frac{1}{2}\Gamma^r_{kj}R^i{}_{rml}dx^m \wedge dx^k \wedge dx^l.
\end{aligned}$$

Thus the Bianchi identity looks like

$$(R^i{}_{jkl,m} + \Gamma^i_{mr}R^r{}_{jkl} + \Gamma^r_{kj}R^i{}_{rml})dx^m \wedge dx^k \wedge dx^l = 0. \tag{9.19}$$

We now try to convert the ordinary derivative of R into a covariant derivative. The second term in the equation above already corresponds to the term in $R^i{}_{jkl;m}$ for the contravariant index i. The next term in the bracket is the term for the covariant index j provided we interchange the names of dummy indices k and m and rearrange $dx^k \wedge dx^m \wedge dx^l$ to $-dx^m \wedge dx^k \wedge dx^l$. The two more terms in $R^i{}_{jkl;m}$, namely $-\Gamma^r_{mk}R^i{}_{jrl} - \Gamma^r_{ml}R^i{}_{jkr}$ which ought to be there but are not, can be trivially written down inside the bracket because they are symmetric in km and lm respectively while the bracket is multiplied by $dx^m \wedge dx^k \wedge dx^l$ which is antisymmetric in these indices. So these terms are effectively zero.

Thus we have

$$R^i{}_{jkl;m}dx^m \wedge dx^k \wedge dx^l = 0. \tag{9.20}$$

Of course, this does not imply that $R^i{}_{jkl;m} = 0$ because $dx^m \wedge dx^k \wedge dx^l$ are linearly independent only for $m < k < l$ and the sum above is over all values of these indices. The simplest way to see what these equations imply is to express the wedge product in terms of tensor products:

$$\begin{aligned}
R^i{}_{jkl;m}dx^m \wedge dx^k \wedge dx^l &= R^i{}_{jkl;m}(dx^m \otimes dx^k \otimes dx^l - dx^m \otimes dx^l \otimes dx^k \\
&\quad - dx^k \otimes dx^m \otimes dx^l + dx^k \otimes dx^l \otimes dx^m \\
&\quad + dx^l \otimes dx^m \otimes dx^k - dx^l \otimes dx^k \otimes dx^m).
\end{aligned}$$

We rearrange these terms as follows. Remember that $R^i{}_{jkl}$ is antisymmetric in the last two indices k and l. The first and second terms in the equation above can be combined together as $2R^i{}_{jkl;m}dx^m \otimes dx^k \otimes dx^l$ if the dummy indices k and l are interchanged in the second term. Similarly third and fourth terms can be combined as $2R^i{}_{jlm;k}dx^m \otimes dx^k \otimes dx^l$ by interchanging m and k and the last two as $2R^i{}_{jmk;l}dx^m \otimes dx^k \otimes dx^l$ by interchanging m and l. Therefore,

$$2(R^i{}_{jkl;m} + R^i{}_{jlm;k} + R^i{}_{jmk;l})dx^m \otimes dx^k \otimes dx^l = 0.$$

We can now say that

$$R^i{}_{jkl;m} + R^i{}_{jlm;k} + R^i{}_{jmk;l} = 0. \tag{9.21}$$

This is the useful form mentioned above.

9.4 Symmetry Properties of the Curvature Tensor

Recall that $R^i{}_{jkl}$ is antisymmetric in the last two indices because these correspond to the curvature two-form matrix $\Omega^i{}_j$,

$$R^i{}_{jkl} = -R^i{}_{jlk}. \tag{9.22}$$

Also, for any symmetric connection $\Gamma^i_{jk} = \Gamma^i_{kj}$ we have the **first Bianchi identity**, written in components

$$R^i{}_{jkl} + R^i{}_{klj} + R^i{}_{ljk} = 0. \tag{9.23}$$

We can check it by just writing down the expressions cyclically and adding. Note that this identity is non-trivial only when all three indices j, k, l are different. When two of these indices are equal, one term drops out and the remaining terms just express the antisymmetry in the last two indices of the curvature tensor.

The curvature tensor as determined by the Levi-Civita connection has additional symmetry properties which are seen when the contravariant index is brought down. Define

$$R_{ijkl} \equiv g_{im} R^m{}_{jkl}. \tag{9.24}$$

Then we see below that this covariant version of the Riemann tensor is antisymmetric in the first two indices,

$$R_{ijkl} = -R_{jikl} \tag{9.25}$$

and the index pair ij and kl can be switched without any change,

$$R_{ijkl} = R_{klij}. \tag{9.26}$$

We write the expression for R_{ijkl},

$$-R_{ijkl} = g_{im}(\Gamma^m_{jk})_{,l} + g_{im}\Gamma^r_{jk}\Gamma^m_{rl} - [k \leftrightarrow l].$$

In the expression for $-R_{ijkl}$ the first term is

$$g_{im}(g^{mn}[jk, n])_{,l} = [jk, i]_{,l} - g^{mn}g_{im,l}[jk, n]$$

where we use the convenient notation

$$[jk, i] = \frac{1}{2}(g_{ji,k} + g_{ki,j} - g_{jk,i})$$

introduced earlier. As $g_{im,l} = [il, m] + [lm, i]$ we can write

$$g_{im}(\Gamma^m_{jk})_{,l} = [jk, i]_{,l} - g^{mn}([il, m][jk, n] + [lm, i][jk, n]).$$

The second term in the expression for R_{ijkl} is

$$g_{im}\Gamma^r_{jk}\Gamma^m_{rl} = g^{rs}[jk,s][rl,i] = g^{mn}[lm,i][jk,n].$$

When we add the first and the second terms, we get

$$
\begin{aligned}
g_{im}(\Gamma^m_{jk})_{,l} + g_{im}\Gamma^r_{jk}\Gamma^m_{rl} &= [jk,i]_{,l} - g^{mn}[il,m][jk,n] \\
&= \frac{1}{2}(g_{ij,kl} + g_{ik,jl} - g_{jk,il}) - g^{mn}[il,m][jk,n].
\end{aligned}
$$

When we subtract $[k \leftrightarrow l]$ terms the very first term $g_{ij,kl}$, being symmetric in kl, vanishes, and we get (using the symmetry of g^{mn})

$$
\begin{aligned}
-R_{ijkl} &= g_{im}(\Gamma^m_{jk})_{,l} + g_{im}\Gamma^r_{jk}\Gamma^m_{rl} - [k \leftrightarrow l] \\
&= \frac{1}{2}(g_{ik,jl} - g_{jk,il} - g_{il,jk} + g_{jl,ik}) \\
&+ g^{mn}([ik,m][jl,n] - [il,m][jk,n] + [ik,n][jl,m] - [il,n][jk,m])/2 \\
&= \frac{1}{2}[\{g_{ik,jl} - i \leftrightarrow j\} - k \leftrightarrow l] \\
&+ g^{mn}[\{[ik,m][jl,n] - i \leftrightarrow j\} - k \leftrightarrow l]/2.
\end{aligned}
$$

In this form all symmetry properties are clearly visible.

To summarise: R_{ijkl} has the following symmetry properties.

1. $R_{ijkl} = -R_{ijlk}$: antisymmetry in last two indices.

2. $R_{ijkl} = -R_{jikl}$: antisymmetry in first two.

3. $R_{ijkl} = R_{klij}$: symmetry in pairs of indices.

4. $R_{ijkl} + R_{iklj} + R_{iljk} = 0$: cyclic sum in last three indices. (First Bianchi Identity.)

In the tutorial to this chapter a more elegant proof of these properties is given using the abstract definition of the covariant form of Riemann tensor $R(W, Z; X, Y)$ in section 8 later in this chapter.

Because of these symmetry properties the number of independent components of the Riemann tensor R_{ijkl} in a space of dimension n is

$$\text{Number of independent } R_{ijkl} = \frac{n^2(n^2 - 1)}{12}.$$

For four spacetime dimensions it is twenty. We discussed this in a tutorial in Chapter 3.

9.5 Ricci, Einstein and Weyl Tensors

9.5.1 Ricci and Einstein Tensors

We define the **Ricci tensor** as the contraction

$$R_{ij} \equiv R^k{}_{ikj}. \tag{9.27}$$

This is a symmetric tensor $R_{ij} = R_{ji}$ because of the symmetry properties of the Riemann curvature tensor

$$R_{ij} = g^{kl} R_{likj} = g^{kl} R_{kjli} = R_{ji}.$$

Further contraction of the Ricci tensor leads to **scalar curvature**

$$R \equiv g^{ij} R_{ij}. \tag{9.28}$$

The tensor defined by

$$G_{ij} = R_{ij} - \frac{1}{2} g_{ij} R \tag{9.29}$$

is called the **Einstein tensor** with a related tensor defined by

$$G^i{}_j \equiv g^{ik} G_{kj} = g^{ik} R_{kj} - \frac{1}{2} \delta^i_j R = R^i{}_j - \frac{1}{2} \delta^i_j R. \tag{9.30}$$

This tensor satisfies the **contracted Bianchi identity**

$$G^i{}_{j;i} = 0. \tag{9.31}$$

The proof of the contracted Bianchi identity is straightforward, we only have to remember that the metric tensor g_{ij} or its contravariant version g^{ij} have their covariant derivatives zero so they can be taken in or out of covariant derivations like constants. Starting with the Bianchi identity

$$R^i{}_{jkl;m} + R^i{}_{jlm;k} + R^i{}_{jmk;l} = 0,$$

write the last term as $-R^i{}_{jkm;l}$ and then contract i and k, that is put $i = k$ and sum. We get

$$R_{jl;m} + R^i{}_{jlm;i} - R_{jm;l} = 0.$$

Now multiply by g^{jm} summing over both indices:

$$R^m{}_{l;m} + (g^{jm} R^i{}_{jlm})_{;i} - (g^{jm} R_{jm})_{;l} = 0.$$

But in the second term

$$g^{jm} R^i{}_{jlm} = g^{jm} g^{is} R_{sjlm} = g^{jm} g^{is} R_{jsml} = g^{is} R_{sl} = R^i{}_l$$

and the last term becomes $R_{;l}$. So $2R^i{}_{l;i} - R_{;l} = 0$ or,

$$G^i{}_{l;i} = R^i{}_{l;i} - \frac{1}{2} R_{;l} = 0. \tag{9.32}$$

9.5.2   Weyl Tensor

The Riemann tensor R_{ijkl} with its symmetry properties is discussed in the previous sections. The Ricci tensor $R_{jl} = g^{ik}R_{ijkl}$ and scalar curvature $R = g^{jl}R_{jl}$ are particular linear combinations of it (with components of the metric tensor as coefficients). Because of the symmetries, there is no non-zero linear combination possible except $g^{ik}R_{ijkl} = R_{jl}$. This is expressed by saying that the Ricci tensor is the "trace" of the Riemann tensor.

What is the rank-four covariant tensor which has the same symmetry properties as the Riemann tensor and is "traceless"?

The answer to this question is the Weyl tensor C_{ijkl}. We construct the Weyl tensor by a combination of R_{ijkl}, and R_{jl} and R.

To get a fourth-rank tensor from R_{jl} we consider $g_{ik}R_{jl}$. To get antisymmetry in i,j construct $g_{ik}R_{jl} - g_{jk}R_{il}$, and to get antisymmetry in k,l, we subtract from it the same expression with k,l interchanged:

$$g_{ik}R_{jl} - g_{jk}R_{il} - g_{il}R_{jk} + g_{jl}R_{ik}.$$

Now this combination has the correct symmetry properties including symmetry under simultaneous interchange of ij and kl.

Next we consider combinations possible with R. The only choice is

$$(g_{ik}g_{jl} - g_{jk}g_{il})R.$$

So we write tentatively,

$$\begin{aligned} C_{ijkl} = \; & R_{ijkl} + a(g_{ik}R_{jl} - g_{jk}R_{il} - g_{il}R_{jk} + g_{jl}R_{ik}) \\ & + b(g_{ik}g_{jl} - g_{jk}g_{il})R. \end{aligned}$$

The trace zero condition $g^{ik}C_{ijkl} = 0$ determines the coefficients using $g^{ij}g_{ij} = n$,

$$\begin{aligned} C_{ijkl} = \; & R_{ijkl} - \frac{1}{n-2}(g_{ik}R_{jl} - g_{jk}R_{il} - g_{il}R_{jk} + g_{jl}R_{ik}) \\ & + \frac{1}{(n-1)(n-2)}(g_{ik}g_{jl} - g_{jk}g_{il})R. \end{aligned} \tag{9.33}$$

9.5.3   Conformally Related Metrics

Suppose on a Riemannian space with metric tensor g_{ij} we choose another metric

$$\overline{g}_{ij} = e^{2\phi}g_{ij} \tag{9.34}$$

which differs only by a positive factor at each point. The result will be that the lengths of tangent vectors will change but the angles will remain the same. Such a transformation is called a conformal transformation, and two metrics related in this way are called **conformally related**. If one of the two conformally related metrics is flat the other one is called **conformally flat**.

The corresponding quantities for the metric $\overline{g}$ can be calculated. And it will be seen that the Weyl tensor has the simple transformation as

$$\overline{C}_{ijkl} = e^{2\phi}C_{ijkl}. \tag{9.35}$$

9.6 Geodesics

9.6.1 Geodesic Equation and Affine Parameter

The 'straightest possible' curve is a curve whose tangent vector when transported along the curve is found to be in the same direction as the tangent vector at the transported point. Let $\lambda \to x(\lambda)$ be a curve, and x^i and $x^i + \Delta x^i$ be coordinates of two points p and q corresponding to λ and $\lambda + \Delta\lambda$ respectively. Then the tangent vector components $dx^i/d\lambda$ at p become, on transporting from p to q,

$$\left.\frac{dx^i}{d\lambda}\right|_p - \Gamma^i_{jk}\Delta x^j \frac{dx^k}{d\lambda}$$

and these should be proportional to $(dx^i/d\lambda)_q$. Therefore

$$\left.\frac{dx^i}{d\lambda}\right|_p - \Gamma^i_{jk}\Delta x^j \frac{dx^k}{d\lambda} = \phi(\lambda) \left.\frac{dx^i}{d\lambda}\right|_q$$

where $\phi(\lambda)$ is the proportionality constant which may change from point to point. In the limit of $\Delta\lambda \to 0$ we can write $\Delta x^i/\Delta\lambda \to (dx^i/d\lambda)_p$ and

$$\lim_{\Delta\lambda\to 0} \frac{1}{\Delta\lambda}\left(\left.\frac{dx^i}{d\lambda}\right|_q - \left.\frac{dx^i}{d\lambda}\right|_p \right) = \left.\frac{d^2 x^i}{d\lambda^2}\right|_p,$$

therefore the equation for the geodesic is

$$\frac{d^2 x^i}{d\lambda^2} + \Gamma^i_{jk}\frac{dx^j}{d\lambda}\frac{dx^k}{d\lambda} = (1 - \phi(\lambda))\frac{dx^i}{d\lambda}.$$

As discussed in section (2.4) we can redefine the parameter λ in terms of another parameter, say, τ in such a manner that the right-hand side of the geodesic equation is zero,

$$\frac{d^2 x^i}{d\tau^2} + \Gamma^i_{jk}\frac{dx^j}{d\tau}\frac{dx^k}{d\tau} = 0.$$

The only freedom still left in the choice of parameter τ is the one-dimensional affine transformation

$$\tau \rightarrow a\tau + b$$

where a and b are constants along the geodesic curve. A parameter chosen in such a way is called an **affine parameter**

From the very definition of the Levi-Civita connection Γ^i_{jk} we know that a parallel transported vector $\mathbf{v}$ preserves its norm $\langle \mathbf{v}, \mathbf{v} \rangle$ under transport. Therefore the tangent vectors of a geodesic cannot change their character along a geodesic. The geodesics are therefore classified as **time-like, space-like** or **null** depending on the nature of its tangent vectors.

In the case of time-like geodesics we can always take the proper time starting from some fixed point p as affine paramenter:

$$\tau = \int_p \sqrt{-g_{ij} \frac{dx^i}{d\lambda} \frac{dx^j}{d\lambda}} d\lambda$$

Tangent vectors T along the geodesic curve have the constant norm squared $\langle T, T \rangle$ for time-like geodesic when the parameter is chosen as proper time. In this case not only the parallel transported tangent vector is proportional to the tangent vector sitting at the transported point (by the condition of the curve being a geodesic), it has, moreover, the same length. Therefore the proportionality constant ϕ is equal to 1 at all points.

This argument does not apply to null geodesics for which proper time (or length) is identically zero along the geodesic and cannot be used as a parameter. Of course, we can always choose an affine parameter for a null geodesic.

9.6.2   Congruence of Geodesics

A **congruence** of curves in an open set O in a manifold is a family of smooth curves such that from each point $p \in O$ only one curve of the family passes. A congruence of curves defines a **tangent vector field** belonging to the congruence. In principle the affine parameters for the curves of the family can be chosen arbitrarily because an affine parameter of one curve of the family has nothing to do with the parameters of any other.

A congruence of geodesic curves is important in general relativity. A congruence of time-like geodesics may represent trajectories of dust particles falling under gravity and a congruence of null curves may similarly represent paths of light rays. A study of these families of geodesic curves gives important clues to the nature of a gravitational field. We will treat congruence of time-like geodesics in the last chapter in the context of the Raychaudhuri equation.

9.7 Calculating Connection Matrix

9.7.1 Ricci Rotation Coefficients

Although the Levi-Civita connection coefficients Γ^k_{ij} ($n^2(n+1)/2$ in number) are straightforward to calculate from the derivatives of g_{ij}, the calculation may sometimes be long. We can calculate the connection matrix somewhat more simply using Cartan's structural equation (or Lie bracket relations) by using a basis in which the metric components are constant. Usually an orthonormal basis is used.

Let $\{X_a\}$ be basis vector fields, $\{\alpha_a\}$ the corresponding dual basis fields and $(\omega_X)^b{}_a$ the connection matrix. Cartan's equation states that for zero torsion (symmetric connection)

$$d\alpha^a = -(\omega_X)^a{}_b \wedge \alpha^b.$$

These equations provide us with an alternative and simpler way to calculate the connection.

The compatibility condition on the Riemannian metric is that the covariant derivative respects the inner product. In the following we omit the basis symbol X from g and ω. The indices a, b on $g_{ab} = \langle X_a, X_b \rangle$ already distinguish them from g_{ij}.

$$
\begin{aligned}
D_Y(g_{ab}) = d(g_{ab})(Y) &= D_Y \langle X_a, X_b \rangle \\
&= \langle D_Y X_a, X_b \rangle + \langle X_a, D_Y X_b \rangle \\
&= \omega^c{}_a(Y) \langle X_c, X_b \rangle + \omega^c{}_b(Y) \langle X_a, X_c \rangle \\
&= g_{cb} \omega^c{}_a(Y) + g_{ac} \omega^c{}_b(Y) \\
&\equiv (\omega_{ba} + \omega_{ab})(Y)
\end{aligned}
$$

or, what is equivalent to eqn. (9.10),

$$d(g_{ab}) = \omega_{ba} + \omega_{ab}.$$

The one-forms

$$\omega_{ab} \equiv g_{ac} \omega^c{}_b \tag{9.36}$$

can be expanded in the basis $\{\alpha^a\}$,

$$\omega_{ab} = \omega_{abc} \alpha^a. \tag{9.37}$$

Written in this way the coefficients ω_{abc} are called **Ricci rotation coefficients**.

When we choose the basis $\{X_a\}$ orthonormal:

$$g_{ab} = \eta_{ab} = \langle X_a, X_b \rangle$$

then $dg_{ab} = d\eta_{ab} = 0$. This implies that $\omega_{ba} = -\omega_{ab}$ and the Ricci rotation coefficients become antisymmetric in the first two indices:

$$\omega_{abc} = -\omega_{bac}.$$

9.7.2   Calculating Ricci Coefficients

There are $n^2(n-1)/2$ Ricci rotation coefficients — of its three indices ω_{abc} is antisymmetric in the first two. This should be compared to $n^2(n+1)/2$ Christoffel symbols Γ^k_{ij}. For four-dimensional spacetime these numbers are 24 and 40. It is therefore easier to calculate the Ricci coefficients and hence the connection matrix. The price to be paid is finding an orthonormal basis, and converting things back to a natual basis if so required.

Let the exterior derivatives $d\alpha^a$ be expanded as

$$d\alpha^a = \sum_{b<c} F^a_{bc}\alpha^b \wedge \alpha^c.$$

The coefficient functions F^a_{bc} are defined only for $b < c$ but we can extend the definition by $F^a_{cb} = -F^a_{bc}$ and $F^a_{bb} = 0$ for each b. The first structural equation is

$$\begin{aligned}
\eta_{da}d\alpha^a &= -\omega_{db} \wedge \alpha^b \\
&= -\omega_{dbc}\alpha^c \wedge \alpha^b \\
&= \sum_{b<c}(\omega_{dbc} - \omega_{dcb})\alpha^b \wedge \alpha^c.
\end{aligned}$$

The left-hand side is

$$\sum_{b<c} \eta_{da}F^a_{bc}\alpha^b \wedge \alpha^c,$$

therefore

$$f_{dbc} = \omega_{dbc} - \omega_{dcb}, \qquad f_{dbc} \equiv \eta_{da}F^a_{bc}.$$

These are $n^2(n-1)/2$ equations for as many unknowns. The f_{abc} are the known functions antisymmetric in the last two indices, whereas the unknown ω_{abc} are antisymmetric in the first two indices. These linear equations can be easily solved. Add all the 3! permutations of ω_{abc} with appropriate permutation signs

$$\omega_{abc} - \omega_{acb} - \omega_{bac} + \omega_{bca} + \omega_{cab} - \omega_{cba} = f_{abc} + f_{bca} + f_{cab}.$$

Use antisymmetry of ω in the first two indices and the left-hand side becomes

$$2(\omega_{abc} + \omega_{cab} - \omega_{cba}) = 2\omega_{abc} + 2f_{cab};$$

put it back to get

$$2\omega_{abc} = f_{abc} + f_{bca} - f_{cab}. \tag{9.38}$$

9.7.3 Bracket Relations for Ricci Coefficients

There is another method to find Ricci coefficients for orthonormal basis fields. The Lie bracket of basis fields, when expanded back in terms of the basis, determines the $n^2(n-1)/2$ structure functions $C_{ab}^c = -C_{ba}^c$:

$$[X_a, X_b] = C_{ab}^c X_c. \tag{9.39}$$

The Lie bracket is related to the torsion tensor (which is identically zero for our case of Riemannian geometry)

$$0 = T(X_a, X_b) = D_{X_a} X_b - D_{X_b} X_a - [X_a, X_b].$$

Therefore, using the definition of the connection matrix,

$$D_{X_a} X_b = i_{X_a} D X_b = \omega^c{}_b(X_a) X_c$$

and expanding ω in the basis $\{\alpha^a\}$ as $\omega^c{}_b = \omega^c{}_{bd}\alpha^d$ we get

$$D_{X_a} X_b = \omega^c{}_{ba} X_c.$$

Thus

$$C_{ab}^c = \omega^c{}_{ba} - \omega^c{}_{ab}.$$

Lowering the first index by $g_{X\,dc} = \eta_{dc}$,

$$h_{dab} \equiv -\eta_{dc} C_{ab}^c = \omega_{dab} - \omega_{dba}. \tag{9.40}$$

Like the last subsection, these are $n^2(n-1)/2$ equations to determine $n^2(n-1)/2$ Ricci coefficients $\omega_{abc} = -\omega bac$ in terms of known quantities $h_{dab} = -h_{dba}$. The solution is

$$2\omega_{abc} = h_{abc} + h_{bca} - h_{cab}. \tag{9.41}$$

9.8 Covariant Riemann Tensor $R(W, Z; X, Y)$

Recall that the covariant derivative of a vector field Y in the direction of X is written as a vector field $D_X Y$. If the connection is symmetric, that is torsion-free, then

$$D_X Y - D_Y X - [X, Y] = 0.$$

We say that the connection is **compatible with the metric** if the rate of change of $\langle Y, Z \rangle$ along X which is equal to

$$D_X(\langle Y, Z \rangle) = X(\langle Y, Z \rangle) = (d\langle Y, Z \rangle)(X)$$

is such that

$$X(\langle Y, Z \rangle) = \langle D_X Y, Z \rangle + \langle Y, D_X Z \rangle. \tag{9.42}$$

This is the same as saying that *the inner product of two vectors is preserved if they are both parallel displaced by the connection along a curve.* In section 9.2.1 earlier in this chapter we gave the proof of the fundamental theorem that the only symmetric connection which is compatible with the metric is the Levi-Civita connection.

The covariant form of the Riemann tensor can be easily defined as a multilinear mapping which takes four-vector fields X, Y, Z, W into a real number $R(W, Z; X, Y)$,

$$R(W, Z; X, Y) \equiv \langle W, R(X, Y)Z \rangle. \tag{9.43}$$

There should be no confusion in using the same symbol R for the two Riemann tensors; one of them takes three arguments $R(X, Y)Z$ and gives a field while the other takes four arguments $R(W, Z; X, Y)$ and gives a number.

The symmetry properties of the Riemann tensor with components R_{ijkl} can be seen as the following properties of the tensor $R(W, Z; X, Y)$:

$$\begin{aligned}
R(W, Z; X, Y) &= -R(W, Z; Y, X), &\tag{9.44} \\
R(W, Z; X, Y) &= -R(Z, W; X, Y), &\tag{9.45} \\
R(W, Z; X, Y) &= R(X, Y; W, Z), &\tag{9.46} \\
R(W, Z; X, Y) &+ R(W, X; Y, Z) + R(W, Y; Z, X) = 0. &\tag{9.47}
\end{aligned}$$

The first and the fourth of these are direct consequences of properties of $R(X, Y)Z$. The other two are treated in a tutorial.

The relation of the tensor with its components when coordinate basis fields $\partial_i \equiv \partial/\partial x^i$ are used is given by

$$\begin{aligned}
D_{\partial_i} \partial_j &= \Gamma^k_{ij} \partial_k, &\tag{9.48} \\
R(\partial_i, \partial_j) \partial_k &= R^l{}_{kij} \partial_l, &\tag{9.49} \\
R(\partial_i, \partial_j; \partial_k, \partial_l) &= R_{ijkl} = g_{im} R^m{}_{jkl}. &\tag{9.50}
\end{aligned}$$

9.9   Isometries and Killing Vector Fields

Let M be a Riemannian manifold with a metric tensor $\mathbf{g}$. Let ϕ be a one-to-one invertible mapping of the manifold. Such a mapping is called a **diffeomorphism** if its representative in terms of coordinates is infinitely differentiable.

Let $c : \tau \to c(\tau) \in M$ be a curve passing through a point $p = c(0)$. For each τ we can map the point $c(\tau)$ by ϕ to $\phi(c(\tau))$ and obtain a new curve. We can say ϕ maps the curve c into another curve $\phi_* c \equiv: \tau \to \phi(c(\tau)) \in M$ passing through $\phi(p)$ at $\tau = 0$.

As a result the tangent vector $\mathbf{v} = dc/d\tau \in T_p$ is mapped by ϕ into the tangent vector $\phi_* \mathbf{v} \equiv d(\phi_* c)/d\tau \in T_{\phi(p)}$. By considering different curves passing through p we can find a $\Phi_* \mathbf{v}$ for every $\mathbf{v} \in T_p$.

The vector $\Phi_* \mathbf{v} \in T_{\phi(p)}$ is called the **pushforward** of $\mathbf{v}$ by ϕ.

If the inner product of two vectors is preserved by the pushforwarding by ϕ, we say that the diffeomorphism ϕ is an **isometry**. This means

$$\langle \phi_* \mathbf{v}, \phi_* \mathbf{w} \rangle_{\phi(p)} = \langle \mathbf{v}, \mathbf{w} \rangle_p$$

where we have indicated the points at which the the inner product on the left and the right-hand side is calculated.

The existence of an isometry is an indication of **symmetry** in the space. Composition of two isometries is again an isometry. The identity mapping and inverse of an isometry are also isometries. Therefore the set of isometries form a group.

A **one-parameter group of isometries** ϕ_t is a family of isometries labelled by a real parameter $t : -\infty < t < \infty$ which satisfies

$$\phi_t \circ \phi_s = \phi_{t+s}, \qquad \phi_0 = id$$

where id is the identity mapping.

Given a one-parameter group of isometries, at each point p the curve $t \to \phi_t(p)$ determines a tangent vector

$$K(p) = \left. \frac{d\phi_t(p)}{dt} \right|_{t=0}$$

giving rise to a vector field. The vector field defined by the one-parameter group of isometries ϕ_t is called the **Killing vector field** corresponding to the one-parameter group. A Killing vector field can be thought of as the infinitesimal form of isometry or symmetry mapping.

Let us choose a coordinate system x. For infinitesimally small values Δt of the parameter, the point which has coordinates x^i is mapped to a point with coordinates $x^i + \Delta t K^i(x)$ by $\phi_{\Delta t}$ where K^i are the components of the Killing vector.

A tangent vector with components

$$a^i = \left. \frac{dx^i(t)}{dt} \right|_{t=0}$$

at point P having coordinates x will have, at the mapped point $Q = \phi_{\Delta t}(P)$ with coordinates $\sim x^i + K^i \Delta t$, components equal to

$$\left. \frac{d(x^i(t) + K^i \Delta t)}{dt} \right|_{t=0} = a^i + K^i{}_{,j} a^j \Delta t.$$

Similarly, a vector with components b^k will become $b^k + K^k{}_{,l} b^l \Delta t$. The condition of isometry is then

$$g_{ik}(x) a^i b^k = g_{ik}(x + K\Delta t)(a^i + K^i{}_{,j} a^j \Delta t)(b^k + K^k{}_{,l} b^l \Delta t).$$

Keeping terms to first order in Δt this gives, for arbitrary a^i and b^k,

$$g_{in}K^i{}_{,m} + g_{mi}K^i{}_{,n} + g_{mn,l}K^l = 0.$$

By lowering the index to $K_i = g_{ij}K^j$ and using the fact that $g_{ij}g^{jk} = \delta_i^k$ is a constant we get, using the definition of covariant derivative,

$$K_{m;n} + K_{n;m} = 0. \tag{9.51}$$

Killing Vector Fields

We have seen that a one-parameter group of isometries determines a Killing vector field K. The covariant components of the Killing vector field satisfy the equation $K_{m;n} + K_{n;m} = 0$.

Conversely, any vector field with components $V^i(x)$ in coordinate system x can be used to define an infinitesimal mapping which maps a point with coordinates x^i into the point with coordinates $x^i + \epsilon V^i(x)$ where ϵ is an infinitesimal parameter. For different values of ϵ this defines a **local one-parameter group of mappings** which preserves the inner product if and only if $V_{m;n} + V_{n;m} = 0$. Therefore any vector field V which satisfies $V_{m;n} + V_{n;m} = 0$ is called a Killing vector field.

Conserved Quantities Along a Geodesic

There is a physically important relationship between a Killing vector field, which represents a symmetry of the metric and conservation of a quantity along a geodesic.

Let $c : \tau \to x(\tau)$ be a geodesic with tangent vector $\mathbf{t}(\tau)$, and $\mathbf{K} = K^i \partial/\partial x^i$ a Killing vector field. Then

$$\frac{d\langle \mathbf{K}, \mathbf{t}(\tau)\rangle}{d\tau} = 0. \tag{9.52}$$

This can be readily seen as follows.

$$
\begin{aligned}
\frac{d\langle \mathbf{K}, \mathbf{t}(\tau)\rangle}{d\tau} &= \frac{d}{d\tau}\left(K_i \frac{dx^i}{d\tau}\right) \\
&= K_{i,j}\frac{dx^j}{d\tau}\frac{dx^i}{d\tau} + K_i \frac{d^2x^i}{d\tau^2} \\
&= K_{k;j}\frac{dx^j}{d\tau}\frac{dx^k}{d\tau}
\end{aligned}
$$

by using the geodesic equation to substitute for $d^2x^i/d\tau^2$, and using the definition of the covariant derivative. This expression is zero because $K_{k;j} = -K_{j;k}$ and this antisymmetric tensor is multiplied by the symmetric product of tangent vector components.

It is important to realise that the Killing fields form a vector space. For physical applications it is important to find all linearly independent Killing fields. We shall see in a later chapter that in a space of dimension n there cannot be more than $n(n+1)/2$ independent Killing fields. A Riemannian space whose metric has all the $n(n+1)/2$ Killing fields present is called **maximally symmetric**.

For physical applications it is important to identify all possible Killing vector fields for a given metric.

It is easy to show that if the metric tensor components g_{ij} are independent of a particular coordinate, say x^{i_0}, then the corresponding natural basis vector field $\partial/\partial x^{i_0}$ is a Killing vector field.

As an example the flat two-dimensional Euclidean space $ds^2 = dx \otimes dx + dy \otimes dy$ is maximally symmetric. The three Killing fields are:

$$\frac{\partial}{\partial x}, \qquad \frac{\partial}{\partial y}, \qquad x\frac{\partial}{\partial y} - y\frac{\partial}{\partial x}.$$

Conserved Energy-Momentum Four-Vector

We will use the following result later. Let $T^{ij} = T^{ji}$ be a symmetric second-rank tensor satisfying 'conservation equation' $T^{ij}{}_{;j} = 0$. Let K^i be a Killing vector field. Then the associated $K_i \equiv g_{ik}K^k$ satisfies $K_{i;j} = -K_{j;i}$. It follows that P^i defined by

$$P^i = T^{ij}K_j$$

satisfies

$$P^i{}_{;i} = 0$$

because in

$$P^i{}_{;i} = (T^{ij}K_j)_{;i} = T^{ij}{}_{;i}K_j + T^{ij}K_{j;i}$$

the first term is zero (given) and the second is zero because T^{ij} is symmetric and $K_{j;i}$ is antisymmetric.

9.10   Tutorial

Exercise 48. Given that a spherically symmetric four-dimensional spacetime has a metric of the form

$$ds^2 = -a(r)dt \otimes dt + b(r)dr \otimes dr + r^2 d\theta \otimes d\theta + r^2 \sin\theta d\phi \otimes d\phi.$$

Find the connection matrix by choosing an orthonormal basis and calculating the **Ricci rotation coefficients**.

Answer 48. Choose the obvious orthonormal basis

$$\alpha^0 = \sqrt{a(r)}dt, \quad \alpha^1 = \sqrt{b(r)}dr, \quad \alpha^2 = rd\theta, \quad \alpha^3 = r\sin\theta d\phi.$$

The derivatives are

$$
\begin{aligned}
d\alpha^0 &= \frac{a'}{2\sqrt{a}}dr \wedge dt = -\frac{a'}{2a\sqrt{b}}\alpha^0 \wedge \alpha^1, \\
d\alpha^1 &= 0, \\
d\alpha^2 &= \frac{1}{r\sqrt{b}}\alpha^1 \wedge \alpha^2, \\
d\alpha^3 &= \frac{1}{r\sqrt{b}}\alpha^1 \wedge \alpha^3 + \frac{\cot\theta}{r}\alpha^2 \wedge \alpha^3.
\end{aligned}
$$

This determines

$$
\begin{aligned}
f_{001} &= \frac{a'}{2a\sqrt{b}}, \\
f_{212} &= \frac{1}{r\sqrt{b}}, \\
f_{313} &= \frac{1}{r\sqrt{b}}, \\
f_{323} &= \frac{\cot\theta}{r},
\end{aligned}
$$

the other f_{abc} being zero. The Ricci rotation coefficients ω_{abc} and $\omega^a{}_b$ are determined by

$$2\omega_{abc} = f_{abc} + f_{bca} - f_{cab},$$

$$\omega^a{}_b = \eta^{ad}\omega_{db} = \eta^{ad}\omega_{dbc}\alpha^c.$$

The non-zero elements of the connection matrix are

$$
\begin{aligned}
\omega^0{}_1 &= \omega^1{}_0 = \frac{a'}{2\sqrt{ab}}dt, \\
\omega^1{}_2 &= -\omega^2{}_1 = -\frac{1}{\sqrt{b}}d\theta, \\
\omega^1{}_3 &= -\omega^3{}_1 = -\frac{\sin\theta}{\sqrt{b}}d\phi, \\
\omega^2{}_3 &= -\omega^3{}_2 = -\cos\theta d\phi.
\end{aligned}
$$

Exercise 49. Calculate the Christoffel symbols in the natural basis using the transformation formula for the connection matrix for the previous exercise.

Answer 49. The dual basis $\{\alpha^a\}$ is related to the natural basis $\{dx^i\}$ by the diagonal matrix U,

$$
\begin{pmatrix} \alpha^0 \\ \alpha^1 \\ \alpha^2 \\ \alpha^3 \end{pmatrix} = \begin{pmatrix} \sqrt{a} & 0 & 0 & 0 \\ 0 & \sqrt{b} & 0 & 0 \\ 0 & 0 & r & 0 \\ 0 & 0 & 0 & r\sin\theta \end{pmatrix} \begin{pmatrix} dt \\ dr \\ d\theta \\ d\phi \end{pmatrix},
$$

therefore the natural dual basis ∂_i is obtained by the same matrix U operating on the dual basis $\{X_a\}$. This means the connection matrix $\Gamma^j_{ki} dx^k$ in natural coordinates is

$$\Gamma^j_{ki} dx^k = U^{-1} dU + U^{-1} \omega U.$$

Thus the non-zero Γ's are

$$
\begin{aligned}
\Gamma^0_{10} &= \Gamma^0_{01} = \frac{a'}{2a}, \\
\Gamma^1_{00} &= \frac{a'}{2b}, \\
\Gamma^1_{11} &= \frac{b'}{2b}, \\
\Gamma^1_{22} &= -\frac{r}{b}, \\
\Gamma^1_{33} &= -\frac{r \sin^2 \theta}{b}, \\
\Gamma^2_{21} &= \Gamma^2_{12} = \Gamma^3_{31} = \Gamma^3_{13} = \frac{1}{r}, \\
\Gamma^2_{33} &= -\sin \theta \cos \theta, \\
\Gamma^3_{32} &= \Gamma^3_{23} = \cot \theta.
\end{aligned}
$$

These are just the coefficients we had calculated in Chapter 4 for obtaining the Schwarzschild solution.

Exercise 50. Prove the symmetry properties of the Riemann tensor:

$$
\begin{aligned}
R(W, Z; X, Y) &+ R(W, X; Y, Z) + R(W, Y; Z, X) = 0, \\
R(W, Z; X, Y) &= -R(W, Z; Y, X), \\
R(W, Z; X, Y) &= -R(Z, W; X, Y), \\
R(W, Z; X, Y) &= R(X, Y; W, Z).
\end{aligned}
$$

Answer 50. We do not have to prove the first two. The first equation is just the first Bianchi identity (proved in section 8.9.3 in Chapter 8)

$$R(X, Y)Z + R(Y, Z)X + R(Z, X)Y = 0.$$

Here it is written for $R(W, Z; X, Y) \equiv \langle W, R(X, Y)Z \rangle$.

The second property (antisymmetry in X and Y) is part of the definition.

For proving $R(W, Z; X, Y) = -R(Z, W; X, Y)$ we show that $R(W, Z; X, Y) + R(Z, W; X, Y)$ adds to zero. We first note that terms in $R(W, Z; X, Y) = \langle W, D_X D_Y Z - D_Y D_X Z - D_{[X,Y]} Z \rangle$ can be expanded using the metric property $[A(\langle B, C \rangle) = \langle D_A B, C \rangle + \langle B, D_A C \rangle]$ as follows:

$$
\begin{aligned}
\langle W, D_X D_Y Z \rangle &= X(\langle W, D_Y Z \rangle) - \langle D_X W, D_Y Z \rangle \\
&= X\big(Y(\langle W, Z \rangle) - \langle D_Y W, Z \rangle\big) - \langle D_X W, D_Y Z \rangle \\
&= X\big(Y(\langle W, Z \rangle)\big) - X\big(\langle D_Y W, Z \rangle\big) - \langle D_X W, D_Y Z \rangle
\end{aligned}
$$

with a similar expression for $\langle W, D_X D_Y Z \rangle$ with X and Y interchanged. Therefore $R(W, Z; X, Y)$ has the first two terms

$$
\begin{aligned}
\langle\, W, D_X D_Y Z &\;-\; D_Y D_X Z\,\rangle \\
&= X\big(Y(\langle W, Z\rangle)\big) - X\big(\langle D_Y W, Z\rangle\big) - \langle D_X W, D_Y Z\rangle \\
&\quad -Y\big(X(\langle W, Z\rangle)\big) + Y\big(\langle D_X W, Z\rangle\big) + \langle D_Y W, D_X Z\rangle \\
&= [X, Y](\langle W, Z\rangle) - X\big(\langle D_Y W, Z\rangle\big) - \langle D_X W, D_Y Z\rangle \\
&\quad +Y\big(\langle D_X W, Z\rangle\big) + \langle D_Y W, D_X Z\rangle.
\end{aligned}
$$

Similarly $R(Z, W; X, Y)$ has two first terms $\langle W, D_X D_Y Z - D_Y D_X Z\rangle$ which can be obtained from the expression for $R(W, Z; X, Y)$ by interchanging W and Z. Adding these together we get

$$
\begin{aligned}
\langle W, D_X D_Y Z - D_Y D_X Z\rangle &\;+\; \langle W, D_X D_Y Z - D_Y D_X Z\rangle \\
&= 2[X, Y](\langle W, Z\rangle) - X\big(\langle D_Y W, Z\rangle + \langle D_Y Z, W\rangle\big) \\
&\quad +Y\big(\langle D_X W, Z\rangle + \langle D_X Z, W\rangle\big) \\
&= [X, Y](\langle W, Z\rangle).
\end{aligned}
$$

On the other hand the third terms in both add to

$$
-\langle W, D_{[X,Y]} Z\rangle - \langle Z, D_{[X,Y]} W\rangle = -[X, Y](\langle W, Z\rangle).
$$

Thus

$$
R(W, Z; X, Y) + R(Z, W; X, Y) = 0
$$

proving the antisymmetry in the first two indices.

To show $R(W, Z; X, Y) = R(X, Y; W, Z)$ write the first Bianchi identity four times cyclically:

$$
\begin{array}{rclcrcl}
R(W, Z; X, Y) &+& R(W, X; Y, Z) &+& R(W, Y; Z, X) &=& 0, \\
R(Z, X; Y, W) &+& R(Z, Y; W, X) &+& R(Z, W; X, Y) &=& 0, \\
R(X, Y; W, Z) &+& R(X, W; Z, Y) &+& R(X, Z; Y, W) &=& 0, \\
R(Y, W; Z, X) &+& R(Y, Z; X, W) &+& R(Y, X; W, Z) &=& 0.
\end{array}
$$

Adding the left-hand sides of these equations and numbering these terms from 1 to 12 we notice that due to antisymmetry properties just proved, term 1 cancels with 6; 3 with 10; 4 with 9 and 7 with 12. Terms 2 and 8 add to $2R(W, X; Y, Z)$ and 5 and 11 add to $-2R(Y, Z; W, X)$. Therefore equating the sum of the twelve terms to zero, we get

$$
R(W, X; Y, Z) = R(Y, Z; W, X).
$$

Exercise 51. GEODESICS ARE EXTREMAL PATHS

Show that the geodesic equation can be obtained also as the extremal for the variational problem

$$
\delta \int_{s_1}^{s_2} ds = 0 \tag{9.53}
$$

where $ds^2 = g_{ij} dx^i dx^j$, and the paths are fixed at two ends at s_1 and s_2.

Answer 51.

$$\delta \int \left(g_{ij} \frac{dx^i}{ds} \frac{dx^j}{ds} \right)^{\frac{1}{2}} ds = \frac{1}{2} \int \left[\delta(g_{ij}) \frac{dx^i}{ds} \frac{dx^j}{ds} + g_{ij} \frac{d\delta x^i}{ds} \frac{dx^j}{ds} + g_{ij} \frac{dx^i}{ds} \frac{d\delta x^j}{ds} \right],$$

so we can interchange the dummy indices i, j in the middle term, and transfer d/ds using integration-by-parts to write the right-hand side

$$
\begin{aligned}
\text{r. h. s.} \quad &= \quad \frac{1}{2} \int \left[g_{ij,k} \frac{dx^i}{ds} \frac{dx^j}{ds} - 2 \frac{d}{ds} \left(g_{jk} \frac{dx^j}{ds} \right) \right] \delta x^k \\
&= \quad \frac{1}{2} \int \left[g_{ij,k} \frac{dx^i}{ds} \frac{dx^j}{ds} - 2 g_{jk,l} \frac{dx^l}{ds} \frac{dx^j}{ds} - 2 g_{jk} \frac{d^2 x^j}{ds^2} \right] \delta x^k \\
&= \quad \frac{1}{2} \int \left[g_{ij,k} \frac{dx^i}{ds} \frac{dx^j}{ds} - g_{jk,l} \frac{dx^l}{ds} \frac{dx^j}{ds} - g_{lk,j} \frac{dx^l}{ds} \frac{dx^j}{ds} - 2 g_{jk} \frac{d^2 x^j}{ds^2} \right] \delta x^k .
\end{aligned}
$$

Setting the quantity in square brackets equal to zero and by multiplying the inverse matrix g^{jm} and summing over j we get the geodesic equation.

Exercise 52. Show that the Weyl tensors of two conformally related metrics

$$\overline{g}_{ij} = e^{2\phi} g_{ij}$$

are very simply related as

$$\overline{C}_{ijkl} = e^{2\phi} C_{ijkl}.$$

Answer 52. Straightforward computation.

Chapter 10

Additional Topics in Geometry

10.1 Mappings Between Manifolds

We have already encountered diffeomorphisms in the previous chapter as one-to-one invertible bothways-differentiable maps of a manifold onto itself. We now define maps between different manifolds.

Let M and N be two differentiable manifolds of dimensions m and n respectively, and ψ a mapping $\psi : M \to N$. Then ψ is called differentiable if (any of the) local coordinates $y^1, \ldots, y^n$ of the point $q = \psi(p)$ are smooth functions of (any of the) local coordinates $x^1, \ldots, x^m$ of p, and this happens for all points of M. We can express this correspondence in terms of coordinates as

$$y^i = \psi^i(x^1, \ldots, x^m) \qquad i = 1, \ldots, n. \tag{10.1}$$

Pullback and Pushforward Maps

The mapping ψ sets a correspondence between smooth functions $\mathcal{F}(N)$ on N and $\mathcal{F}(M)$ on M. For any $f \in \mathcal{F}(N)$ we can define a **pullback** function $\psi^* f = f \circ \psi \in \mathcal{F}(M)$ which simply assigns to point $p \in M$ the same value that f does to $\psi(p)$.

Let $c : I \to M$ be a curve in M. The mapping ψ then defines a curve on N by mapping $t \in I$ to $\psi(c(t)) = (\psi \circ c)(t)$. A tangent vector $\mathbf{v} \in T_p(M)$ to the curve c at a point $p = c(t_0) \in M$ corresponding to $t = t_0 \in I$ determines a tangent vector $\mathbf{w} \in T_{\psi(p)}(N)$ tangent to the curve $\psi \circ c$ at the same t_0.

Let the curve c be represented in coordinates by functions $x^i(t), i = 1, \ldots, m$. With ψ these points are mapped to a curve in N whose coordinates can be given by functions $y^j(t) = \psi^j(x(t)), j = 1, \ldots, n$. Therefore the components w^j of the tangent vector $\mathbf{w} = w^j \frac{\partial}{\partial y^j}$ to the curve at point $\psi(p)$ are given in terms of components of the vector $\mathbf{v} = v^i \frac{\partial}{\partial x^i}$ as

$$w^j = \frac{dy^j}{dt}\bigg|_{t_0} = \frac{\partial \psi^j}{\partial x^i} \frac{dx^i}{dt}\bigg|_{t_0} = \frac{\partial \psi^j}{\partial x^i} v^i.$$

This correspondence of a tangent vector $\mathbf{v}$ at a point p determining a tangent vector $\mathbf{w}$ via ψ is called a **pushforward** by ψ and written

$$\mathbf{w} = \psi_*(\mathbf{v}). \tag{10.2}$$

The above discussion of the tangent vector is not dependent on the existence of a curve, although a concrete curve makes the concept very transparent. We can discuss the tangent vectors in terms of their action on smooth functions defined on the manifolds. If $f \in \mathcal{F}(N)$ is a smooth function on N, then the action of $\mathbf{w}$ on f is

$$\mathbf{w}(f) = \mathbf{v}(\psi^* f) = \mathbf{v}(f \circ \psi). \tag{10.3}$$

Notice the use of '*' in the superscript ψ^* for pullback, that is in the direction opposite to that of ψ and in the subscript for pushforward. This is the standard notation.

ψ-Related Vector Fields

It might appear that if X is a vector field on M then it will determine a vector field on N by pushing forward every vector to the mapped point. But this is not so in general, because two different points of M may map to the same point in N and the two pushed-forward vectors may not match.

A vector field X which does, in fact, determine a vector field on N unambiguously is said to be ψ-**related**.

We write $\psi_* X$ for the pushed-forward vector field.

The Lie bracket of two ψ-related vector fields is also ψ-related and equal to the bracket of pushed-forward fields,

$$\psi_*[X, Y] = [\psi_* X, \psi_* Y].$$

Pullback of a 1-Form

A 1-form or cotangent vector $\alpha \in T^*_{\psi(p)}$ at $\psi(p)$ is pulled back to point $p \in M$ to $\psi^* \alpha \in T^*_p$ which acts on members of T_p as

$$\psi^* \alpha(\mathbf{v}) = \alpha(\psi_* \mathbf{v}). \tag{10.4}$$

There is no ambiguity in defining the pullback of the cotangent vector field, unlike the case for tangent vector fields.

Pullback and Pushforward

For higher rank tensors the push-forward or pullback is defined in a straightforward manner on decomposable vectors as

$$\psi_*(\mathbf{v} \otimes \cdots \otimes \mathbf{w}) \equiv (\psi_* \mathbf{v}) \otimes \cdots \otimes (\psi_* \mathbf{w}), \tag{10.5}$$

$$\psi^*(\alpha \otimes \cdots \otimes \beta) \equiv (\psi^* \alpha) \otimes \cdots \otimes (\psi^* \beta). \tag{10.6}$$

On general vectors the mapping is extended by using linearity. It is useful to write down the formulas for basis elements:

$$\psi_* \left(\frac{\partial}{\partial x^i} \right) = \frac{\partial \psi^j}{\partial x^i} \frac{\partial}{\partial y^j}, \tag{10.7}$$

$$\psi^* (dy^j) = \frac{\partial \psi^j}{\partial x^i} dx^i. \tag{10.8}$$

In general all *contravariant* tensor fields can be pushed *forward* by $\psi : M \to N$ and all *covariant* tensor fields which are multilinear functionals on contravariant quantities are pulled *back* from N to M.

The only caution to be exercised is that pushed-forward fields must be ψ-related. But there is no restriction on the pullbacks, even in those cases where the range of ψ is a proper subset of N.

Pullback of Forms and Exterior Derivative

For pulled-back differential forms the following result is fundamental.

The pullback operation commutes with that of taking the exterior derivative,

$$\psi^* (d\alpha) = d(\psi^* \alpha). \tag{10.9}$$

This is a very important result although it is straightforward to prove. It shows the significance of the notion of exterior derivative in differential geometry. The definition of integration on manifolds depends crucially on this result.

Diffeomorphisms

A smooth mapping $\psi : M \to N$ from a manifold M to a manifold N of the same dimension is called a **diffeomorphism** if it is invertible and the inverse mapping $\psi^{-1} : N \to M$ is also differentiable. In this case all geometrical quantities can be pushed to or fro using $\psi^*, \psi_*, \psi^{-1*}, \psi_*^{-1}$. In particular there exist diffeomorphisms of a manifold to itself. Two manifolds between which there exists a diffeomorphism, are called **diffeomorphic** and they are practically identical as far as their differential structure is concerned.

The set $Diff(M)$ of all diffeomorphisms of a manifold onto itself forms a group with composition of mappings as the group operation. The identity of the group is the mapping which maps every point of M to itself. This is an *infinite* group in the sense that different members of the group cannot be labelled by a finite number of parameters. Still, we can talk about diffeomorphisms which are **close to identity** if every point is mapped to a point in its neighbourhood in some definite sense.

Let us consider a diffeomorphism Φ close to identity. Choose a coordinate chart and let a point p with coordinates $x(p)$ be mapped to a point $q = \Phi(p)$; then we can assume that the point q also lies in the same chart with coordinates $x(q)$. $x(q)$ as numbers are close to $x(p)$ and we write $x(q) = x(p) + \xi$. As the point p varies in a region within the chart, we get a functional relationship

$$y^i = x^i(q) = x^i(p) + \xi^i(x).$$

A vector $\mathbf{v} \in T_p$ with components v^i in a coordinate basis at the point p is pushed forward to $\phi_* \mathbf{v} \in T_{\phi(p)}$ whose action on a function f is

$$\phi_* \mathbf{v}(f) = \mathbf{v}(f \circ \phi).$$

If the coordinate representation of the function f is $f(x)$ then the coordinate representation of the function $f \circ \phi$ is $f(y) = f(x + \xi)$. Therefore

$$\phi_* \mathbf{v}(f) \;=\; v^i \left.\frac{\partial}{\partial x^i}\right|_p (f \circ \phi)$$

$$=\; v^i \frac{\partial y^j}{\partial x^i} \left.\frac{\partial}{\partial x^j}\right|_q f.$$

This gives the transformation formula for the pushed-forward vector components

$$(\phi_* \mathbf{v})^j = v^i \frac{\partial y^j}{\partial x^i}.$$

Similarly components of other geometric quantities can be calculated.

A Word of Warning

One should never, *never* confuse a diffeomorphism with a coordinate transformation. A point in a manifold may be described by two charts defined in its neighbourhood. The coordinates in these respective charts may be, say, x^i and y^i. These numbers *refer to the same point* p. A diffeomorphism Φ maps *all* points of the manifold into other points of the manifold. And barring exception a point p is mapped to a *different* point $q = \Phi(p)$. The points q and p may happen to lie in the same chart but their coordinates refer to two different points. The relationship $y^i = x^i + \xi^i$ above is therefore not a coordinate transformation but just a local coordinate expression of the diffeomorphism ϕ when it happens to be close to identity.

This caveat is necessary because in many texts this distinction is not emphasized enough. Physicists define vectors or tensors as quantities which 'transform' in a certain way. The formula which gives a change in the components of a vector when coordinates are changed and the formula above which gives the components of a pushed-forward vector at q in terms components of the original vector components at p are similar. Maybe that is why this confusion is prevalent.

10.2 Integral Curves of a Vector Field

We know that a smooth curve $c : I \to M$ from an interval I of the real line into a manifold M defines tangent vectors all along itself. Suppose $X = v^i(x)\partial/\partial x^i$ is a smooth vector field. Is it possible to find curves such that vectors $X(p)$ are just the tangent vectors to the curve passing through the point p?

Let us assume such a curve exists, $c : I \to M$. Without any loss of generality, we can take I to be an interval containing the point $t = 0$. Let a^i be the coordinates of the point $c(0)$ and $x^i(t)$ that of points $c(t)$. Therefore

$$\frac{dx^i}{dt} = v^i(x), \qquad x^i(0) = a^i.$$

Solutions to this differential equation exist not only for the initial conditions $x^i = a^i$ but also for initial conditions in a neighbourhood of a^i. This family of solutions corresponds (by the coordinate charts) to a family of curves in a neighbourhood U of the original point, $c[p] : I_p \to M; p \in U$ such that

$$c[p](0) = p, \qquad \frac{dc[p]}{dt} = X(c(t)). \tag{10.10}$$

We say that the curve $c[p]$ is an **integral curve** of the vector field **centered at** p. In the following we consider all these curves $c[p]$ in their totality, and we will assume that there is a common minimum interval I which is contained in all the intervals I_p, $p \in U$.

Let us fix t to be small so that $p_1 = c[p](t)$ is still in U. Now, there exists a curve $c[p_1]$ centered at p_1 such that if s is the parameter of this curve, $c[p_1](s = 0) = p_1 = c[p](t)$ and

$$\frac{dc[p_1]}{ds} = X(c[p_1](s)).$$

Let us define a new curve as $C(s) \equiv c[p](t + s)$; then C is centered at p_1 and satisfies the same differential equation and initial conditions as the curve $c[p_1]$. Calling $T = t + s$,

$$C(0) = p_1, \qquad \frac{dC}{ds} = \frac{dc[p]}{dT} = X(c[p](T)) = X(C(s)).$$

The solution to the differential equation is unique, therefore it follows that C and $c[p_1]$ are the same curve and they coincide on the union of the intervals I_p and I_{p_1},

$$c[p](t + s) = C(s) = c[p_1](s) = c[c[p](t)](s).$$

Let us define a mapping $\phi_t : U \to U$ for small enough values of t, which maps the initial points p to the points $c[p](t)$ further along the curve through them, $\phi_t(p) \equiv c[p](t)$. We can apply ϕ_s on the image $\phi_t(p)$ and get

$$(\phi_s \circ \phi_t)(p) = \phi_s(\phi_t(p)) = c[c[p](t)](s) = c[p](t + s) = \phi_{t+s}(p).$$

This set of mappings defined for a small interval containing $t = 0$ follows the composition law of an additive group of real numbers. These mappings form a **local one-parameter group of diffeomorphisms** determined by the vector field X. It is a group because ϕ_0 is the identity mapping and $\phi_t^{-1} = \phi_{-t}$. But in general, the composition law holds only for t and s close to zero such that $c[p](t), c[p](s)$ and $c[p](t + s)$ are all defined. Therefore these mappings represent only a "local" additive group of real numbers close to zero, and not the group of all real numbers on the real line.

In terms of the mappings ϕ_t we can write the equation for the curves $c[p]$ symbolically as

$$\frac{d\phi_t}{dt} = X \circ \phi_t. \tag{10.11}$$

10.3   Lie Derivative

Given a diffeomorphism ϕ on a manifold we can pushforward all contravariant vectors and tensors from a point p to $q = \phi(p)$ by the mapping ϕ_* and pull back all covariant tensors from q to p by ϕ^*. Moreover since ϕ is invertible we can use ϕ^{-1} to transport these vectors and tensors in the opposite direction as well. This means we can pushforward and pullback any mixed tensors as needed by suitably using the mappings determined by ϕ or ϕ^{-1}. We shall denote all push-forwards (in the direction of ϕ) by ϕ_* even though for covariant quantities it is actually pullback ϕ^{-1*} of ϕ^{-1}. Similarly all pullbacks will be denoted by the general symbol ϕ^*.

Let X be a vector field and, in some region, ϕ_t be the one-parameter group of diffeomorphisms determined by the integral curves $c[p]$ of X. Let α be a tensor field defined in the region. Our aim is to find out how α varies *along the integral curve* $c[p]$, that is, how does α at $q = c[p](t) = \phi_t(p)$ compare with α at p as $t \to 0$. Note that these quantities cannot be directly compared because they belong to different tangent spaces (or tensor products of tangent spaces) at two different points p and q. The idea behind the Lie derivative is to use the mapping ϕ_t which transports p to q to pullback α from q to p for a finite t, find the difference with α at p, divide by t and take the limit $t \to 0$. As we subtract quantities in the same tangent space, for all values of t, the derivative is well defined. This Lie derivative is written L_X:

$$L_X \alpha \equiv \lim_{t \to 0} \frac{\phi_t^* \alpha - \alpha}{t}. \tag{10.12}$$

Let us calculate the Lie derivative of the basic quantities. To make the discussion simple we choose a set of coordinates in the region. In these coordinates let the vector field have components $v^i(x)$. Let the coordinates of the point p be x^i and that of $q = \phi_t(p)$ be $y^i(t)$. Then, for sufficiently small t,

$$y^i(t) \simeq x^i + t v^i(x). \tag{10.13}$$

Lie Derivative of a Function f

We first calculate the Lie derivative of a function f.

The pullback $\phi_t^* f$ of a function f is simply $f \circ \phi_t$. Therefore, calling the coordinate representation of f also by f,

$$
\begin{aligned}
\phi_t^* f(x) = (f \circ \phi_t)(x) \ &= \ f(y) \\
&= \ f(x^i + tv^i + \cdots) \\
&= \ f(x) + tv^i \frac{\partial f}{\partial x^i} + \cdots
\end{aligned}
$$

which gives,

$$
L_X f = \lim_{t \to 0} \frac{\phi_t^* f - f}{t} = v^i \frac{\partial f}{\partial x^i} = X(f). \tag{10.14}
$$

Now to calculate the Lie derivative of a vector field Y.

Lie Derivative of a Vector Field

Let $Y = w^i(x)\partial/\partial x^i$. In section 10.1 we saw that if, under a mapping coordinates x^i are mapped to y^i, then the pushforward of a vector components v^i is given by

$$
(\psi_* v)^i(y) = \frac{\partial y^i}{\partial x^j} v^j(x).
$$

Here we have to pushforward the point y to x by ϕ_t^{-1}, therefore we must express x as functions of y. We have,

$$
x^i = y^i - tv^i(y) + \cdots .
$$

Therefore, the components w^i brought back to x from y are

$$
\begin{aligned}
(\phi_t^{-1}{}_* w^i)(x) \ &= \ w^j(y)\frac{\partial}{\partial y^j}\left(y^i - tv^i(y)\right) \\
&= \ w^i(x + tv) - tw^j(x + tv)\frac{\partial v^i}{\partial y^j} \\
&= \ w^i(x) + tv^j(x)\frac{\partial w^i}{\partial x^j} - tw^j(x)\frac{\partial v^i}{\partial x^j} + \cdots .
\end{aligned}
$$

The Lie derivative is thus the vector with components

$$
v^j(x)\frac{\partial w^i}{\partial x^j} - w^j(x)\frac{\partial v^i}{\partial x^j}
$$

which is recognised as those of the Lie bracket $[X, Y]$. We write

$$
L_X Y = [X, Y]. \tag{10.15}
$$

Lie Derivative of a One-Form

Let $\alpha = a_i dx^i$ be a covariant vector field or one-form. By a reasoning similar to the contravariant case in the previous subsection, the pullback components are (for the same vector field $X = v^i \partial/\partial x^i$)

$$
\begin{aligned}
\phi_t^* a_j(x) &= a_i(y) \frac{\partial y^i}{\partial x^j} \\
&= a_i(x + tv) \left(\delta_j^i + t \frac{\partial v^i}{\partial x^j} \right) \\
&= a_j(x) + t v^i \frac{\partial a_j}{\partial x^i} + t a_i(x) \frac{\partial v^i}{\partial x^j} + \cdots .
\end{aligned}
$$

Therefore, the Lie derivative is a one-form with components

$$
v^i \frac{\partial a_j}{\partial x^i} + a_i \frac{\partial v^i}{\partial x^j}.
$$

Or, we can write

$$
L_X \alpha = \left(v^i \frac{\partial a_j}{\partial x^i} + a_i \frac{\partial v^i}{\partial x^j} \right) dx^j. \tag{10.16}
$$

There are two standard ways to write this result in component-free from. We can rearrange

$$
\left(v^i \frac{\partial a_j}{\partial x^i} + a_i \frac{\partial v^i}{\partial x^j} \right) dx^j = v^i \left(\frac{\partial a_j}{\partial x^i} - \frac{\partial a_i}{\partial x^j} \right) dx^j + \frac{\partial}{\partial x^j} (a_i v^i) dx^j
$$

which gives

$$
L_X \alpha = i_X d\alpha + d(i_X \alpha) \tag{10.17}
$$

where i_X is the contraction operator (or the interior product operator) defined on any covariant tensor $\phi \in T_k^0$ and giving the covariant tensor $i_X \phi \in T_{k-1}^0$ of rank one lower,

$$
(i_X \phi)(X_1, \ldots, X_{k-1}) = \phi(X, X_1, \ldots, X_{k-1}). \tag{10.18}
$$

The other way to write $L_X \alpha$ is to see its effect on a vector field $Y = w^k \partial/\partial x^k$,

$$
\begin{aligned}
(L_X \alpha)(Y) &= v^i \frac{\partial a_j}{\partial x^i} w^j + a_i w^j \frac{\partial v^i}{\partial x^j} \\
&= v^i \frac{\partial}{\partial x^i} (a_j w^j) - v^i a_j \frac{\partial w^j}{\partial x^i} + a_j w^i \frac{\partial v^j}{\partial x^i}
\end{aligned}
$$

where indices i and j are interchanged in the last term. This can be written now as

$$
(L_X \alpha)(Y) = X(\alpha(Y)) - \alpha([X, Y]). \tag{10.19}
$$

Lie Derivative and Tensor Products

It is important to note that pullback or pushforward generated by a diffeomorphism, by definition, respects tensor products and contractions,

$$\phi^*(\alpha \otimes \beta) = \phi^*(\alpha) \otimes \phi^*(\beta),$$
$$\phi^*(\alpha(X)) = (\phi^*(\alpha))(\phi^* X),$$

therefore the Lie derivative satisfies the Leibnitz rule:

$$L_X(\alpha \otimes \beta) = (L_X \alpha) \otimes \beta + \alpha \otimes (L_X \beta),$$
$$L_X(\alpha(Y)) = (L_X \alpha)(Y) + \alpha(L_X Y). \tag{10.20}$$

Therefore also

$$L_X(\alpha \wedge \beta) = (L_X \alpha) \wedge \beta + \alpha \wedge (L_X \beta)$$

for any two differential forms. The formula for Lie derivative of a 1-form given above (10.19)is an example of using the Leibnitz property under tensor product and contraction.

$$X(\alpha(Y)) = L_X(\alpha(Y)) = (L_X \alpha)(Y) + \alpha(L_X Y) = (L_X \alpha)(Y) + \alpha([X, Y]).$$

Lie Derivative of any Tensor Field

If we know the Lie derivative for a function f, for $\partial/\partial x^i$ and for dx^i then the Lie derivative for all the tensors can be easily calculated.

From the calculations above, we know that for a vector field X,

$$X = v^i \frac{\partial}{\partial x^i},$$
$$L_X f = v^i \frac{\partial f}{\partial x^i}, \tag{10.21}$$
$$L_X \left(\frac{\partial}{\partial x^i} \right) = -\frac{\partial v^j}{\partial x^i} \frac{\partial}{\partial x^j}, \tag{10.22}$$
$$L_X(dx^i) = \frac{\partial v^i}{\partial x^j} dx^j. \tag{10.23}$$

Starting from these as given and using the Leibnitz property we can rederive the formulas for vector fields and for one-forms.

A very important formula for the derivative of the metric tensor can be written using the above rules:

$$L_X(g_{ij} dx^i \otimes dx^j) = [v_{i;j} + v_{j;i}] dx^i \otimes dx^j. \tag{10.24}$$

It is given as a tutorial problem.

We can also show that if X and Y are two vector fields, the Lie derivative satisfies

$$L_X \circ L_Y - L_Y \circ L_X = L_{[X,Y]}. \tag{10.25}$$

We can show this first on the basic quantities: functions f, tangent space basis vectors $\partial/\partial x^i$ and cotangent basis vectors dx^i. Then prove that

$$\begin{aligned}(L_X \circ L_Y - L_Y \circ L_X)(\alpha \otimes \beta) &= ((L_X \circ L_Y - L_Y \circ L_X)\alpha) \otimes \beta \\ &\quad + \alpha \otimes (L_X \circ L_Y - L_Y \circ L_X)\beta\end{aligned}$$

for any two tensors α and β.

General Formula for the Lie Derivative of an r-Form

One can show by induction

$$\begin{aligned}&(L_X \alpha)(X_1, \ldots, X_r) \\ &= X(\alpha(X_1, \ldots, X_r)) \quad - \quad \sum_{i=1}^{r} \alpha(X_1, \ldots, [X, X_i], \ldots, X_r). \end{aligned} \tag{10.26}$$

10.4 Submanifolds

We have discussed two-dimensional surfaces embedded in three-dimensional ambient space quite early in this book.

A one-one map $\psi : S \to M$ from an $r < n$-dimensional manifold into an n-dimensional manifold M is called a **submanifold** if the pushforward map ψ_* takes the r linearly independent vectors in the tangent space $T_p(S)$ at every point $p \in S$ into r linearly independent vectors in $T_q(M), q = \psi(p)$. Actually, we will assume the submanifold to satisfy a further condition: open sets in S are mapped by ψ into sets which can be obtained as intersections of an open set in M with $\psi(S)$. This is technically called an **embedding**, and we shall assume this further condition as well. This allows us to identify the submanifold S with its image $\psi(S)$ with the differential structure of coordinates and tangent spaces identified with the one-one map ψ.

Still, one must not confuse the subset $\psi(S) \subset M$, *just as a subset*, with the submanifold $\psi(S)$ equipped with its own coordinates and tangent spaces. This is clarified by using the **inclusion map** $i : \psi(S) \to M$ which maps a point p regarded as a point of $\psi(S)$ to $p \in M$.

In physics our manifold M is spacetime, and there is nothing outside it. There we have to imagine, perforce, a submanifold to be a subset $\psi(S) \subset M$ which is the image of S under the map ψ defining the submanifold. In practice this creates no difficulties as the following discussion shows.

If $u^1, \ldots, u^r$ are coordinates around a point $p \in S$ and $x^1, \ldots, x^n$ around $\psi(p) \in M$, then the mapping ψ is determined by n functions $\psi^i : R^r \to R$,

$$x^i = \psi^i(u), \qquad i = 1, \ldots, n. \tag{10.27}$$

The condition about linearly independent vectors mapping into linearly independent vectors means that the Jacobian matrix $\partial x^i / \partial u^a$ where $i = 1, \ldots, n$ and $a = 1, \ldots, r$, has the maximal rank r. By eliminating the r u's from n equations $x^i = \psi^i(u)$ we get $n - r$ equations

$$\chi^A(x) = 0, \ A = 1, \ldots, n - r$$

which characterize the submanifold.

10.5 Frobenius Theorem

We have seen that when a smooth vector field is given, we can find integral curves, that is, curves whose tangent vector at a point coincides with the value of the vector field at the point.

Now suppose we are given a smooth **distribution** of r-dimensional subspaces of the tangent spaces of a manifold M. By smooth distribution of subspaces in a region we mean that there exist smooth vector fields $X_1, \ldots, X_r$ such that they are linearly independent at all points and span r-dimensional subspaces of tangent spaces at those points.

The question we can ask is: Does there exist an r-dimensional submanifold such that vectors tangent to the submanifolds constitute precisely the subspaces given by the distribution? Frobenius theorem answers this question.

We cannot prove the theorem here (not because it is difficult but because we have limited space) and quote the result.

For a smooth distribution determined by the vector fields $X_1, \ldots, X_r$ there exists a submanifold whose tangent spaces coincide with the subspaces of the distribution if and only if the vector fields $X_1, \ldots, X_r$ are **involutive**, that is, they are such that $[X_i, X_j]$ is again a linear combination of these fields,

$$[X_i, X_j] = \sum_{k=1}^{r} C_{ij}^k X_k \tag{10.28}$$

where C_{ij}^k are smooth functions.

This condition can also be written in terms of the dual basis of forms.

We take vector fields $X_1, \ldots, X_r$ which are independent at each point and complete the basis by adding $n - r$ additional linearly independent fields $Y_1, \ldots, Y_{n-r}$. Let $\{\gamma_i\} = \{\alpha^1, \ldots, \alpha^r, \omega^1, \ldots, \omega^{n-r}\}$ be dual to the basis $\{X_1, \ldots, X_r, Y_1, \ldots, Y_{n-r}\}$.

By definition, $\omega^1, \ldots, \omega^{n-r}$ *annihilate* the vector fields of the distribution:

$$\omega^j(X_i) = 0, \qquad j = 1, \ldots, n-r \qquad i = 1, \ldots, r.$$

Let ω be any of the ω^j, $j = 1, \ldots, n-r$ and X, X' be fields of the distribution. Calculate $d\omega$ by the standard formula

$$d\omega(X, X') = X(\omega(X')) - X'(\omega(X)) - \omega([X, X'])$$

where the right-hand side is zero because ω annihilates fields of the distribution. This means that if the two-form $d\omega$ is expanded in the basis $\gamma^i \wedge \gamma^j$, every term has at least one factor corresponding to the annihilators $\omega^1, \ldots, \omega^{n-r}$.

$$d\omega^i = \sum a^i_{jk} \omega^j \wedge \gamma^k, \qquad i, j = 1, \ldots, n-r, \ \ k = 1, \ldots, n,$$

where a^i_{jk} are coefficient functions.

10.6 Induced Metric

If there is a metric $\langle \ , \ \rangle$ defined on the manifold M, then we can define metric $\langle \ , \ \rangle_S$, called the **induced metric** on a submanifold S, by pulling back the 2-form which defines $\langle \ , \ \rangle$. It amounts to saying that the inner product of two vectors $\mathbf{v}, \mathbf{u} \in T_p(S)$ is the same as that between the pushed-forward vectors $\psi_* \mathbf{v}, \psi_* \mathbf{u} \in T_q(M)$,

$$\langle \mathbf{v}, \mathbf{u} \rangle_S = \langle \psi_* \mathbf{v}, \psi_* \mathbf{u} \rangle. \tag{10.29}$$

But one must remember that the induced metric can be degenerate (i.e., a non-zero vector can be orthogonal to all vectors of the tangent space) if the metric in the space M is non-degenerate but not positive definite.

10.7 Hypersurface

An $n-1$ dimensional submanifold $\psi : S \to M$ of an n-dimensional manifold M is called a **hypersurface**. In this case there is one function $\chi(x) = 0$ which characterizes the hypersurface. If coordinates x^i are determined by coordinate function $\phi : M \to R^n$, then $\chi(x)$ is the coordinate representative of the smooth function $f \equiv \chi \circ \phi$. The image of S near point $\psi(p)$ is the constant f 'surface', $f = 0$.

The set of vectors $H_q(M) = \psi_*(T_p(S)), q = \psi(p)$ is an $(n-1)$-dimensional subspace of vectors in $T_q(M)$ tangent to the surface.

Let there be a curve in M passing through the point $q = \psi(p)$ and lying entirely in the surface $f = 0$. Then the tangent vector $\mathbf{v} \in T_q$ to the curve will give zero rate of change for f:

$$\mathbf{v}(f) = df(\mathbf{v}) = 0.$$

Thus the tangent subspace $H_q(M)$ is determined by the condition

$$H_q(M) = \{\mathbf{v} \in T_q(M)| df(\mathbf{v}) = 0\}. \tag{10.30}$$

We can similarly construct subspaces tangent to the surface at all points of the submanifold.

Let $\langle \, , \, \rangle$ be the metric in M. The correspondence set up by the metric between forms and tangent vectors then singles out a vector field obtained by raising the 1-form df as $\mathbf{N} \equiv (df)^\sharp$ such that for any vector $\mathbf{v} \in H_q$,

$$df(\mathbf{v}) = \langle \mathbf{N}, \mathbf{v} \rangle. \tag{10.31}$$

The vector $\mathbf{N}$ is called the **normal** to the surface because it is orthogonal to all vectors $\mathbf{v} \in H_q$ tangent to the hypersurface: $\langle \mathbf{N}, \mathbf{v} \rangle = 0$.

We only deal with manifolds M which are equipped with a Lorentzian metric, whose form in an orthonormal basis is $\eta_{ij} = \text{diagonal}\{-1, +1, \ldots, +1\}$.

When $\langle \mathbf{N}, \mathbf{N} \rangle < 0$ everywhere (that is the normal is time-like) the hypersurface is called a **space-like hypersurface**, when $\langle \mathbf{N}, \mathbf{N} \rangle > 0$ (the normal is space-like) the hypersurface is called **time-like** and if $\langle \mathbf{N}, \mathbf{N} \rangle = 0$ the hypersurface is called a **null hypersurface**.

10.8 Homogeneous and Isotropic Spaces

Let M be a Riemannian manifold with metric tensor $\mathbf{g}$. In the most general case the metric may have no isometries or Killing vector fields at all. But in physics and cosmology we do encounter Riemannian spaces which have a fair degree of symmetry, and Killing vector fields do exist.

It is clear from the definition that if X and Y are two Killing fields, then any linear combination (with constant coefficients) is again a Killing vector field. Moreover from the fact that $L_X \circ L_Y - L_Y \circ L_X = L_{[X,Y]}$ it follows that $L_X \mathbf{g} = 0$ and $L_Y \mathbf{g} = 0$ imply that the bracket $[X, Y]$ is also a Killing vector field.

Thus the set of all Killing vector fields forms a **Lie algebra**. What is even more interesting is that there is a limit to which a space can be symmetric: there are no more than $n(n + 1)/2$ independent Killing vector fields and the space which has them all is called **maximally symmetric**.

We now outline the arguments which show how this is so. For more details the student should see for example Weinberg's *Gravitation and Cosmology*.

Recall that a Killing vector field with components $v^i(x)$ must satisfy

$$v_{i;j} = -v_{j;i}.$$

Using the Ricci identity

$$v_{i;j;k} - v_{i;k;j} = R^l{}_{ijk} v_l$$

and adding to it its cyclic versions, and recalling that

$$R^l{}_{ijk} + R^l{}_{jki} + R^l{}_{kij} = 0$$

we get

$$v_{i;j;k} - v_{i;k;j} + v_{j;k;i} - v_{j;i;k} + v_{k;i;j} - v_{k;j;i} = 0.$$

Using the defining property $v_{j;i;k} = -v_{i;j;k}$ and $v_{i;k;j} = -v_{k;i;j}$ for the Killing field, we get

$$2v_{i;j;k} + 2v_{k;i;j} - 2v_{k;j;i} = 0$$

or, using the above formula for repeated covariant derivatives again,

$$v_{i;j;k} = -R^l{}_{kij}v_l. \tag{10.32}$$

Now suppose we specify the numbers $b_i = v_i$ at a fixed point. We also specify the antisymmetric matrix numbers $c_{ij} = v_{i;j} = -v_{j;i}$ at the same point. As $c_{ij} = v_{i;j} = v_{i,j} - \Gamma^k_{ij}b_k$ this gives us the first derivatives $v_{i,j}$. Next, the above equation determines the second derivatives from

$$\begin{aligned} v_{i;j;k} &= (v_{i,j} - \Gamma^l_{ij}v_l)_{,k} - \text{combinations of } \Gamma \text{ and } c_{ij} \\ &= R^l{}_{kij}b_l. \end{aligned}$$

The third partial derivatives, similarly, will be linear combinations of b_i and c_{ij} and lower partial derivatives. And so, by repeating this process all higher partial derivatives of v_i can be constructed out of given n numbers b_i, and $n(n-1)/2$ numbers c_{ij} (in addition to known quantites like the curvature tensor and its covariant derivatives). The Killing field is therefore determined by its Taylor expansion in a neighbourhood of the point at which the values v_i and $v_{i;j}$ are specified.

Of course, choosing some numbers v_i and $v_{i;j}$ at a point will not necessarily ensure a solution because the values of v_i determined by the Taylor expansion at a neighbouring point, will in their turn, determine v_i at other points including the original point, and they have to match. The important point is that in the best possible case, when all possible arbitrary values of v_i and $v_{i;j}$ can be chosen, there will only be at most $n + n(n-1)/2 = n(n+1)/2$ independent Killing fields. That is the maximal dimension of the Lie algebra of Killing fields.

One should look at a Killing vector field as an infinitesimal symmetry transformation which compares the metric tensor at two neighbouring points x and $y = x + tv$ for small t.

If it is possible to choose Killing fields such that at a fixed point the n components v^i of the field take all possible values, then every point in the neighbourhood can be approached by the infinitesimal symmetry transformation starting from the given point x. The space is called **homogeneous** around the point x. If Killing fields of this type are defined everywhere we say the space is homogeneous.

If, on the other hand, at a point, Killing fields exist which have $v_i = 0$ for all i, then the infinitesimal symmetry transformations determined by these fields keep the point fixed and instead the non-zero values of $v_{i;j}$ determine "rotations" of neighbouring points about the given point. If it is possible to choose all the $n(n-1)/2$ independent values for $v_{i;j}$ we say that the space is **isotropic** about the point.

A space which is homogeneous as well as isotropic about a single point is actually isotropic about all points and is maximally symmetric.

On the other hand a space which is isotropic about every point is homogeneous and therefore again maximally symmetric. The proof goes somewhat along these lines. Choose Killing fields for which $v_i(x) = 0$ but $v_{i;j}(x)$ take all possible values. As there is isotropy at a neighbouring point y as well, there exist Killing fields with compnents w^i such that $w_i(y) = 0$ but $w_{i;j}(y)$ take all possible values. The proof consists in showing that suitable combinations u^i of Killing fields w^i can be taken which will permit all possible values for $u_i(x)$ at x, proving the homogeneous nature at x. And this argument can be given for all points.

10.9 Maximally Symmetric Spaces

Existence of symmetry puts restrictions on g_{ij} and $R^i{}_{jkl}$. A maximally symmetric space is therefore possible only for very special values of these quantities. It is therefore not surprising that a maximally symmetric Riemannian space is essentially unique.

For a maximally symmetric space the following hold:

1. The scalar curvature R is a constant.

2. The Ricci tensor is proportional to the metric tensor

$$R_{ij} = \frac{1}{n} R g_{ij}.$$

3. The Riemann curvature tensor is

$$R_{ijkl} = \frac{R}{n(n-1)} (g_{ik} g_{jl} - g_{jk} g_{il}).$$

Any maximally symmetric space can be shown to be diffeomorphic to a certain standard space which is characterized by the scalar curvature constant R and the **signature** of matrix g_{ij}, that is the numbers n_+ and n_- of its positive and negative eigenvalues. (There are no zero eigenvalues of g_{ij} because it is non-degenerate.)

These standard spaces are constructed by starting with an $(n+1)$-dimensional Eucliden or Minkowskian space and restricting to the sphere or hyperboloid.

Let us first consider an $(n+1)$-dimensional Euclidean space with coordinates $x^1, \ldots, x^n, z$ and metric

$$(ds)^2|_{(n+1)} = \sum_{i=1}^{n} dx^i \, dx^i + \frac{\epsilon}{k^2}(dz)^2$$

where ϵ is equal to $+1$ or -1 and k is a positive constant.

Our maximally symmetric n-dimensional space is the subspace of this Euclidean space given by points whose coordinates satisfy

$$\sum_i \epsilon k^2 (x^i)^2 + z^2 = 1$$

which, for $\epsilon = +1$ is a spheroid and for $\epsilon = -1$, a hyperboloid.

We consider the case $\epsilon = +1$ in detail to illustrate.

The induced metric is calculated by retaining x^i as coordinates and expressing dz in terms of them:

$$2k^2 \sum x^i dx^i + 2z dz = 0$$

or

$$dz = -k^2 \frac{\sum x^i dx^i}{\sqrt{1 - k^2 \sum (x^i)^2}}.$$

Thus the n-dimensional metric $\mathbf{g}_{(n)}$ induced on the sphere due to the Euclidean $(n+1)$-dimensional metric is

$$(ds)^2|_{(n)} = \sum_i dx^i \, dx^i + k^2 \sum_{i,j} \frac{x^i x^j dx^i dx^j}{1 - k^2 \sum_i (x^i)^2}. \qquad (10.33)$$

We can also write this in vector notation as

$$(ds)^2|_{(n)} = (d\mathbf{x}).(d\mathbf{x}) + k^2 \frac{(\mathbf{x}.d\mathbf{x})^2}{1 - k^2 |\mathbf{x}|^2}.$$

Isometries of the metric are easy to see. Any $n \times n$ rotation (that is an orthogonal matrix) R which takes x^i to $R_{ij} x^j$ is an isometry in the larger $n+1$ dimensional space and therefore also for the induced metric. There are $n(n-1)/2$ independent rotations corresponding to as many planes in which the motion takes place. For example, a one-parameter group of isometry depending on a parameter t which maps $x \to x(t)$ (corresponding to a rotation in the (12)-plane) is

$$
\begin{aligned}
x^1(t) &= x^1 \cos t + x^2 \sin t, \\
x^2(t) &= x^2 \cos t - x^1 \sin t, \\
x^i(t) &= x^i \qquad i = 3, \ldots, n.
\end{aligned}
$$

The tangent vectors to these curves at $t = 0$ give the Killing fields of the form

$$\left.\frac{dx^i(t)}{dt}\right|_{t=0} = M^{(12)}_{ij} x^j$$

where $M^{(12)}_{ij}$ is the antisymmetric matrix whose (12) element is 1, (21) element -1 and all the rest zero.

The remaining n Killing fields (there should be $n(n+1)/2$ for a maximally symmetric space) come from the isometries involving rotation in the plane of z and one of the coordinates x^i. For $i = 1$ choose

$$\begin{aligned}
kx^1(t) &= kx^1 \cos t + z \sin t, \\
z(t) &= z \cos t - kx^1 \sin t, \\
x^i(t) &= x^i \qquad i = 2, \ldots, n,
\end{aligned}$$

with the understanding that z is to be replaced in terms of the x using the equation of the sphere. (The second of these equations is not really needed because our space does not have z as a coordinate.)

The case $\epsilon = -1$ can be discussed similarly with the metric given by

$$(ds)^2|_{(n)} = \sum_i dx^i\, dx^i - k^2 \sum_{i,j} \frac{x^i x^j\, dx^i\, dx^j}{1 + k^2 \sum_i (x^i)^2} \tag{10.34}$$

and in vector notation as

$$(ds)^2|_{(n)} = (d\mathbf{x}).(d\mathbf{x}) - k^2 \frac{(\mathbf{x}.d\mathbf{x})^2}{1 + k^2 \mathbf{x}.\mathbf{x}}.$$

de Sitter Space

We start with the physical four-dimensional Minkowski space, with additional fifth w coordinate and metric

$$ds^2|_5 = -(dt)^2 + (dx)^2 + (dy)^2 + (dz)^2 + (dw)^2$$

and restrict to the four-dimensional subspace determined by the hyperboloid

$$-t^2 + x^2 + y^2 + z^2 + w^2 = \rho^2.$$

Choose coordinates on the surface compatible with the geometry

$$\begin{aligned}
t &= \rho \sinh \tau, \\
w &= (\rho \cosh \tau) \cos \chi, \\
z &= (\rho \cosh \tau \sin \chi) \cos \theta, \\
x &= (\rho \cosh \tau \sin \chi \sin \theta) \cos \phi, \\
y &= (\rho \cosh \tau \sin \chi \sin \theta) \sin \phi;
\end{aligned}$$

then we get the metric of the **de Sitter space**

$$(ds)^2|_{\text{dS}} = -\rho^2(d\tau)^2 + (\rho\cosh\tau)^2[(d\chi)^2 + \sin^2\chi((d\theta)^2 + \sin^2\theta(d\phi)^2)].$$

There is another form of de Sitter metric which is useful in cosmology. If we take the outgoing and ingoing null coordinates in the direction w and define u by $w + t = u$ and $v = t - w$, then the equation of the surface is

$$-uv + x^2 + y^2 + z^2 = \rho^2$$

which allows us to eliminate v,

$$v = \frac{1}{u}[x^2 + y^2 + z^2 - \rho^2],$$

$$dv = -\frac{du}{u^2}[x^2 + y^2 + z^2 - \rho^2] + 2\frac{xdx + ydy + zdz}{u}.$$

The metric becomes

$$(ds)^2|_{\text{dS}} = -dudv + (dx)^2 + (dy)^2 + (dz)^2$$

$$= [x^2 + y^2 + z^2 - \rho^2]\frac{(du)^2}{u^2}$$

$$-2[xdx + ydy + zdz]\frac{du}{u} + (dx)^2 + (dy)^2 + (dz)^2.$$

This suggests that we define $dU = du/u$ which integrates to $u = \rho\exp(U)$ if we chose the constant of integration to get the factor ρ. So

$$(ds)^2|_{\text{dS}} = -\rho^2(dU)^2 + (dx - xdU)^2 + (dy - ydU)^2 + (dz - zdU)^2.$$

It is easy to verify that if we now define

$$X = x/u = xe^{-U}/\rho, \qquad Y = y/u = ye^{-U}/\rho, \qquad Z = z/u = ze^{-U}/\rho$$

so that $(\rho e^U)dX = dx - xdU$ etc., this will give us the final form

$$\frac{1}{\rho^2}(ds)^2|_{\text{dS}} = -(dU)^2 + e^{2U}[(dX)^2 + (dY)^2 + (dZ)^2] \qquad (10.35)$$

which shows the homogeniety in the space coordinates X, Y, Z explicitly.

Maximally Symmetric Subspaces

Physically interesting spaces may not be maximally symmetric themselves but may admit a family of maximally symmetric subspaces which fill up the entire space. In such a case it can be shown that if there are m-dimensional subspaces which are maximally symmetric as subspaces, then there exist coordinates (v, u) with $v^a, a = 1, \ldots, n - m$ and $u^r, r = 1, \ldots, m$ such that the metric is

$$(ds)^2|_{(n)} = g_{ab}dv^a\, dv^b + f(v)\tilde{g}_{rs}du^r\, du^s \tag{10.36}$$

where the subspaces are defined by $v^a =$ constant. It follows that there are $m(m+1)/2$ Killing vector fields of the general form

$$X = w^r(u, v)\frac{\partial}{\partial u^r}. \tag{10.37}$$

The integral curves of these Killing fields do not cut across subspaces and lie entirely in the subspaces.

Two examples of physical importance of metrics whose subspaces are maximally symmetirc are given below.

Spherically Symmetric Spacetime

In this case the v-coordinates are r, t and the u-coordinates are θ, ϕ. The metric is given by,

$$\begin{aligned}(ds)^2 &= g_{tt}(r, t)(dt)^2 + 2g_{rt}(r, t)(drdt) + g_{rr}(r, t)(dr)^2\\ &+ f(r, t)[(d\theta)^2 + \sin^2\theta(d\phi)^2].\end{aligned} \tag{10.38}$$

Friedman-Robertson-Walker Spacetime

Here the three-dimensional space-like surfaces are maximally symmetric, that is homegeneous and isotropic. This assumption about the maximal symmetry of the universe at large scales is called the cosmological principle. If x^1, x^2, x^3 are the three homogeneous space coordinates, then

$$(ds)^2 = g(v)(dv)^2 + f(v)\left[d\mathbf{x}.d\mathbf{x} \pm k^2\frac{(\mathbf{x}.d\mathbf{x})^2}{1 \mp k^2\mathbf{x}.\mathbf{x}}\right] \tag{10.39}$$

where k is the constant of maximal symmetry which can be normalized to take values 0 or ± 1. It is customary to redefine $dt = (g(v))^{1/2}dv$ and use $\mathbf{x}$ related to r, θ, ϕ in the standard way. Then $\mathbf{x}.d\mathbf{x} = rdr$ and the metric is in the standard Friedman-Robertson-Walker or FRW form

$$(ds)^2 = (dt)^2 - R(t)^2\left[\frac{(dr)^2}{1 \mp k^2r^2} + r^2(d\theta)^2 + r^2\sin^2\theta(d\phi)^2\right]. \tag{10.40}$$

In this form the isotropy of the space (about the point $r = 0$) is manifest but homogeneity is not.

10.10   Integration

Integration is a limiting process of summation. It is not, contrary to what one learns in school, a process opposite to differentiation. Integration of a function on a Euclidean space is done by dividing the region of integration into convenient non-overlapping elementary subsets (for example by a rectangular grid). Each of these elementary subsets is equipped with a "measure" or volume. The value of the function (assumed to be regular in some well-defined sense) in each subset is approximated and is multiplied by the volume of the subset. The sum of these products over all subsets approximates the integral. Choosing finer and finer divisions of the region of integration into elementary subsets, the sum approaches a limit, called the integral of the function.

In the case of integration over a one-dimensional interval, the integral of a function depends on the two boundary points (limits of integration). The fundamental theorem of calculus assures us that this integral is dependent on the limits in a differentiable manner and the derivative with respect to the upper limit is just the function which was integrated. This shows that the differentiation of an integral gives the original function. This is the concept of 'indefinite' integral. On the other hand, if we differentiate a function first and then integrate, the result is just the difference of the values of the function at the two boundary points of the region of integration.

In higher dimensions there is no anologue of the indefinite integral. The integral of a 'function' depends on a region and its boundary in a more complicated way. There is a very important result known as **Stokes theorem**. It relates the integral of the exterior derivative of an r-form over an $r + 1$-dimensional region to the value of the form on the r-dimensional boundary of the region.

Integration on Euclidean Space

Differential forms have a deep and intimate relation to integration of functions of several variables over regions in the space of those variables.

Recall that integration of a function on an interval can be defined by dividing that interval into many smaller intervals by choosing points (a partition) $x_0 = a, x_1, x_2, \ldots, x_N = b$. The value of a function in the interval is approximated by the constant value at some point $\xi_i \in x_{i+1} - x_i$. The "measure" or volume of the i-th infinitesimal interval is taken to be just its length $x_{i+1} - x_i$. The integral of the function is then approximated by

$$\int_a^b f(x)dx = \sum_i f(\xi_i)(x_{i+1} - x_i).$$

As the limit of the partitioning is made finer and finer, the sums converge to a limit called the integral of the function.

Now consider an integral over some compact region in two-dimensional Euclidean space. If we chose Cartesian coordinates we can divide the region of integration by a rectangular grid with lines parallel to the two (say x and y axes). Then the "measure" of a typical cell is the area $\Delta x \Delta y$. We can construct the integral of a function $f(x, y)$ of two variables in a manner similar to the one-dimensional case by approximating the function in some standard way, multiplying it with the area of the cell and summing over all cells. In the limit of cell size going to zero the integral is obtained. But it is not necessary to use rectanguar coordinates. Suppose we use instead $\xi(x, y)$ and $\eta(x, y)$ as the variables in place of x, y. The grid of lines corresponding to $\xi(x, y) =$const and $\eta(x, y) =$const can now be used to define the integral, but the area of the infinitesimal cell with sides whose (ξ, η) coordinates differ by $\Delta \xi$ and $\Delta \eta$ is not $\Delta \xi \Delta \eta$ anymore. The infinitesimal vector in the directions of $\eta = b =$const and ξ changing from $\xi = a$ to $\xi + \Delta \xi$ has components (in the x, y rectangular coordinates) equal to

$$\mathbf{n}_\xi = \left(\frac{\partial x}{\partial \xi} \Delta \xi, \frac{\partial y}{\partial \xi} \Delta \xi \right).$$

Similarly the infinitesimal vector in the directions of $\xi = a =$const and η changing from $\eta = b$ to $b + \Delta \eta$ has components equal to

$$\mathbf{n}_\eta = \left(\frac{\partial x}{\partial \eta} \Delta \eta, \frac{\partial y}{\partial \eta} \Delta \eta \right).$$

Thus the area of the little parallelogram contained within the lines $\xi = a, \xi = a + \Delta \xi$ and $\eta = b, b + \Delta \eta$ is

$$\mathbf{n}_\xi \times \mathbf{n}_\eta = \Delta \xi \Delta \eta \left(\frac{\partial x}{\partial \xi} \frac{\partial y}{\partial \eta} - \frac{\partial x}{\partial \eta} \frac{\partial y}{\partial \xi} \right)$$

$$= \Delta \xi \Delta \eta \det \left| \frac{\partial(x, y)}{\partial(\xi, \eta)} \right|.$$

Therefore the integral of the same function in $(\xi \eta)$ coordinates looks like

$$\int f(x, y) \det \left| \frac{\partial(x, y)}{\partial(\xi, \eta)} \right| d\xi d\eta$$

where x, y are supposed to have been expressed as functions of ξ and η.

We rephrase this by saying that we are actually integrating a differential two-form $\alpha = f(x, y) dx \wedge dy$ over a region. Under a change of coordinates

$$\int f(x, y) dx \wedge dy = \int f(x, y) \left(\frac{\partial x}{\partial \xi} d\xi + \frac{\partial x}{\partial \eta} d\eta \right) \wedge \left(\frac{\partial y}{\partial \xi} d\xi + \frac{\partial y}{\partial \eta} d\eta \right)$$

$$= \int f(x, y) \det \left| \frac{\partial(x, y)}{\partial(\xi, \eta)} \right| d\xi \wedge d\eta.$$

The infinitesimal cells can be defined by infinitesimal tangent vectors which constitute a grid. They can be chosen in any arbitrary way but usually they are specific to some coordinate system as above.

The process of evaluation of the integral simply means evaluating the two-form α on the two vectors of the grid $\alpha(e_1, e_2)$ and adding this number to similar numbers obtained from other cells. The student can convince herself that the integral obtained will be the same because of the natural way in which the Jacobian matrix appears under change of coordinates. This is the deep-rooted connection of differential forms to integration.

One can similarly check that integration over a three-dimensional region involves a three-form evaluated on a grid of infinitesimal tangent vectors, and so on.

Integration on Manifolds

Integration of a 1-form on a one-dimensional submanifold of an n-dimensional manifold M is defined as follows. Let the one-dimensional submanifold be obtained as a mapping $\psi : T \to M$. If t is a coordinate function on T and x^i on M, let this mapping ψ be represented in coordinates by $t \to x^i(t)$.

A one-form $\alpha = A_i(x)dx^i$ is then integrated as

$$\int \alpha = \int_{t_1}^{t_2} A_i(x) \frac{dx^i}{dt} dt. \tag{10.41}$$

It is obvious that what we are really integrating is the pullback $\psi^*(\alpha)$ on the real line and using the definition of integration on the one-dimensional Euclidean space. Note that the value of the integral is independent of the choice of coordinates t on T or x on M.

Similarly, one can integrate an r-form β on M on an r-dimensional submanifold $\psi : S \to M$. Let ψ be represented in coordinates by $x^i(u)$ where $u^1, \ldots, u^r$ are coordinates on the r-dimensional manifold S. We pull back the form

$$\beta = \sum_{i_1 < \cdots < i_r} B_{i_1 \ldots i_r}(x) dx^{i_1} \wedge \cdots \wedge dx^{i_r}$$

to S as

$$\psi^*(\beta) = \sum_{i_1 < \cdots < i_r; j_1, \ldots, j_r} B_{i_1 \ldots i_r}(x(u)) \frac{\partial x^{i_1}}{\partial u^{j_1}} \cdots \frac{\partial x^{i_r}}{\partial u^{j_r}} du^{j_1} \wedge \cdots \wedge du^{j_r}.$$

This is just the usual r-form on an r-dimensional manifold. Again the integral is independent of the choice of coordinates x or u.

Stokes Theorem

A region D contained in a submanifold of dimension r has a boundary ∂D which is a submanifold of dimension $r - 1$. The Stokes theorem simply says that if the r-form α which is to be integrated on a region D happens to be an exact form, that is, $\alpha = d\beta$ where β is an $(r-1)$-form, then, the integral of α over D is equal to the integral of β over ∂D,

$$\int_D d\beta = \int_{\partial D} \beta. \tag{10.42}$$

The student can look up the proof of the theorem in the book by Frank Warner, *Foundations of Differential Geometry and Lie Groups*.

Summary and Remarks

One-forms are integrated on one-dimensional submanifolds, two-forms on two-dimensional submanifolds and so on. The integral is independent of any coordinate system because the change in coordinate system is compensated by the Jacobian determinant.

   If we integrate an r-form α which is obtained by exterior derivative of an $(r-1)$-form β, that is, $\alpha = d\beta$, over a region D the result is the same as the integral of β over the boundary ∂D of the region.

10.11   Integration on a Riemannian Manifold

We have seen how r-forms can be integrated over an r-dimensional region. The integration proceeds with the assumption that the value of the form on the grid of infinitesimal vectors gives the "content" or "measure" of the quantity to be summed. There is no preferred form which can be taken as standard. But a Riemannian structure gives us a definition of "length" and orthogonality. Then there is a preferred "volume" form corresponding to an orthonormal basis.

Orientation

Let $x^1, \ldots, x^n$ and $y^1, \ldots, y^n$ be two coordinate systems or charts with overlapping domain of definition around a point p in a Riemannian space. The space of n-forms is one-dimensional and all n-forms are proportional to each other. In particular we know that

$$dy^1 \wedge \cdots \wedge dy^n = \det \left| \frac{\partial y}{\partial x} \right| dx^1 \wedge \cdots \wedge dx^n. \tag{10.43}$$

We say the two coordinate systems have the **same orientation** if the determinant $\det |\partial y/\partial x|$ is positive.

   A Riemannian manifold is called **orientable** if there exists an atlas (that is charts covering the whole space) so that all charts have the same orientation. We deal with only such spaces. When we choose an atlas of this type we say we have chosen an orientation.

It may occur to the student that one can always get the same orientation for the two charts by switching the name of two coordinates (say x^1 and x^2 in one chart to make a determinant sign change) but that is not always possible. The reason is that the two coordinate systems may overlap in two disjoint places and the det $|\partial y/\partial x|$ may be positive in one and negative in the other. A prime example of a non-orientable manifold is the Mobius band.

Construct an orthonormal basis of form fields $\alpha^1, \ldots, \alpha^n$. Let

$$\Omega \equiv \alpha^1 \wedge \cdots \wedge \alpha^n.$$

It is an n-form basis element of the one-dimensional space of n-forms and therefore proportional to any other n-form We say that Ω has the same orientation as the coordinate basis if

$$\Omega = (\text{a positive number})dx^1 \wedge \cdots \wedge dx^n$$

at each point with local chart $x^1, \ldots, x^n$. We call Ω the **volume form** for the given orientation.

The 'positive number' can be easily calculated. Let $\mathbf{e}_1, \ldots, \mathbf{e}_n$ be the orthonormal basis dual to $\alpha^1, \ldots, \alpha^n$ with

$$\langle \mathbf{e}_i, \mathbf{e}_j \rangle = \eta_{ij} = \begin{pmatrix} \epsilon_1 & & & \\ & \epsilon_2 & & \\ & & \ddots & \\ & & & \epsilon_n \end{pmatrix}$$

where ϵ_i are fixed constants equal to 1 or -1. Recall that the number of vectors with norm square 1 or -1 in an o.n. basis is fixed. Let

$$\mathbf{e}_i = a_i{}^j \frac{\partial}{\partial x^j}, \tag{10.44}$$

then

$$\alpha^i = b^i{}_j dx^j \tag{10.45}$$

where matrices a and b are inverse-transpose of each other because $\{e_i\}$ and $\{\alpha^i\}$ are dual bases. Therefore

$$a_i{}^k b^i{}_j = \delta^k_j = b^k{}_i a_j{}^i. \tag{10.46}$$

Now,

$$\eta_{ij} = a_i{}^k a_j{}^l \left\langle \frac{\partial}{\partial x^k}, \frac{\partial}{\partial x^l} \right\rangle = a_i{}^k a_j{}^l g_{kl} \tag{10.47}$$

which, on taking the determinant, gives

$$\epsilon = \epsilon_1 \epsilon_2 \ldots \epsilon_n = (\det \|a\|)^2 \det \|g\|. \tag{10.48}$$

Since $\det \|b\|$ is the inverse of $\det \|a\|$, we find that

$$\det \|b\| = \sqrt{\epsilon g} \tag{10.49}$$

where according to usual custom we write g for $\det \|g\|$. Thus

$$\begin{aligned}
\alpha^1 \wedge \cdots \wedge \alpha^n &= b^1{}_i \ldots b^n{}_j dx^i \wedge \cdots \wedge dx^j \\
&= \det \|b\| dx^1 \wedge \cdots \wedge dx^n \\
&= \sqrt{\epsilon g} dx^1 \wedge \cdots \wedge dx^n.
\end{aligned} \tag{10.50}$$

Gauss' Divergence Theorem

This is an important application of the Stokes theorem. We need it for discussing conservation of a physical quantity in general relativity.

First, we define the divergence of a vector field. Let X be a vector field on an orientable Riemann manifold M which has Ω as the volume form. The Lie derivative $L_X \Omega$ of the volume form with respect to the vector field is another n-form proportional to Ω. The proportionality factor is a scalar field called the divergence $\operatorname{div}X$ of the field. Thus

$$L_X \Omega = (\operatorname{div}X)\Omega. \tag{10.51}$$

Since L_X acts like $L_X = d \circ i_X + i_X \circ d$ on any form (where i_X is the interior product or contraction by X),

$$L_X \Omega = d(i_X \Omega) \tag{10.52}$$

because the exterior derivative acting on the n-form gives zero. Integrating on a region D we get

$$\int_D (\operatorname{div}X)\Omega = \int_D d(i_X \Omega) = \int_{\partial D} (i_X \Omega). \tag{10.53}$$

Now we choose an orthonormal basis field on points of ∂D in such a manner that $\mathbf{e}_1 = \mathbf{n}$ points in the direction normal to the hypersurface ∂D and $\mathbf{e}_2, \ldots, \mathbf{e}_n$ span the tangent space of ∂D. Let $\alpha^1, \cdots, \alpha^n$ be the basis dual to $\mathbf{e}_1, \ldots, \mathbf{e}_n$ so that $\Omega = \alpha^1 \wedge \cdots \wedge \alpha^n$. Out of the basis for $(n-1)$-forms, only $\alpha^2 \wedge \cdots \wedge \alpha^n \equiv (\Omega)_{\partial D}$ is non-zero on ∂D and is the volume $(n-1)$-form on ∂D. Thus,

$$\begin{aligned}
i_X(\Omega) &= i_X(\alpha^1 \wedge \cdots \wedge \alpha^n) \\
&= \alpha^1(X)\alpha^2 \wedge \cdots \wedge \alpha^n - \alpha^2(X)\alpha^1 \wedge \alpha^3 \wedge \cdots \wedge \alpha^n + \cdots
\end{aligned}$$

and therefore, on ∂D

$$\begin{aligned}
i_X(\Omega)|_{\partial D} &= \epsilon_1 \langle \mathbf{e}_1, X \rangle \alpha^2 \wedge \cdots \wedge \alpha^n \\
&= \langle \mathbf{n}, \mathbf{n} \rangle \langle X, \mathbf{n} \rangle (\Omega)_{\partial D}.
\end{aligned}$$

Here we have used eqn. (6.8), so that $\alpha^1(X) = \langle(\alpha^1)^\sharp, X\rangle = \epsilon_1\langle\mathbf{e}_1, X\rangle$ which is just $\langle\mathbf{n}, \mathbf{n}\rangle\langle X, \mathbf{n}\rangle$. Therefore,

$$\int_D (\mathrm{div}X)\Omega = \int_{\partial D} \langle\mathbf{n}, \mathbf{n}\rangle\langle X, \mathbf{n}\rangle(\Omega)_{\partial D}. \tag{10.54}$$

The orientation on ∂D is fixed by the Stokes theorem in terms of the choice of the unit normal vector $\mathbf{n}$. We are omitting some technical subtleties here when $\mathbf{n}$ becomes a null vector. But these do not arise in discussing conservation laws.

Before we leave this topic we calculate the divergence $\mathrm{div}X$ in terms of coordinate components.

$$\begin{aligned}
L_X\Omega &= L_X(\sqrt{-g}dx^1 \wedge \cdots \wedge dx^n) \\
&= X^i \frac{(\sqrt{-g})_{,i}}{\sqrt{-g}}\Omega + \sqrt{-g}(X^1)_{,j}dx^j \wedge dx^2 \wedge \cdots \wedge dx^n \\
&\quad + \sqrt{-g}dx^1 \wedge [(X^2)_{,j}dx^j] \wedge dx^3 \wedge \cdots \wedge dx^n + \cdots \\
&= \left((X^i)_{,i} + X^i\Gamma^j_{ij}\right)\Omega \\
&= (X^i)_{;i}\Omega.
\end{aligned}$$

Therefore

$$\mathrm{div}X = (X^i)_{;i}. \tag{10.55}$$

10.12   Tutorial

Exercise 53. Show that the map $\phi : R \to R$ such that $x \to x^3$ is one-one, invertible and infinitely differentiable. But it is not a diffeomorphism. Why?

Answer 53. The inverse map is not differentiable at $x = 0$.

Exercise 54. Two differential 1-forms α^1, α^2 are given to be linearly independent in a region U of a two-dimensional manifold. Show that there exist local coordinates x^1, x^2 such that $\alpha^1 = dx^1$ and $\alpha^2 = dx^2$. Generalize this result to n-dimensions.

Answer 54. OUTLINE OF SOLUTION.
Construct the dual basis X_1, X_2 vector fields. That is, find the dual to the basis $\alpha^1(p), \alpha^2(p) \in T_p^*$ in each tangent space T_p. Then X_1, X_2 are differentiable vector fields. Choose a point and find an integral curve for X_1 passing through the point at parameter $t_1 = 0$. Label each point on this integral curve by $t_1 \in I_1$ where I_1 is an interval about $t_1 = 0$. Find the integral curve of X_2 from each point labelled with fixed t_1 and similarly label points with values (t_1, t_2). Then coordinates (t_1, t_2) are defined in a neighbourhood of the chosen point, the point itself at $(0,0)$ and $X_1 = \partial/\partial t_1$ and $\partial/\partial t_2$. The dual basis condition $\alpha^i(X_j) = \delta^i_j$ requires that $\alpha^1 = dt_1$ and $\alpha^2 = dt_2$.

Exercise 55. Show that the Lie derivative of the metric tensor $\mathbf{g} = g_{ij}dx^i \otimes dx^j$ with respect to a vector field $K = K^i\partial/\partial x^i$ is equal to

$$L_K(\mathbf{g}) = (K_{i;j} + K_{j;i})dx^i \otimes dx^j \tag{10.56}$$

where $K_i = g_{ij}K^j$ and the semicolon denotes the covariant derivative.

Answer 55.

$$
\begin{aligned}
L_K(g_{ij}dx^i \otimes dx^j) &= K(g_{ij})dx^i \otimes dx^j + g_{ij}L_K(dx^i) \otimes dx^j + g_{ij}dx^i \otimes L_K(dx^j) \\
&= [g_{ij,k}K^k + g_{kj}K^k_{,i} + g_{ik}K^k_{,j}]dx^i \otimes dx^j
\end{aligned}
$$

converting the ordinary derivatives into covariant derivatives using a formula such as

$$
K^k_{,i} = K^k_{;i} - \Gamma^k_{mi}K^m
$$

and taking the metric components inside the covariant derivatives (because they act like constants under covariant differentiation) the expression simplifies to

$$
L_K(\mathbf{g}) = [K_{i;j} + K_{j;i}]dx^i \otimes dx^j. \tag{10.57}
$$

If the Lie derivative of the metric with respect to a vector field K vanishes, then that field is called a Killing vector field.

Part III

Gravitation

Chapter 11

The Einstein Equation

We have learnt something about the left-hand side of the Einstein equation

$$G_{\mu\nu} \equiv R_{\mu\nu} - \frac{1}{2}g_{\mu\nu}R = \frac{8\pi G}{c^4}T_{\mu\nu}$$

in chapters on geometry. It is time now to have a look at the equation itself.

Einstein guessed this equation by a difficult process of intution, trial and error. A detailed record of these efforts is contained in Abraham Pais' book *Subtle is the Lord...* Einstein knew that $g_{\mu\nu}$ which determine the geometry of spacetime are supposed to decide the motion of freely falling bodies and light in a gravitational field by the equivalence principle. The question was what determines $g_{\mu\nu}$ themselves? As we know, a gravitational field is determined by matter distribution, therefore stress-energy-momentum tensor $T_{\mu\nu}$ which represents the density and flow of matter should somehow be equated to a tensor dependent on $g_{\mu\nu}$. After a long and tortuous search ('superhuman exertions' in his own words) Einstein was able to converge on the tensor $G_{\mu\nu}$ (now named after him) whose distinguishing feature is that it satisfies the contracted Bianchi identities (four equations) $(g^{\mu\nu}G_{\nu\sigma})_{;\mu} = 0$. But it was not as straightforward as it sounds. He was unaware of the Bianchi identities. The usual argument given is that $G^{\mu\nu}$ should be proportional to the stress-energy tensor $T^{\mu\nu}$ because then $T^{\mu\nu}{}_{;\nu} = 0$ follows automatically, which represents the **local** conservation of energy and momentum. Energy momentum conservation laws in general relativity are a delicate matter because in a spacetime with no symmetries there are no quantities whose conservation (that is, depletion of quantity inside a closed region related to the flowing out of that quantity from the boundary of the region) can be guaranteed. Another way to say this is that there is no local energy density of a gravitational field because a gravitational field can be transformed away by choosing a local inertial frame at any point.

The identification of the stress energy tensor with $G_{\mu\nu}$ (up to a constant factor) gives a theory of gravitation. We can calculate how spacetime geometry is determined by matter distribution. Spacetime in general relativity is a dynamical quantity not a passive arena in which things happen. What is more, spacetime affects itself and determines its own development.

We take the Einstein equation as a basic axiom. If we can write the stress-energy-momentum of the source of a gravitational field, the equation can be solved (in principle) for the gravitational field represented by metric components $g_{\mu\nu}$. What is more, the motion of matter in this gravitational field is also determined by the equation.

The Einstein equation can also be derived from an action principle as we see later in this chapter. In that case the choice of the Lagrangian becomes the basic axiom. But the action principle has an advantage. It turns out that gravity interacts with all other fields in a *minimal way*. The Lagrangian for any field is first written in its special relativistic form in a local inertial frame and then the Minkowski metric $\eta_{\mu\nu}$ is converted to $g_{\mu\nu}$ and the ordinary derivative into a covariant derivative. There are no separate "interaction terms" to be added to the total Lagrangian. Variation of the metric $g_{\mu\nu}$ produces the stress tensor $T^{\mu\nu}$ which governs the interaction of the gravitational field with the given field representing matter. The action principle allows us to calculate the stress-energy tensor for any field if we can write the Lagrangian of that field in general relativistic fashion. It is this stress tensor which causes gravity and responds to it. As an aside it is the most convenient way to calculate the stress tensor of any field even in Minkowski space.

After these preliminary remarks we start with a definition of the stress-tensor. We review the Newtonian concept first for pedagogical reasons.

11.1 Stress-Energy-Momentum Tensor

Classical description of matter involves the mass or energy density and its flow. The flow of mass is momentum and the flow of momentum (momentum per unit time per unit area = force per unit area) is called stress. The stress-energy-momentum tensor (called "stress-energy tensor' for short) is represented mathematically by a second-rank symmetric tensor whose components contain matter or energy density.

Before we discuss how the distribution of matter density and matter currents are described in relativity, we first deal with the non-relativistic case.

11.1.1 Newton's Second Law for Fluids

We use a Cartesian system of coordinates and call the three coordinates $x = x^i, i = 1, 2, 3$.

The motion of continuous matter, a fluid, is characterized by two quantities: a mass density $\rho(x, t)$, which may change from place to place and may depend on time, and the velocity $v^i(x, t)$ of the fluid which, too, changes from place to place and with time.

Let the total force per unit volume around point x at time t be denoted by $f^i(x, t)$

Take a small volume dV about x at time t and fix attention on the matter contained inside it. The velocity of this portion of matter is $v^i(x, t)$. A little later, at time $t + \Delta t$, all this matter has moved to a new point $x + \Delta x$ with $\Delta x^i \simeq v^i(x, t)\Delta t$. The acceleration of this mass packet is

$$a^i = \lim_{\Delta t \to 0} \frac{v^i(x + \Delta x, t + \Delta t) - v^i(x, t)}{\Delta t} = \frac{\partial v^i}{\partial t} + v^j \frac{\partial v^i}{\partial x^j}.$$

The force on this packet which has mass $\rho(x, t)dV$ is $f^i(x, t)dV$. Therefore Newton's second law can be written down as

$$\rho a^i = \rho \left(\frac{\partial v^i}{\partial t} + v^j \frac{\partial v^i}{\partial x^j} \right) = f^i. \tag{11.1}$$

11.1.2 Continuity Equation

Since mass is neither created nor destroyed in Newtonian mechanics, the flowing out of mass from a volume is equal to the loss of mass contained in it. This is expressed in the continuity equation,

$$\frac{\partial \rho}{\partial t} + \frac{\partial (\rho v^j)}{\partial x^j} = 0. \tag{11.2}$$

11.1.3 Stress Tensor

Let us assume for the moment there are no forces other than that of fluid on itself. These internal forces are the 'area forces' like pressure and friction (or viscosity). It is a good assumption that these forces act only between neighbouring portions of matter.

Consider a small area da centered at a point x at time t with a normal unit vector $\mathbf{n}$.

Let $\mathbf{F}$ be the force acting on matter immediately in the frontside of this area element due to matter behind the area element. That is, matter on the $-\mathbf{n}$ side of the area element pushes the matter on the $\mathbf{n}$ side.

In a **perfect fluid** which on the whole is at rest, the force is in the same direction as $\mathbf{n}$ and proportional to area, so $\mathbf{F} = p(x, t)\mathbf{n}da$. The constant of proportionality $p(x, t)$ is called **pressure** at the point x at time t.

In the general case the tangential or 'shear' forces due to tendency of layers of fluid moving with differing velocities to drag adjacent matter have an only slightly more complicated form. The relationhip is linear, that is, each component of force is a linear combination of components of the normal vector

$$F^i(x, t) = t^{ik}(x, t)n_k da; \tag{11.3}$$

here (F^1, F^2, F^3) are components of the force $\mathbf{F}$ and (n_1, n_2, n_3) are components of $\mathbf{n}$.

The coefficients t^{ik} are called components of the **stress tensor**. We can say that t^{ik} is the force component (acting near x) in the i-direction on matter in front of a unit area element chosen perpendicular to the k-direction by matter behind that element. Across every area element there is equilibrium of forces by Newton's third law. Therefore the force component in the i-direction on the same area element by matter immediately in the front of the element on matter just behind is $-t^{ik}$.

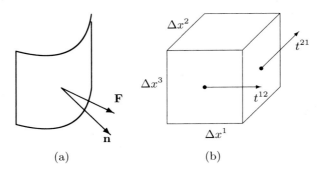

(a)   (b)

Fig. 11.1: (a) Force $\mathbf{F}$ acts on fluid matter immediately in the front of an area element,(normal $\mathbf{n}$ determines which side is front) by matter behind. Components of the stress tensor relate $\mathbf{F}$ to $\mathbf{n}$ by $F^i(x,t) = t^{ik}(x,t)n_k\,da$ where da is the area. (b) The condition that torque exerted by matter just outside a volume should be zero if there are no external forces gives the symmetry of the stress tensor.

For a stationary perfect fluid the components of the stress-tensor have the simplest possible form, $t^{ik} = p(x,t)\delta^{ik}$ where $p(x,t)$ is called pressure.

11.1.4 Symmetry of the Stress Tensor

We now show that the stress tensor components t^{ik} are symmetric, that is $t^{ik} = t^{ki}$.

Let us consider a small cube-like region (Figure 11.1(b)) formed by points with coordinates $(x^1 \pm \Delta x^1/2, x^2 \pm \Delta x^2/2, x^3 \pm \Delta x^3/2)$. The 3-component of the moment of forces (that is torque) with respect to point (x^1, x^2, x^3) acting on the cubical region by the matter outside it is

$$-t^{12}\Delta x^1 \Delta x^3 \left(-\frac{\Delta x^2}{2}\right) + t^{12}\Delta x^1 \Delta x^3 \left(+\frac{\Delta x^2}{2}\right)$$

$$-t^{21}\Delta x^2 \Delta x^3 \left(+\frac{\Delta x^1}{2}\right) + t^{21}\Delta x^2 \Delta x^3 \left(-\frac{\Delta x^1}{2}\right)$$

$$= (t^{12} - t^{21})dV.$$

This shows that in the absence of external forces the angular momentum of the volume element will keep on changing unless $t^{12} = t^{21}$. We can show in a similar way the symmetry of other components, $t^{13} = t^{31}, t^{23} = t^{32}$.

11.1.5 Euler Equation

We now switch on the external (volume) forces assumed to be equal to $g^i(x, t)$ per unit volume.

Consider a small volume V with surface area S. By Gauss' theorem, the sum of forces components in i-th direction acting on its infinitesimal surface elements is

$$- \int_S t^{ik} n_k da$$

where the negative sign appears because we are considering the force on matter contained inside V by matter outside, that is by matter on the *same* side as **n**, and not behind. By Gauss' theorem this surface integral can be written as a volume integral

$$- \int_S t^{ik} n_k da = - \int_V (\partial_k t^{ik}) dV.$$

By making the volume V small we see that the force per unit volume due to internal forces is equal to $-\partial_k t^{ik}$. Therefore, the total force per unit volume on continuous matter is

$$f^i(x, t) = g^i(x, t) - \partial_k t^{ik}$$

where $g^i(x, t)$ is the external force per unit volume.

We can now combine all the ingredients together. The equation representing Newton's second law can be written as

$$\rho \left(\frac{\partial v^i}{\partial t} + v^j \frac{\partial v^i}{\partial x^j} \right) = g^i(x, t) - \partial_k t^{ik}.$$

Using the equation of continuity, we can write

$$\rho \frac{\partial v^i}{\partial t} \equiv \frac{\partial(\rho v^i)}{\partial t} - v^i \frac{\partial \rho}{\partial t} = \frac{\partial(\rho v^i)}{\partial t} + v^i \frac{\partial(\rho v^j)}{\partial x^j}.$$

Therefore, the equation of motion can be written

$$\frac{\partial(\rho v^i)}{\partial t} + \frac{\partial}{\partial x^j} \left(\rho v^i v^j + t^{ij} \right) = g^i. \tag{11.4}$$

This equation, called **Euler's Equation**, expresses Newton's second law for continuous matter. Together with the equation of continuity it determines the dynamics of continuous matter.

11.2   Relativistic Perfect Fluid

11.2.1   Continuous Matter in Special Relativity

The stress-energy tensor for continuous matter in special relativity can be written by generalising the non-relativistic equations treated in the last section.

The mechanics of continuous matter is determined by two equations: the continuity equation which represents conservation of mass and the Euler equation which is just Newton's second law. We shall see that these two equations are time and space parts of a single equation.

In the absence of external forces, the two equations are,

$$\frac{\partial \rho}{\partial t} + \frac{\partial (\rho v^j)}{\partial x^j} = 0,$$

$$\frac{\partial (\rho v^i)}{\partial t} + \frac{\partial}{\partial x^j} \left(\rho v^i v^j + t^{ij} \right) = 0.$$

The non-relativistic velocity v^i is contained in the limit of the four-vector

$$U^\mu = (U^0, U^i) = (c\gamma, v^i \gamma)$$

where $\gamma = 1/\sqrt{1 - \mathbf{v}^2/c^2}$. In the limit of low velocities $U^\mu \approx (c, v^i)$. We can write the two equations by defining $x^0 = ct$,

$$\frac{\partial (\rho U^0 U^0)}{\partial x^0} + \frac{\partial (\rho U^0 U^j)}{\partial x^j} = 0,$$

$$\frac{\partial (\rho U^i U^0)}{\partial x^0} + \frac{\partial}{\partial x^j} \left(\rho U^i U^j + t^{ij} \right) = 0$$

which are parts of the single equation

$$\frac{\partial T^{\mu\nu}}{\partial x^\nu} = 0, \qquad T^{\mu\nu} = \rho U^\mu U^\nu + t^{\mu\nu} \tag{11.5}$$

where

$$t^{\mu\nu} = \begin{pmatrix} 0 & 0 & 0 & 0 \\ 0 & & & \\ 0 & & t^{ij} & \\ 0 & & & \end{pmatrix}. \tag{11.6}$$

We see that one part $(U^\mu U^\nu)$ of $T^{\mu\nu}$ is already in a relativistically covariant form if U^μ is allowed to have the full relativistic status. The second part can be generalised too, provided we can give meaning to the tensor $t^{\mu\nu}$ in a relativistic, frame-independent way.

We will do that not for the most general relativistic fluid, but for the most simple type of fluid known as a **perfect fluid**.

Relativistic dust is defined to be matter for which $t^{\mu\nu} = 0$. This is the same as a perfect fluid with zero pressure.

A **relativistic perfect fluid** is defined to be one for which the $t^{\mu\nu}$ part of the stress energy tensor *when seen in a local frame moving along with the fluid* has the same form as the non-relativistic perfect fluid. We denote the components of the stress tensor in this **comoving frame** by $t^{\mu\nu}_{(0)}$. Then

$$t^{ij}_{(0)} = p(x)\delta^{ij}$$

Here $p(x)$ is a scalar quantity equal to the physical fluid pressure in the comoving frame.

To generalise the continuity and Euler equations to the relativistic case we assume that the equations written above at a given point are written for the local frame which is moving with the fluid. Then $U^\mu = U^\mu_{(0)} = (c, \mathbf{0})$ and then *these equations are exact equations relativistically as well*. To get the general form we only have to transform the tensor to the frame moving with arbitrary velocity using an appropriate Lorentz transformation.

The part $t^{\mu\nu}$ has the form given by the matrix above with $t^{ij} = p(x)\delta^{ij}$ in the comoving frame for a perfect fluid.

We can calculate how $t^{\mu\nu}$ looks in a general frame by transforming it as a second-rank tensor with the same Lorentz transformation which transforms the velocity four-vector $U^\mu_{(0)} = (c, \mathbf{0})$ to $U^\mu = (U^0, U^i) = (c\gamma, v^i\gamma)$. We have already noted this transformation before (in a tutorial in Chapter 1),

$$\Lambda^\mu{}_\nu = \begin{pmatrix} U^0/c & U^i/c \\ U^i/c & \delta^{ij} + U^iU^j/(c^2(\gamma+1)) \end{pmatrix}.$$

The answer is

$$t^{\mu\nu} = p\left(\eta^{\mu\nu} + U^\mu U^\nu/c^2\right) \tag{11.7}$$

as we can see by the following steps.

(1) $t^{\mu\nu}_{(0)}$ in the comoving frame is

$$t^{\mu\nu}_{(0)} = \begin{pmatrix} 0 & 0 & 0 & 0 \\ 0 & p & 0 & 0 \\ 0 & 0 & p & 0 \\ 0 & 0 & 0 & p \end{pmatrix}$$

(2) In the general frame therefore

$$t^{\mu\nu} = \Lambda^\mu{}_\sigma \Lambda^\nu{}_\tau t^{\sigma\tau}_{(0)}.$$

We can write these equations as

$$t^{00} = p \sum_k \Lambda^0{}_k \Lambda^0{}_k = p(\gamma^2 - 1),$$

$$t^{0i} = p \sum_k \Lambda^0{}_k \Lambda^i{}_k = p\gamma U^i/c,$$

$$t^{ij} = p \sum_k \Lambda^i{}_k \Lambda^j{}_k = p(\delta^{ij} + U^i U^j/c^2),$$

and so, after arranging the factors a little, $t^{\mu\nu} = p(\eta^{\mu\nu} + U^\mu U^\nu/c^2)$.

The full stress-energy tensor is thus

$$T^{\mu\nu} = p\eta^{\mu\nu} + \left(\frac{p}{c^2} + \rho\right) U^\mu U^\nu; \tag{11.8}$$

$\rho(x)$ is the **proper mass density** as a function of spacetime. This means that at a fixed point at a given time if we choose a frame of reference (comoving with the fluid) ρc^2 gives the energy per unit volume in a small neighbourhood of the point.

The Lorentz transformation Λ which transforms the four-vector $U^\mu_{(0)} = (c, \mathbf{0})$ into $U^\mu = (c\gamma, v^i\gamma)$ is not determined uniquely. We can show without much difficulty that any other transformation which serves the same purpose is of the form $B\Lambda A$ where A and B are Lorentz transformations which belong to the 'stability groups' of $U^\mu_{(0)}$ and U^μ respectively, that is, $AU_{(0)} = U_{(0)}$ and $BU = U$. Then the form of the stress-energy tensor $T^{\mu\nu} = p\eta^{\mu\nu} + (p/c^2 + \rho)U^\mu U^\nu$ does not change even for the general $B\Lambda A$.

To summarise, we say that the mechanics of a relativistic perfect fluid is described by a stress-energy tensor of the above type. The continuity and Euler equations are combined into a single four-dimensional law

$$T^{\mu\nu}{}_{,\nu} = 0. \tag{11.9}$$

11.2.2   Perfect Fluid in General Relativity

Perfect fluid in general relativity is defined to have the same form for the stress tensor as in special relativity except that $\eta^{\mu\nu}$ is replaced by $g^{\mu\nu}$:

$$T^{\mu\nu} = pg^{\mu\nu} + \left(\frac{p}{c^2} + \rho\right) U^\mu U^\nu. \tag{11.10}$$

This is the quantity that acts as the source of a gravitational field in the Einstein equation.

11.3 Interpretation of $T^{\mu\nu}{}_{;\nu} = 0$

The Einstein equation $G^{\mu\nu} = 8\pi G T^{\mu\nu}/c^4$ determines the gravitational field given the source of gravity, that is, the stress-energy tensor of matter (or of radiation). But there is more to it.

Because of the contracted Bianchi identity $G^{\mu\nu}_{;\nu} \equiv 0$, the stress-energy tensor satisfies

$$T^{\mu\nu}{}_{;\nu} = 0. \tag{11.11}$$

Note that this equation involves the gravitational field through the covariant derivative. It expresses not the conservation of energy momentum but rather *the exchange of energy and momentum of matter with the gravitational field*.

$T^{\mu\nu}{}_{,\nu} = 0$ gives Newton's equations of motion for a fluid moving under no other forces except its own stresses. The generalised equation $T^{\mu\nu}{}_{;\nu} = 0$ gives the equation of motion of the fluid in the presence of gravity.

Thus the Einstein equation plays both roles: of determining the gravitational field from matter distribution as well as determining the motion of matter in that field. Contrast this with electrodynamics in special relativity: Maxwell's equations determine electromagnetic fields from charge-current distributions, but the equation of motion for a charged particle, the Lorentz force equation, has to be postulated separately.

The electromagnetic fields and their interaction with gravity can be formulated entirely in terms of the field tensor $F_{\mu\nu}$. But to discuss the interaction of electromagnetic fields with charged matter we must either postulate the Lorentz force equation separately or introduce the electromagnetic potential A_μ in the Lagrangian. In quantum theory the interaction of electromagnetic fields with charged matter fields comes out naturally from the principle of gauge invariance by the introduction of the connection 1-form $A_\mu dx^\mu$ which is just the potential.

For a small body which is away from the main matter distribution causing the gravitational field, the equation $T^{\mu\nu}{}_{,\nu} = 0$ implies that the body follows a geodesic path. This is not easy to prove and we shall take this fact (that particles falling freely in gravity follow geodesic paths) as an independent postulate.

11.4 Electromagnetic Fields

11.4.1 Maxwell's Equations in Minkowski Space

Electromagnetic fields are described by a differential 1-form $A = A_\mu dx^\mu$ called the four-potential. It is related to the electric and magnetic potentials by

$$A_0 = -\phi/c, \qquad (A_1, A_2, A_3) = \mathbf{A}. \tag{11.12}$$

We are using the SI system of units where the potential 1-form $A = A_\mu dx^\mu$ when multiplied by the electric charge has the dimensions of action and ϕ, the electrostatic potential, has dimensions of energy per unit charge.

The exterior derivative of A is a 2-form $F = dA$ given by

$$
\begin{aligned}
F &= dA = (\partial_\nu A_\mu) dx^\nu \wedge dx^\mu = \sum_{\nu < \mu} F_{\nu\mu} dx^\nu \wedge dx^\mu \\
&= \frac{1}{2} (\partial_\nu A_\mu - \partial_\mu A_\nu) dx^\nu \wedge dx^\mu
\end{aligned}
$$

where the antisymmetric components $F_{\mu\nu}$ of F are related to the electric and magnetic fields by

$$
(F_{23}, F_{31}, F_{12}) = \mathbf{B}, \qquad (F_{01}, F_{02}, F_{03}) = -\mathbf{E}/c.
$$

The exterior derivative of F is zero $dF = d(dA) \equiv 0$ because F is exact.

$$
dF = \frac{1}{2} (\partial_\sigma F_{\nu\mu}) dx^\sigma \wedge dx^\nu \wedge dx^\mu = 0. \tag{11.13}
$$

In order to infer what these equations imply, convert into the independent basis elements $dx^\sigma \otimes dx^\nu \otimes dx^\mu$, then we get

$$
\partial_\sigma F_{\nu\mu} + \partial_\nu F_{\mu\sigma} + \partial_\mu F_{\sigma\nu} = 0.
$$

These are equivalent to the two Maxwell equations

$$
\nabla \cdot \mathbf{B} = 0, \qquad \nabla \times \mathbf{E} = -\frac{\partial}{\partial t} \mathbf{B}.
$$

To express the remaining two equations we need to define the charge current density. Let the contravariant four-vector current have components

$$
J^\mu \equiv (c\rho, \mathbf{j}). \tag{11.14}
$$

Construct the 1-form $J \equiv J_\mu dx^\mu$ by lowering the index $J_\mu = \eta_{\mu\nu} J^\nu = (-c\rho, \mathbf{j})$. Then the Hodge star applied to J gives the 3-form $*J$ whose components are

$$
\begin{aligned}
*J &= (c\rho) dx^1 \wedge dx^2 \wedge dx^3 - (j_1) dx^0 \wedge dx^2 \wedge dx^3 \\
&\quad - (j_2) dx^0 \wedge dx^3 \wedge dx^1 - (j_3) dx^0 \wedge dx^1 \wedge dx^2.
\end{aligned}
$$

The physical dimension of the 3-form $*J$ is charge times velocity.

Similarly let $*F$ be the Hodge star-dual to F. Then the remaining Maxwell equations can be written compactly as

$$
d(*F) = \frac{*J}{\epsilon_0 c^2}. \tag{11.15}
$$

The Lagrangian for the electromagenetic field can be written as the 4-form

$$\mathsf{L}_{\text{em}} = -\frac{1}{2}(\epsilon_0 c^2)\, F \wedge (*F). \tag{11.16}$$

Interaction with a charged distribution is determined by the 4-form

$$\mathsf{L}_{\text{int}} = \frac{1}{c}(*J) \wedge A. \tag{11.17}$$

11.4.2 Electromagnetic Fields in General Relativity

Electromagnetic fields in general relativity are determined by the same Lagrangian interpreted in general coordinates instead of Minkowki space coordinates. Thus there is a one-form $A = A_\mu dx^\mu$ and $F = dA$. The Lagrangian is again

$$\mathsf{L}_{\text{em}} = -\frac{1}{2}(\epsilon_0 c^2)\, F \wedge (*F) \tag{11.18}$$

except that *the gravitational field $g_{\mu\nu}$ has sneaked in through the star operator*. We can show that (see tutorial in this chapter)

$$\mathsf{L}_{\text{em}} = -\frac{1}{4}(\epsilon_0 c^2)\, F_{\mu\nu} F^{\mu\nu} \sqrt{-g}\, d^4 x \tag{11.19}$$

where the contravariant antisymmetric tensor with components is defined by raising indices

$$F^{\mu\nu} \equiv g^{\mu\sigma} g^{\nu\tau} F_{\sigma\tau}. \tag{11.20}$$

11.5 Action Principle

The action principle is a *variational principle*. It is based on an extremum or maximum-minimum problem. A quantity called action is dependent on 'trajectories' or 'histories of evolution of configurations' (called simply the 'configurations') of a physical system in a given region of spacetime. The principle says that only those configurations are realized in nature for which the action is stationary when compared to the (suitably defined) 'neighbouring' configurations.

Let M be the spacetime manifold with metric components $g_{\mu\nu}$ in some coordinate system x. Let $\phi_r(x), r = 1, \ldots, N$ be classical fields corresponding to matter (or radiation) defined on M. We will call all these fields "matter fields" even though in some other contexts it may be necessary to distinguish matter from radiation.

We may assume the fields ϕ_r to be real. At each point with coordinates x, $\phi_r(x)$ have (covariant) derivatives $\phi_{r;\mu}$. Some of the indices r may be spacetime indices (like μ, ν, etc.) indicating the vector or tensor character of the fields. The metric tensor components $g_{\mu\nu}$ themselves are fields, but their covariant derivatives are zero. Their ordinary derivatives, however, are present in the Γ's which occur in covariant derivatives $\phi_{r;\mu}$ and in the curvature tensors.

Action A on a 'domain' D (that is a subset of spacetime with a regular boundary ∂D) depends on fields $\phi_r(x), g_{\mu\nu}(x)$ and their derivatives. In all the cases the action is assumed to be of the form of an integral

$$A = \int_D \mathsf{L} \tag{11.21}$$

where L is a four-form.

The action four-form can be written as a scalar field $\mathcal{L}$ multiplied by the volume form

$$\eta = \sqrt{-g}\, d^4x \equiv \sqrt{-g}\, dx^0 \wedge \cdots \wedge dx^3. \tag{11.22}$$

The scalar field $\mathcal{L}$ is called the **action density** or the **Lagrangian**. Thus

$$\mathsf{L} = \mathcal{L}\, \eta \equiv \mathcal{L}(x)\, \sqrt{-g}\, d^4x \tag{11.23}$$

and action is

$$A = \int_D \mathsf{L} = \int_D \mathcal{L}(x)\, \sqrt{-g}\, d^4x. \tag{11.24}$$

The action depends on the fields $g_{\mu\nu}(x), \phi_r(x)$ through the Lagrangian. In all cases of application the Lagrangian (which is a scalar), is constructed from fields and their derivatives as a local product. The scalar character is obtained by balancing tensor indices in a suitable manner.

Variation

Action depends on the domain D and on the fields defined on it. We choose a region on spacetime D and keep it fixed throughout. Then we calculate the action for some field configuration. Next we calculate the action by replacing the fields by a different configuration inside D but keeping the same values on the boundary ∂D.

To vary fields $g_{\mu\nu}$ we choose a family of metrics $g_{\mu\nu}^{(\lambda)}$ depending on a continuous parameter λ. $g_{\mu\nu}^{(\lambda)}$ has the same value as $g_{\mu\nu}$ on and outside the boundary ∂D of the integration region D. Inside D, $g_{\mu\nu}^{(\lambda)}$ differs from $g_{\mu\nu}$ but tends to $g_{\mu\nu}$ as $\lambda \to 0$. We define

$$\delta g_{\mu\nu}(x) = \left.\frac{d}{d\lambda}\right|_{\lambda=0} g_{\mu\nu}^{(\lambda)}(x) \tag{11.25}$$

where we use the symbol δ:

$$\delta \equiv \left.\frac{d}{d\lambda}\right|_{\lambda=0}. \tag{11.26}$$

On the boundary ∂D the $g_{\mu\nu}$ remain fixed. Therefore we assume that the derivatives of $g_{\mu\nu}$ are all zero. This is, strictly speaking, true only for derivatives tangential to the boundary surface but we assume variations become zero in all directions.

By a variation of fields ϕ_r we similarly mean a family of field configurations $\phi_r^{\lambda'}(x)$ such that on and outside the boundary of D there is no variation: $\phi_r^{\lambda'}(x) = \phi_r(x)$ for all λ' whereas inside $\phi_r^{\lambda'}(x) \to \phi_r(x)$ as $\lambda' \to 0$.

$$\delta\phi_r(x) = \frac{d}{d\lambda'}\bigg|_{\lambda'=0} \phi_r^{\lambda'}(x). \tag{11.27}$$

We use the same symbol δ to denote variation of ϕ_r as well as of $g_{\mu\nu}$.

The variation of action with respect to $g_{\mu\nu}$ is the difference of action evaluated for the varied and original fields in the limit of $\lambda \to 0$:

$$\delta A = \frac{d}{d\lambda}\bigg|_{\lambda=0} A[g^{(\lambda)}, \phi_r]. \tag{11.28}$$

The variation of A with respect to matter fields ϕ_r is similarly defined.

The variation of action can be calculated knowing the dependence of $\mathcal{L}$ on fields. By doing integration by parts we can write the variation in the form

$$\delta A = \int_D d^4x \sqrt{-g} \left[\frac{\delta A}{\delta g_{\mu\nu}(x)} \right] \delta g_{\mu\nu}(x) + \int_D d^4x \sqrt{-g} \left[\frac{\delta A}{\delta \phi_r(x)} \right] \delta \phi_r(x); \tag{11.29}$$

the quantities in brackets are called the **variational derivatives** of the action with respect to the variations of fields $g_{\mu\nu}$ and ϕ_r.

The action principle can now be stated as:

Only those field configurations are realised in nature for which the action is an extremum. This means

$$\frac{\delta A}{\delta g_{\mu\nu}(x)} = 0, \tag{11.30}$$

$$\frac{\delta A}{\delta \phi_r(x)} = 0. \tag{11.31}$$

Lagrangian

We assume that the Lagrangian can be written as a sum of two terms

$$\mathcal{L} = \mathcal{L}_{\text{gravity}} + \mathcal{L}_{\text{matter}}. \tag{11.32}$$

These are, as indicated, for the gravitational and for the matter fields ϕ_r respectively. The gravity Lagrangian is chosen to be

$$\mathcal{L}_{\text{gravity}} = \frac{c^4}{16\pi G} R \tag{11.33}$$

where R is the Ricci scalar curvature. The Lagrangian for a gravitational field was written down by David Hilbert in 1916 within a few weeks of Einstein's discovery of his equation in November 1915. This action is therefore called the **Einstein-Hilbert** action.

The matter Lagrangian $\mathcal{L}_{\text{matter}}$ is constructed as follows: (i) we first write the Lagrangian density as it is written in special relativity, (ii) we replace the Minkowski metric $\eta_{\mu\nu}$ wherever it occurs in the Lagrangian by $g_{\mu\nu}$, (iii) then we replace the ordinary derivatives with respect to coordinates by *covariant* derivatives: that is we use the "comma" to "semicolon" rule.

It is remarkable there is no "interaction Lagrangian" for interactions of gravity with fields. Whatever interactions between gravity and matter do exist, they are supposed to be already contained in the matter Lagrangian through $g_{\mu\nu}$ or its derivatives occuring in the covariant derivatives and in balancing the indices to make a scalar $\mathcal{L}_{\text{matter}}$. This is called the principle of *minimal coupling* of gravity. This is exactly analogous to the coupling of matter fields with gauge fields.

Remark

The method of constructing $\mathcal{L}_{\text{matter}}$ has ambiguities. In translating from a special relativistic formula to the general relativistic formula there may be terms proportional to curvature, but invisible in special relativistic formulas. How do we know those are not present in reality? This ambiguity is similar to the one which occurs in quantization of a classical system. When a classical formula is changed to its quantum mechanical counterpart, different orderings of factors of coordinates and momenta give different operator expressions differing by terms proportional to powers of $\hbar$. In general relativity the commuting ordinary derivatives on flat space are raised to the status of non-commuting covariant derivatives on curved space.

11.5.1   Variation of A_{gravity} with Respect to $g_{\mu\nu}$

First we discuss the variational derivative of the action when $g_{\mu\nu}$ .are varied.

The scalar R depends on $g_{\mu\nu}$, their first as well as second derivatives (because R involves derivatives of Γ's). We shall see that despite this we do not get Euler-Lagrange equations of higher than second-order.

The action is calculated for infinitesimally differing values $\lambda = 0$ and $\lambda = \Delta\lambda$ and subtracted. The variational principle says that in nature only that (or those) metric(s) exist for which the difference goes to zero as $\Delta\lambda \to 0$. In other words

$$\delta A[g^{(\lambda)}] \equiv \left.\frac{d}{d\lambda}\right|_{\lambda=0} A[g^{(\lambda)}] = 0.$$

Omitting the constant $c^4/16\pi G$ (to be restored later):

$$
\begin{aligned}
\delta A_{\text{gravity}} &= \int_D \left[(\delta R)\sqrt{-g} + R(\delta\sqrt{-g})\right] dx^0 \wedge \cdots \wedge dx^3 \\
&= \int_D [(\delta g^{\mu\nu})R_{\mu\nu}\sqrt{-g} + g^{\mu\nu}(\delta R_{\mu\nu})\sqrt{-g} \\
&\quad + R(\delta\sqrt{-g})] \, dx^0 \wedge \cdots \wedge dx^3.
\end{aligned}
\tag{11.34}
$$

Since $g^{\mu\nu}$ are completely determined by $g_{\mu\nu}$ and vice-versa, it is a little more convenient to make variations in $g^{\mu\nu}$ than in $g_{\mu\nu}$. In any case a relation between variations of the two is obtained by differentiating $g^{\mu\nu}g_{\nu\sigma} = \delta^{\mu}_{\sigma}$ and equating to zero,

$$(\delta g^{\mu\nu})g_{\nu\sigma} + g^{\mu\nu}(\delta g_{\nu\sigma}) = 0 \tag{11.35}$$

so that we can always change one from the other when required.

The variational terms can be calculated as follows.

Calculate $\delta\sqrt{-g}$

$\delta\sqrt{-g}$ is easy to calculate. We know from Chapter 9 ("a useful formula") that

$$\delta \det \|g\| = \det \|g\| \mathrm{Tr}(\|g\|^{-1}\delta\|g\|),$$

therefore

$$\delta\sqrt{-g} = \frac{1}{2}\sqrt{-g}g^{\mu\nu}\delta g_{\mu\nu} = -\frac{1}{2}\sqrt{-g}(\delta g^{\mu\nu})g_{\mu\nu}. \tag{11.36}$$

Calculate $\delta R_{\alpha\beta}$

This one requires special care. $R_{\alpha\beta}$ is a sum of components of the Riemann tensor which involves products of two Γ's and derivatives of Γ's. We first calculate $\delta R_{\alpha\beta}$ at a fixed point by choosing a special coordinate system in the neighbourhood of the point. After the calculation the result is in a tensor form and we can relax the choice of special coordinate system and go back to the original system.

Special Coordinate System

At the given point we choose a coordinate system such that the connection coefficients are zero at the point. We will prove this result in the next chapter. Therefore $\delta(\Gamma\Gamma)$ which give $\Gamma\delta\Gamma$ are all zero. So $\delta R_{\alpha\beta}$ depends only on the variation of $-\Gamma^{\sigma}_{\alpha\sigma,\beta} + \Gamma^{\sigma}_{\alpha\beta,\sigma}$ terms.

$\delta(\Gamma^{\sigma}_{\alpha\sigma,\beta})$

$$
\begin{aligned}
\delta(\Gamma^{\sigma}_{\alpha\sigma,\beta}) &= \lim \frac{1}{\Delta\lambda}\left(\Gamma^{\sigma}_{\alpha\sigma,\beta}[g^{(\Delta\lambda)}] - \Gamma^{\sigma}_{\alpha\sigma,\beta}\right) \\
&= \lim \frac{1}{\Delta\lambda}\left(\Gamma^{\sigma}_{\alpha\sigma}[g^{(\Delta\lambda)}] - \Gamma^{\sigma}_{\alpha\sigma}\right)_{,\beta} \\
&= (\delta\Gamma^{\sigma}_{\alpha\sigma})_{,\beta}. \tag{11.37}
\end{aligned}
$$

$\delta\Gamma^{\sigma}_{\alpha\sigma}$ is a Tensor

Use the fact that the difference of two connections is a tensor. For the tensor $\delta\Gamma^{\sigma}_{\alpha\sigma}$ we can write the ordinary derivative as a covariant derivative because in the chosen coordinate system the connection coefficients are zero:

$$(\delta\Gamma^{\sigma}_{\alpha\sigma})_{,\beta} = (\delta\Gamma^{\sigma}_{\alpha\sigma})_{;\beta}. \tag{11.38}$$

After the expression is put in a tensor form the condition of the special coordinate system can be relaxed. This same argument is repeated for all points. When multiplied by $g^{\alpha\beta}$ this term gives

$$(g^{\alpha\beta}\delta\Gamma^{\sigma}_{\alpha\sigma})_{;\beta} = A^{\beta}_{;\beta}, \qquad A^{\beta} \equiv g^{\alpha\beta}\delta\Gamma^{\sigma}_{\alpha\sigma}$$

because $g^{\alpha\beta}$ is a constant as far as covariant derivative is concerned and can move freely in or out of covariant derivatives.

$\delta\Gamma^{\sigma}_{\alpha\beta,\sigma}$

Similarly,

$$g^{\alpha\beta}(\delta\Gamma^{\sigma}_{\alpha\beta,\sigma}) = (g^{\alpha\sigma}\delta\Gamma^{\beta}_{\alpha\sigma})_{;\beta} = B^{\beta}_{;\beta}, \qquad B^{\beta} \equiv g^{\alpha\sigma}\delta\Gamma^{\beta}_{\alpha\sigma}.$$

$\delta A_{\text{gravity}}$

Therefore, the variation of the gravitational action gives,

$$
\begin{aligned}
\delta A_{\text{gravity}} &= \frac{c^4}{16\pi G}\int_D d^4x[(\delta g^{\mu\nu})R_{\mu\nu}\sqrt{-g} + g^{\mu\nu}(\delta R_{\mu\nu})\sqrt{-g} \\
&\quad + R(\delta\sqrt{-g})] \\
&= \frac{c^4}{16\pi G}\int_D d^4x\sqrt{-g}\left[R_{\mu\nu} - \frac{1}{2}g_{\mu\nu}R\right]\delta g^{\mu\nu} \\
&\quad + \int_D d^4x\sqrt{-g}(B^{\nu} - A^{\nu})_{;\nu}.
\end{aligned}
$$

The last term being a divergence becomes a surface integral on ∂D and since we have assumed the derivatives (of $g_{\mu\nu}$) to have zero variation there, these terms are zero.

11.5.2   Variation of A_{matter} with Respect to $g_{\mu\nu}$

We can calculate the variation of matter action only if we know the form of the Lagrangian. But what is remarkable is that the variation of matter action with respect to the minimally coupled gravitational field $g_{\mu\nu}$ is always the symmetric tensor we traditionally associate with stress-energy of that field. Therefore, we define the second-rank symmetric **stress-energy tensor of matter** $T_{\mu\nu}$ as the quantity obtained by the variation of $g^{\mu\nu}$ in the matter fields action,

$$\frac{\delta A_{\text{matter}}}{\delta g^{\mu\nu}} \equiv -\frac{1}{2}T_{\mu\nu}. \tag{11.39}$$

Therefore if the action is given by $\int_D \mathcal{L}_{\text{matter}}\sqrt{-g}d^4x$ and if the Lagrangian density depends only on $g^{\mu\nu}$ and not its derivatives, then

$$T_{\mu\nu} = -2\frac{\partial \mathcal{L}_{\text{matter}}}{\partial g^{\mu\nu}} + g_{\mu\nu}\mathcal{L}_{\text{matter}}. \tag{11.40}$$

This identification of the stress-energy tensor is based on examples of stress energy tensors of classical fields. A prime example is electrodynamics. See examples in the Tutorial exercises.

11.5.3 Einstein's Equation

Einstein's equation is the Euler-Lagrange equation for variation with respect to $g_{\mu\nu}$. We get from previous subsection $\delta A \equiv \delta(A_{\text{gravity}} + A_{\text{matter}}) = 0$,

$$\delta A = \int_D d^4x\sqrt{-g}\left[\frac{c^4}{16\pi G}\left(R_{\mu\nu} - \frac{1}{2}g_{\mu\nu}R\right) - \frac{1}{2}T_{\mu\nu}\right]\delta g^{\mu\nu}$$

which gives the Einstein equation for arbitrary variations.

$$R_{\mu\nu} - \frac{1}{2}g_{\mu\nu}R = \frac{8\pi G}{c^4}T_{\mu\nu}. \tag{11.41}$$

11.5.4 Variation with Respect to Matter Fields

The matter fields appear only in the matter Lagrangian. The variation of the matter action can be written quite generally as

$$\delta A_{\text{matter}} = \int_D d^4x\sqrt{-g}\left[\frac{\delta A_{\text{matter}}}{\delta \phi_r}\right]\delta\phi_r. \tag{11.42}$$

The exact form of the variational derivative will depend on the specific Lagrangian. But we know that under arbitrary variations the total action will be stationary if the variational derivative vanishes:

$$\frac{\delta A_{\text{matter}}}{\delta \phi_r} = 0. \tag{11.43}$$

11.6   Diffeomorphic Invariance

We have seen that the variation of action for gravitational and matter field

$$\delta A[\phi, g^{\mu\nu}] = \delta A_{\text{gravity}}[g^{\mu\nu}] + \delta A_{\text{matter}}[\phi_r, g^{\mu\nu}] \tag{11.44}$$

$$= \delta \int \mathcal{L}_{\text{gravity}} d^4 x + \delta \int \mathcal{L}_{\text{matter}} d^4 x \tag{11.45}$$

where

$$\delta A_{\text{gravity}} = \int_D d^4 x \sqrt{-g} \left[\frac{\delta A_{\text{gravity}}}{\delta g^{\mu\nu}} \right] \delta g^{\mu\nu}$$

$$\equiv \int_D d^4 x \sqrt{-g} \left[\frac{c^4}{16\pi G} \left(R_{\mu\nu} - \frac{1}{2} g_{\mu\nu} R \right) \right] \delta g^{\mu\nu} \tag{11.46}$$

and

$$\delta A_{\text{matter}} = \int_D d^4 x \sqrt{-g} \left[\frac{\delta A_{\text{matter}}}{\delta g^{\mu\nu}} \right] \delta g^{\mu\nu}$$

$$\equiv \int_D d^4 x \sqrt{-g} \left[-\frac{1}{2} T_{\mu\nu} \right] \delta g^{\mu\nu}. \tag{11.47}$$

Our action functionals are integrals of 4-forms

$$A_{\text{gravity}} = \int_D \mathsf{L}_{\text{gravity}}, \qquad A_{\text{matter}} = \int_D \mathsf{L}_{\text{matter}}.$$

Let there be a family of diffeomorphisms Φ_t depending on a continuous parameter t with $\Phi_0 = $ identity. We can choose a diffeomorphism such that under Φ_t points p of the domain D are mapped to points q within the domain D. For any diffeomorphism the action of the pullback form $(\Phi^{-1})^* \mathsf{L}$ over a grid of vectors pushed-forward to a point q by Φ_* is the same as that of L on those vectors at the original point p. Therefore, an **action integral does not change under such a diffeomorphism**.

$$A_{\text{gravity}} = \int_D \mathsf{L}_{\text{gravity}} = \int_{\Phi(D)=D} (\Phi^{-1})^* \mathsf{L}_{\text{gravity}},$$

$$A_{\text{matter}} = \int_D \mathsf{L}_{\text{matter}} = \int_{\Phi(D)=D} (\Phi^{-1})^* \mathsf{L}_{\text{matter}}.$$

In the limit of $t \to 0$ the diffeomorphism is generated by the vector field X (given by the tangent vectors $d\Phi_t(p)/dt$ at p).By definition of the Lie derivative,

$$\int_D L_X L_{\text{gravity}} = 0, \qquad \text{as well as} \qquad \int_D L_X L_{\text{matter}} = 0.$$

We shall now work out the consequences of the invariance of the gravitational and the matter actions under the diffeomorphism generated by the vector field X.

Since we have already calculated the variation for arbitrary $\delta g^{\mu\nu}$ all we need to do is to replace $\delta g_{\mu\nu}$ by $L_X g_{\mu\nu}$. We know that

$$\delta g_{\mu\nu} = L_X g_{\mu\nu} = X_{\mu;\nu} + X_{\nu;\mu}$$

so using

$$\delta g^{\mu\nu} = -g^{\mu\sigma} g^{\nu\tau} \delta g_{\sigma\tau}$$

we obtain

$$
\begin{aligned}
0 &= -\int_D d^4x \sqrt{-g} \left[\frac{\delta A_{\text{gravity}}}{\delta g^{\mu\nu}} \right] g^{\mu\sigma} g^{\nu\tau} (X_{\sigma;\tau} + X_{\tau;\sigma}) \\
&= -2 \int_D d^4x \sqrt{-g} \left[\frac{c^4}{16\pi G} \left(R^{\mu\nu} - \frac{1}{2} g^{\mu\nu} R \right) \right] X_{\mu;\nu}
\end{aligned}
$$

where $R^{\mu\nu} = g^{\mu\sigma} g^{\nu\tau} R_{\sigma\tau}$ etc. We can transfer the covariant derivative from X to the other factor,

$$(\ldots)^{\mu\nu} X_{\mu;\nu} = ((\ldots)^{\mu\nu} X_\mu)_{;\nu} - ((\ldots)^{\mu\nu})_{;\nu} X_\mu.$$

The first term (which is a total divergence) can be converted by Gauss' divergence theorem as the surface integral where we have chosen X to be zero. The other term gives

$$\left(\left[\frac{c^4}{16\pi G} \left(R^{\mu\nu} - \frac{1}{2} g^{\mu\nu} R \right) \right] \right)_{;\nu} = 0$$

because the vector field X can be arbitrarily chosen inside D. Thus we get the contracted Bianchi identities

$$\left(R^{\mu\nu} - \frac{1}{2} g^{\mu\nu} R \right)_{;\nu} = 0. \tag{11.48}$$

We must emphasize that these equations are just a consequence of the diffeomorphic invariance of the gravitational action. It is not necessary that the Euler-Lagrange equations (that is Einstein equations) be satisfied. (In Part II of this book we have already verified that the Bianchi identities follow just from the definition of the curvature tensor.)

We now see the consequence of the matter action being a diffeomorphic invariant.

The matter Lagrangian contains both the gravitational field $g_{\mu\nu}$ as well as the matter fields ϕ_r. The diffeomorphism on spacetime will change *both the fields*. Let $\delta\phi_r \equiv L_X\phi_r$ and $\delta g_{\mu\nu} \equiv L_X g_{\mu\nu} = X_{\mu;\nu} + X_{\nu;\mu}$. Then the variation is

$$\delta A_{\text{matter}} = \int_D d^4x\sqrt{-g}\left[\frac{\delta A_{\text{matter}}}{\delta\phi_r}\right]\delta\phi_r - \frac{1}{2}\int_D d^4x\sqrt{-g}\,T_{\mu\nu}\delta g^{\mu\nu}.$$

We immediately see that *if the matter field equations are satisfied*, the first term is zero, and the second term, by an argument similar to that for the gravity Lagrangian gives

$$T^{\mu\nu}{}_{;\nu} = 0.$$

This is just the expression of the local conservation of energy-momentum. Here it follows from the fact that the field equations for the matter field are satisfied.

Remark

Note that one does not necessarily *have* to assume that the field equations for matter are satisfied. One can equally say that $T^{\mu\nu}{}_{;\nu} = 0$ follows from Einsteins equations and Bianchi's identities $(R^{\mu\nu} - g^{\mu\nu}R/2)_{;\nu} = 0$. In the early days of general relativity there was some confusion as neither Einstein nor Hilbert were aware of the Bianchi identities in 1915. The situation was clarified by Emma Noether in 1918 when she showed that for *any* variational problem in which the action is invariant under a group of transformations whose elements depend on arbitrary functions, there would exist identities involving the variational derivatives (treated as functions) and their ordinary derivatives. In the case of general theory of relativity the group in question is the group of diffeomorphisms. In gauge theories too there are such identities. These identities are now called generalised Bianchi identities. If there are gauge fields there will be these generalised Bianchi identities as well. This result is called **Emma Noether's second theorem**. The first theorem is about invariance of the action functional under transformations which depend on a finite number of parameters.

Incidently, Bianchi identities were discovered by A. Voss in 1880, then independently by G. Ricci in 1889 and again, independently by L. Bianchi in 1902. See Abraham Pais' book for historical details. An accessible account of Noether's first and second theorems is given in N. P. Konopleva and V. N. Popov, *Gauge Fields*.

11.7   Tutorial

Exercise 56. Verify that the equation $d(*F) = *J/\epsilon_0 c^2$ does indeed give the Maxwell equations with sources for electromagnetic fields in Minkowski space.

Answer 56. Recall the definition of the star operator for Minkowski space (sections 6.3 and 6.4). The *-dual to the field tensor F is

$$
\begin{aligned}
*F &= \sum_{\mu<\nu} F_{\mu\nu} * (dx^\mu \wedge dx^\nu) \\
&= -F_{01}dx^2 \wedge dx^3 + \cdots - F_{12}dx^3 \wedge dx^0 + \cdots \\
&= \frac{E_1}{c}dx^2 \wedge dx^3 + \cdots + B_3 dx^0 \wedge dx^3 - \cdots,
\end{aligned}
$$

therefore

$$
\begin{aligned}
d(*F) &= \frac{1}{c}\frac{\partial E_1}{\partial x^1}dx^1 \wedge dx^2 \wedge dx^3 + \frac{1}{c}\frac{\partial E_1}{\partial x^0}dx^0 \wedge dx^2 \wedge dx^3 + \cdots \\
&\quad + \frac{\partial B_3}{\partial x^1}dx^1 \wedge dx^0 \wedge dx^3 + \frac{\partial B_3}{\partial x^2}dx^2 \wedge dx^0 \wedge dx^3 + \cdots \\
&= \frac{1}{c}\frac{\partial E_1}{\partial x^1}dx^1 \wedge dx^2 \wedge dx^3 + \cdots \\
&\quad + \left[\frac{1}{c^2}\frac{\partial E_1}{\partial t} - \left(\frac{\partial B_3}{\partial x^2} + \frac{\partial B_2}{\partial x^3}\right)\right]dx^0 \wedge dx^2 \wedge dx^3 + \cdots ;
\end{aligned}
$$

comparing it with the expression for the three-form $*J/\epsilon_0 c^2$ and equating components we get the result

$$
\nabla.\mathbf{E} = \rho/\epsilon_0, \qquad c^2\nabla \times \mathbf{B} = \frac{\mathbf{j}}{\epsilon_0} + \frac{\partial \mathbf{E}}{\partial t}.
$$

Exercise 57. ALTERNATIVE FORMS FOR THE ELECTROMAGNETIC LAGRANGIANS
Calculate L_{em} and L_{int} in components and put them in more traditional forms.

Answer 57.

$$
\begin{aligned}
(*J) \wedge A &= [(c\rho)dx^1 \wedge dx^2 \wedge dx^3 - (j_1)dx^0 \wedge dx^2 \wedge dx^3 \\
&\quad -(j_2)dx^0 \wedge dx^3 \wedge dx^1 - (j_3)dx^0 \wedge dx^1 \wedge dx^2] \wedge A_\mu dx^\mu \\
&= -[(c\rho)A_0 + (j_1)A_1 + (j_2)A_2 + (j_3)A_3]dx^0 \wedge dx^1 \wedge dx^2 \wedge dx^3 \\
&= -(J^\mu A_\mu)d^4x.
\end{aligned}
$$

The electromagnetic 4-form is similarly

$$
\begin{aligned}
F \wedge *F &= \sum_{\mu<\nu,\sigma<\tau} F_{\mu\nu}(*F)_{\sigma\tau}dx^\mu \wedge dx^\nu \wedge dx^\sigma \wedge dx^\tau \\
&= -(\mathbf{E}^2/c^2 - \mathbf{B}^2).
\end{aligned}
$$

If we define $F^{\mu\nu} \equiv \eta^{\mu\alpha}\eta^{\nu\beta}F_{\alpha\beta}$, then

$$
(F^{23}, F^{31}, F^{12}) = \mathbf{B}, \qquad (F^{01}, F^{02}, F^{03}) = \mathbf{E}/c
$$

and we find

$$
F_{\mu\nu}F^{\mu\nu} = -2(\mathbf{E}^2/c^2 - \mathbf{B}^2)
$$

so we can write the Lagrangian

$$
L_{em} = -\frac{1}{2}\epsilon_0 c^2 F \wedge (*F) = -\frac{1}{4}\epsilon_0 c^2 F_{\mu\nu}F^{\mu\nu}d^4x.
$$

Exercise 58. MAXWELL ELECTROMAGNETIC STRESS-ENERGY TENSOR
Calculate the stress-energy tensor of the electromagnetic field.

Answer 58. Use the formula

$$T_{\mu\nu} = -2\frac{\partial\mathcal{L}_{\text{matter}}}{\partial g^{\mu\nu}} + g_{\mu\nu}\mathcal{L}_{\text{matter}}$$

for the electromagnetic Lagrangian of the previous exercise with $g_{\mu\nu}$ replacing $\eta_{\mu\nu}$ for variational purposes. The result is

$$T_{\mu\nu} = \epsilon_0 c^2 \left[g^{\sigma\tau} F_{\mu\sigma} F_{\nu\tau} - \frac{1}{4} g_{\mu\nu} F_{\sigma\tau} F^{\sigma\tau} \right]. \tag{11.49}$$

Exercise 59. ELECTROMAGNETIC STRESS-ENERGY TENSOR IN SPECIAL CASES
Calculate the electromagnetic stress-energy tensor $T^{\mu\nu}$ in Minkowski space for the following types of fields: (1) plane wave in the z-direction and (2)diffused radiation, (that is superposition of electromagnetic waves so that the average field everywhere is zero).

Answer 59. Using the formulas given in the previous three excercises, we easily obtain

$$T^{\mu\nu} = \begin{pmatrix} \rho & \mathbf{P}^i \\ \mathbf{P}^i & \rho\delta_{ij} - \epsilon_0 E^i E^j - B^i B^j/\mu_0 \end{pmatrix} \tag{11.50}$$

where ρ is the energy density and $\mathbf{P}$ is the Poynting vector ;

$$\rho = \frac{1}{2}\epsilon_0\mathbf{E}^2 + \frac{1}{2\mu_0}\mathbf{B}^2, \tag{11.51}$$

$$\mu_0 = \frac{1}{\epsilon_0 c^2}, \tag{11.52}$$

$$\mathbf{P} = \epsilon_0 c\,\mathbf{E} \times \mathbf{B}. \tag{11.53}$$

For (1) plane wave choose $E^2, E^3 = 0, E^1 = E\cos(\omega t - kx^3)$, then $B^1, B^3 = 0, B^2 = E^1/c$ and

$$T^{\mu\nu} = \epsilon_0\,(E^1)^2 \begin{pmatrix} 1 & 0 & 0 & 1 \\ 0 & 0 & 0 & 0 \\ 0 & 0 & 0 & 0 \\ 1 & 0 & 0 & 1 \end{pmatrix}. \tag{11.54}$$

Similarly for (2) if we take averages (denote by $\langle\ \rangle$) for diffuse radiation, then

$$\langle E^1\rangle = \langle E^2\rangle = \langle E^3\rangle = \langle B^1\rangle = \langle B^2\rangle = \langle B^3\rangle = 0$$

and

$$\langle (E^1)^2\rangle = \langle (E^2)^2\rangle = \langle (E^3)^2\rangle = \langle(\mathbf{E})^2/3\rangle,$$
$$\langle (B^1)^2\rangle = \langle (B^2)^2\rangle = \langle (B^3)^2\rangle = \langle(\mathbf{B})^2/3\rangle.$$

Therefore

$$T^{\mu\nu} = \langle\rho\rangle \begin{pmatrix} 1 & 0 & 0 & 0 \\ 0 & 1/3 & 0 & 0 \\ 0 & 0 & 1/3 & 0 \\ 0 & 0 & 0 & 1/3 \end{pmatrix} \tag{11.55}$$

where $\langle \rho \rangle$ is the average energy density

$$\langle \rho \rangle = \frac{1}{2}\epsilon_0 \langle \mathbf{E}^2 \rangle + \frac{1}{2\mu_0} \langle \mathbf{B}^2 \rangle.$$

This means that the diffused radiation (for example in thermal equilibrium) behaves like a perfect fluid with a pressure $p = \langle \rho \rangle /3$. We shall briefly use this result in the chapter on cosmology.

Exercise 60. Calculate the stress-energy tensor for the scalar field ϕ whose Lagrangian density is given by

$$\mathcal{L} = -\frac{1}{2}g^{\mu\nu}\phi_{;\mu}\phi_{;\nu} - \frac{1}{2}m^2\phi^2. \tag{11.56}$$

Answer 60. The covariant derivatives of the scalar field are ordinary derivatives and do not involve components of the metric tensor. Thus

$$T_{\mu\nu} = \frac{1}{2}\phi_{;\mu}\phi_{;\nu} - \frac{1}{2}g_{\mu\nu}m^2\phi^2. \tag{11.57}$$

Project on Reissner-Nordstrom Metric

A neutral particle creates the Schwarzschild field. A charged particle creates the Reissner-Nordstrom field. One must solve the Einstein equations not for vacuum but with the Coulomb field.

Exercise 61. Calculate the static gravitational and electromagnetic field of a point mass with electric charge Q and show that they are given in Schwarzschild-like coordinates ct, r, θ, ϕ by

$$\begin{aligned}
ds^2 &= -\left(1 - \frac{2GM}{c^2 r} + \frac{GQ^2}{c^2(4\pi\epsilon_0)r^2}\right)c^2(dt)^2 \\
&\quad + \left(1 - \frac{2GM}{c^2 r} + \frac{GQ^2}{c^2(4\pi\epsilon_0)r^2}\right)^{-1}(dr)^2 \\
&\quad + r^2(d\theta)^2 + r^2\sin^2\theta(d\phi)^2,
\end{aligned} \tag{11.58}$$

$$F = -E\,dt \wedge dr, \qquad E = \frac{Q}{4\pi\epsilon_0 r^2}. \tag{11.59}$$

Answer 61. We locate the particle at $r = 0$. For static fields, we assume the Schwarzschild form of Chapter 4,

$$ds^2 = -a(r)c^2(dt)^2 + b(r)dr^2 + r^2(d\theta)^2 + r^2\sin^2\theta(d\phi)^2$$

where $a(r), b(r) \to 1$ as $r \to \infty$ The geometry of the arrangement demands that there is only electrostatic potential

$$A = -\phi(r)dt, \qquad F = dA = -E\,dt \wedge dr.$$

The dual tensor is

$$*F = -\frac{Er^2\sin\theta}{2c\sqrt{ab}}d\theta \wedge d\phi.$$

The charge is confined to the point $r = 0$, so we need solve only the vacuum Maxwell equation

$$d(*F) = 0$$

which implies

$$\frac{d}{dr}\left(\frac{Er^2}{\sqrt{ab}}\right) = 0$$

giving a solution

$$E = \frac{Q\sqrt{ab}}{4\pi\epsilon_0 r^2}$$

where the constant has been fixed by demanding that as $r \to \infty$ we must get the usual coulomb field. The electromagnetic stress tensor is

$$T_{\mu\nu} = \left(\frac{Q}{4\pi\epsilon_0}\right)^2 \frac{\epsilon_0 c^2}{2r^4} \begin{bmatrix} a & & & \\ & -b & & \\ & & r^2 & \\ & & & r^2\sin^2\theta \end{bmatrix}.$$

The Einstein equations

$$R_{\mu\nu} - \frac{1}{2}g_{\mu\nu}R = \frac{8\pi G}{c^4}T_{\mu\nu}$$

can be rewritten as

$$R_{\mu\nu} = \frac{8\pi G}{c^4}\left(T_{\mu\nu} - \frac{1}{2}g_{\mu\nu}T\right)$$

where

$$T \equiv g^{\mu\nu}T_{\mu\nu}.$$

The electromagnetic stress tensor always has $T = 0$, so we have to solve just for

$$R_{\mu\nu} = \frac{8\pi G}{c^4}T_{\mu\nu}.$$

The Ricci tensor has been calculated in the tutorial in Chapter 4 for the Schwarzschild form of metric. The Einstein equations thus become ($A = \ln a$, $B = \ln b$ and prime denotes differentiation with respect to r),

$$A'' + \frac{A'(A' - B')}{2} + \frac{2A'}{r} = \frac{2G}{c^2}\frac{Q^2}{4\pi\epsilon_0}\frac{b}{r^4},$$

$$A'' + \frac{A'(A' - B')}{2} - \frac{2B'}{r} = \frac{2G}{c^2}\frac{Q^2}{4\pi\epsilon_0}\frac{b}{r^4},$$

$$\frac{r}{2}(A' - B') + (1 - b) = -\frac{G}{c^2}\frac{Q^2}{4\pi\epsilon_0}\frac{b}{r^2}.$$

The first two equations give $A' + B' = 0$ which implies $ab = $ constant and the constant should be 1 because both a and b tend to 1 as $r \to 0$. Substituting for b gives the equation for a as

$$a'' + \frac{2}{r}a' = \frac{2}{r^4}\frac{GQ^2}{c^2(4\pi\epsilon_0)}$$

which can be immediately solved for a.

Chapter 12

General Features of Spacetime

12.1 Signature and Time Orientability

A fundamental assumption of the general theory of relativity is that the physical spacetime is a four-dimensional manifold with a non-degenerate metric $\langle\ ,\ \rangle$ of **signature** $+2$. Signature of the metric is the number of positive eigenvalues of metric components matrix $g_{\mu\nu}$ minus the number of its negative eigenvalues. As the metric is non-degenerate, there can be no zero eigenvalues. This statement is independent of coordinate system chosen. The actual numerical values of eigenvalues of the tensor matrix may vary from one coordinate system to another but the number of eigenvalues of positive sign and of negative sign is the same.

This follows from **Sylvester's law of inertia**. See for example R.Bellman, *Introduction to Matrix Analysis*. The law states that if two real symmetric matrices A and B are related by a real non-singular matrix S such that $B = SAS^T$, then A has as many zero eigenvalues and as many eigenvalues of positive and negative signs as B. When S is orthogonal ($S^T = S^{-1}$) then, of course, the eigenvalues of A and B are also numerically equal. Matrices of the metric tensor in a two-coordinate system are related by the formula

$$g'_{\mu\nu}(x') = \frac{\partial x^\sigma}{\partial x'^\mu} g_{\sigma\tau}(x) \frac{\partial x^\tau}{\partial x'^\nu}$$

to which Sylvester's law applies. Since $g_{\mu\nu}(x)$ change continuously with x, and the matrix is non-degenerate everywhere, the sign of an eigenvalue cannot change from point to point.

We can choose an orthonormal basis $\mathbf{e}_a, a = 0, 1, 2, 3$ in any tangent space, such that the tensor components are

$$\eta_{ab} = \langle \mathbf{e}_a, \mathbf{e}_b \rangle$$

where η_{ab} is a diagonal matrix with $\eta_{00} = -1$ and $\eta_{11} = \eta_{22} = \eta_{33} = +1$.

The metric divides the tangent space vectors $\mathbf{v}$ at any point into **space-like**, **time-like** or **null** according as $\langle \mathbf{v}, \mathbf{v} \rangle$ is > 0, < 0 or 0. The **lightcone** consists of null vectors and separates the time-like vectors into two classes, those pointing to the future and those pointing into the past.

The spacetime is supposed to be **time orientable**, that is, it should be possible to pick out the future (or past) pointing cone of time-like vectors unambiguously at all points and this choice should vary smoothly from point to point. This means in particular that there cannot exist any closed time-like curves.

12.2   Local Flatness

In a suitable neighbourhood of every point we can choose coordinates X^μ such that the first derivatives of the metric (and therefore also connection components) vanish at the point.

This is a property of the Riemannian geometry. We show that we can put a one-to-one correspondence between an open neighbourhood about the zero vector in the tangent space of the given point and a certain open neighbourhood of the point itself.

The idea is simple. Let $\phi : M \to R^4 : q \to x^\mu(q)$ be some coordinate system in a neighbourhood of the point p, and let $x^\mu(p) = x^\mu_{(0)}$.

Choose an orthonormal basis

$$\{\mathbf{e}_a\} = \{\mathbf{e}_0, \mathbf{e}_1, \mathbf{e}_2, \mathbf{e}_3\}$$

in the tangent space of the point p. Every vector $\mathbf{v} \in T_p$ can then be expanded in the coordinate basis as well as the orthonormal basis:

$$\mathbf{v} = \xi^a \mathbf{e}_a = v^\mu \frac{\partial}{\partial x^\mu}.$$

This shows that v^μ are linear functions of ξ^a and vice versa.

Construct a geodesic starting from p with affine parameter s and with vector $v^\mu \partial/\partial x^\mu$ as the tangent to the geodesic at the point.

$$\frac{d^2 x^\mu}{ds^2} + \Gamma^\mu_{\nu\sigma} \frac{dx^\nu}{ds} \frac{dx^\sigma}{ds} = 0, \qquad x^\mu|_{s=0} = x^\mu_{(0)}, \qquad \frac{dx^\mu}{ds}\bigg|_{s=0} = v^\mu.$$

The geodesic equations are second-order ordinary differential equations. From Cauchy's theorem on the existence of solutions of ordinary differential equations we know there exists a solution

$$x^\mu(s) = \psi^\mu(s; \xi^a)$$

where we use ξ^a determined by v^μ so that the new coordinates we are going to asssign can be easily identified.

The point q which corresponds to parameter value $s = 1$ is now assigned new coordinates $X^a(q) = \xi^a$:

$$q \qquad \longleftrightarrow \qquad \psi^\mu(s = 1; \xi^a) \qquad \longleftrightarrow \qquad \xi^a.$$

By choosing components ξ^a sufficiently close to zero we can always ensure that the point corresponding to $s = 1$ exists. The point p itself has coordinates $X^a(p) = 0$.

The construction above for assigning coordinates shows that if the affine parameter s in the geodesic equation is replaced by $s' = \lambda s$ where λ is a constant, the geodesic equation remains unchanged and the same identical function (of s') will be obtained as the solution to the geodesic equation, provided the initial tangent vector is now chosen with components ξ^a/λ:

$$\psi^\mu(s; \xi^a) = \psi^\mu(\lambda s; \xi^a/\lambda).$$

Therefore if q is assigned coordinates $X^a(q) = \xi^a$, then a point q' lying on the same geodesic at $s = 1/2$ with x coordinates $x^\mu(q') = \psi^\mu(s = 1/2; \xi^a) = \psi^\mu(s = 1; \xi^a/2)$, would be asigned X coordinates $X^a(q') = \xi^a/2$.

We see that if we restrict attention to a small enough neighbourhood of the point p, a point q in the neighbourhood can be connected by a geodesic starting from p at $s = 0$ with appropriate initial tangent vector ξ^a so as to reach q with affine parameter equal to 1. The point q is assigned coordinates $X^a(q) = \xi^a$ and points along the geodesic from p to q which correspond to the parameter values $0 < s < 1$, are assigned coordinates $X^a(s) = s\xi^a$.

The geodesic in these new coordinates is expressed simply by

$$X^a(s) = s\xi^a. \tag{12.1}$$

The equation for the geodesic in these coordinates *at the point p* becomes

$$\frac{d^2 X^a}{ds^2}\bigg|_p + \Gamma^a_{bc}\big|_p \frac{dX^b}{ds}\frac{dX^c}{ds} = \Gamma^a_{bc}\big|_p \xi^b\xi^c = 0.$$

As this holds for all possible initial values ξ^a, it follows that all the Christoffel symbols are zero at p,

$$\Gamma^a_{bc}\big|_p = 0. \tag{12.2}$$

Note that we can make the statement only for the point p because it is common to all the geodesics corresponding to different values of ξ^a.

It is easy to see that in these coordinates

$$\frac{\partial}{\partial X^a}\bigg|_p = \mathbf{e}_a \tag{12.3}$$

because, for example, the curve $s \to X^a(s) = (s, 0, 0, 0)$ has a tangent vector at p with components $(1, 0, 0, 0)$ in the basis $\{\mathbf{e}_a\}$. The metric components in these coordinates are just η_{ab}.

These coordinates are called the **Riemann normal coordinates**.

12.3   Static and Stationary Spacetimes

Symmetries of a spacetime are determined by its isometries. An isometry preserves the inner product of two vectors under a push-forward map and therefore as far as Riemannian geometry is concerned the new point looks the same as the old one. As we have seen in the mathematical chapters (sections 9.9 and 10.8) a continuous one-parameter group of isometries generates a family of curves whose tangent vectors are Killing vector fields.

A **stationary** gravitational field is described by a spacetime on which a time-like Killing vector field K exists. An observer following the (time-like) integral curve of the Killing field will find the metric tensor independent of time.

If, moreover, the Killing field is orthogonal to a hypersurface, then the space-time is called **static**.

For a static spacetime let the given surface orthogonal to the Killing field be called S_0. Start from every point p of the surface S_0 an integral curve of K. Let the local 1-parameter group of diffeomorphisms ϕ_t determined by K at p map p to q. For fixed t the set of all points q as p varies on S_0 determines another surface S_t. The Killing vectors are orthogonal to the surface S_t as well because vectors tangential to S_0 are mapped to vectors tangential to S_t by construction and ϕ_t is an isometry.

Therefore, at least for t in a small interval around zero, we determine a family of surfaces S_t given by a function t which has the same constant value for points of the surfaces.

Let u^1, u^2, u^3 be a set of coordinates for the surface S_0 with three-dimensional metric

$$\gamma_{ij}(u) = \left\langle \frac{\partial}{\partial u^i}, \frac{\partial}{\partial u^j} \right\rangle.$$

By assigning the same coordinates to the mapped point $q = \phi_t(p)$ we set up a coordinate system t, u^1, u^2, u^3 in the neighbourhood of the surface S_0 such that

$$K = \frac{\partial}{\partial t},$$

$$\left\langle \frac{\partial}{\partial u^i}, \frac{\partial}{\partial t} \right\rangle = 0.$$

The metric in these coordinates is therefore of the form

$$(ds)^2 = -a(u)(dt)^2 + \gamma_{ij}(u)(du^i)(du^j) \tag{12.4}$$

where a is a positive function of u determined by the Killing vector

$$-a(u) = \langle K, K \rangle = \left\langle \frac{\partial}{\partial t}, \frac{\partial}{\partial t} \right\rangle.$$

12.4   Fermi Transport

Construction of a Parallel Vector Field Starting from a Point

Let $C : I \to M$ be a curve, that is, a differentiable mapping from an interval I on the real line into spacetime M. Choose natural coordinates and basis. Let $T(s) = T^\mu \partial/\partial x^\mu|_{C(s)}$ be the tangent vector to the curve.

Let $p = C(s)$ be a point on the curve and let $q = C(s+\Delta s)$ be a neighbouring point on it. Let v^μ be four real numbers chosen as the components of a vector at p. If x^μ and $x^\mu + \Delta x^\mu$ are coordinates of p and q respectively, define

$$v^\mu_\parallel = v^\mu - \Gamma^\mu_{\nu\sigma} \Delta x^\nu v^\sigma$$

as the components of the vector parallel transported to q.

By repeating this process, starting with a single vector at a point p on the curve we can define vectors all along the points on the image of the curve by parallel transporting. Let us write $v^\mu(s)$ as components of the parallel displaced vectors. Then as $\Delta s \to 0$ the ratio $\Delta x^\mu/\Delta s$ tends to $T^\mu(s)$ so

$$\frac{dv^\mu}{ds} + \Gamma^\mu_{\nu\sigma} T^\nu(s) v^\sigma(s) = 0.$$

Thus the parallel transport components are determined by the ordinary differential equation which has a unique solution given the initial value of v^μ at a fixed value of s.

The vector fields $V = v^\mu(s)\partial/\partial x^\mu|_{C(s)}$ and $T = T^\mu(s)\partial/\partial x^\mu|_{C(s)}$ are defined only on the point of the curve. If the fields were defined everywhere in the neighbourhood we would have written the left-hand side of the above equation as

$$
\begin{aligned}
D_T V &\equiv \frac{dv^\mu}{ds} + \Gamma^\mu_{\nu\sigma} T^\nu(s) v^\sigma(s) \\
&= T^\nu \left(\frac{\partial}{\partial x^\nu} v^\mu(x(s)) + \Gamma^\mu_{\nu\sigma} v^\sigma(s) \right) \\
&= T^\nu V^\mu_{;\nu}.
\end{aligned}
$$

We will write the above equation for parallel transport as $D_T V = 0$ with this understanding, that the fields in question are defined only along the curve.

A **geodesic curve** is one for which the parallel transport of the tangent vector is proportional to itself

$$D_T T = f(s)T,$$

and if we choose the parameter s defining the curve as an **affine** parameter, then

$$D_T T = 0 \qquad \text{for a geodesic with an affine parameter.}$$

Restriction of a Vector Field to a Curve

Let X be a vector field in a neighbourhood of the range of a given curve. Let $X(s)$ be the assignment of vectors of the field X at the points corresponding to parameter s of the curve. If there is a coordinate basis in the region, then the components $X^\mu(x(s))$ are functions of s. We can see from the discussion in the previous paragraph and the definition of the covariant derivative that

$$
\begin{aligned}
\Delta s (D_T X)^\mu &= \Delta s T^\nu \left(\frac{\partial X^\mu}{\partial x^\nu} + \Gamma^\mu_{\nu\sigma} T^\sigma \right) \\
&= X^\mu(s + \Delta s) - X^\mu_\parallel(s + \Delta s)
\end{aligned}
$$

where $X^\mu_\parallel(s + \Delta s) = X^\mu(s) - \Gamma^\mu_{\nu\sigma}\Delta x^\nu X^\sigma$ are components of the parallel displaced vector $X(s)$ to $s + \Delta s$). This equation determines the condition for parallelism of the restriction of the field on the curve.

We define a vector field X as **parallel along the curve** C if $D_T X = 0$ at all points on the curve.

Two vector fields X and Y, both parallel along the same curve, may be different elsewhere but if they agree at one point on the curve, then they agree all along the curve because of this uniqueness.

Orthonormal Frames Along a Curve

An observer (confined to a small region of spacetime) will describe physics in his frame of reference using an orthonormal set of tangent vectors as instantaneous axes at each point of his history. An observer moves along a time-like curve $s \to x(s)$ in spacetime. The instantaneous time axis of the observer's frame of reference is along the tangent vector $T(s)$ to the curve. If we choose $s = c\tau$ (proper-time multiplied by light velocity constant c) as the curve parameter, then moreover,

$$
\langle T(s), T(s) \rangle = -1.
$$

Therefore in order to describe physics from the observer's point of view we must further specify the remaining three space-like vectors of the orthonormal set all along the curve.

Observer Along a Geodesic

An observer falling freely in spacetime moves along a time-like geodesic. In this case we are given a time-like geodesic curve, $C : I \to M$, the trajectory of the observer. The unit tangent vectors $T(s)$ to the curve with $\langle T(s), T(s) \rangle = -1$ are in the direction of the instantaneous time axis for the observer. We would like to choose the remaining three space-like unit vectors such that along with $T(s)$ they provide an orthonormal frame of reference for the observer. The number of ways this can be done is infinite. If $X(s), Y(s), Z(s)$ are three vectors orthogonal to each other and to $T(s)$ in the tangent space at $C(s)$, then any rotation of these three can give another possible choice.

As the curve C is a geodesic $D_T T = 0$. We can choose any three vectors $X(0), Y(0), Z(0)$ orthogonal to $T(0)$ and to each other. Then, they can be parallel transported along C from the point $C(0)$ to the point $C(s)$ and defined as $X(s), Y(s), Z(s)$.

Under parallel transport, the inner products do not change, and the vector $T(0)$ is parallel transported to $T(s)$ because the curve is a geodesic. Therefore we have the orthonormal frames $T(s), X(s), Y(s), Z(s)$ at points $C(s)$.

Fermi Transport

When the curve of the observer's history is not a geodesic, that is, the observer is not in free fall but accelerated, the tangents $T(s)$ are still time-like vectors which can be normalised to $\langle T(s), T(s) \rangle = -1$ by choosing $s = c\tau$ as the parameter along the curve.

As the curve is not a geodesic, we do not have $D_T T$ equal to zero. But $D_T T$ is orthogonal to T all along the curve because of the constant norm

$$
\begin{aligned}
\frac{d}{ds} \langle T(s), T(s) \rangle &= D_T\big(\langle T(s), T(s) \rangle\big) \\
&= \langle D_T T, T \rangle + \langle T, D_T T \rangle = 2\langle D_T T, T \rangle = 0.
\end{aligned}
$$

Let $V^\mu(0)$ be four numbers giving the components of a vector in T_p at point p ($s = 0$) orthognal to $T(0)$. If we parallel translate this vector to q corresponding to Δs, its components will be

$$
V_{\|}^\mu = V^\mu(0) - \Gamma^\mu_{\nu\sigma} \Delta x^\nu V^\sigma(0).
$$

When we construct the vector

$$
V_{\|} = V_{\|}^\mu \left. \frac{\partial}{\partial x^\mu} \right|_q,
$$

then $V_{\|}$ is *not orthogonal* to $T(\Delta s)$ but rather to the parallel translate $T_{\|}$ of $T(0)$ to q because the parallel translation respects the inner product. The components $T_{\|}$ are

$$
\begin{aligned}
T_{\|}^\mu &= T^\mu(0) - \Gamma^\mu_{\nu\sigma} \Delta x^\nu T^\sigma(0) \\
&= T^\mu(0) - \Delta s \Gamma^\mu_{\nu\sigma} T^\nu(0) T^\sigma(0) \\
&= T^\mu(\Delta s) - \Delta s (D_T T)^\mu.
\end{aligned}
$$

We must make a correction to $V_{\|}^\mu$ so that it is orthogonal to the time axis $T(\Delta s)$. Let the correction (which has to be of the order Δs) be W, then

$$
\langle V_{\|} + W, T(\Delta s) \rangle = \langle V_{\|} + W, T_{\|} + \Delta s D_T T \rangle = 0;
$$

this gives up to order Δs,

$$\langle W, T_\| \rangle \sim \langle W, T \rangle = -\Delta s \langle V_\|, D_T T \rangle \sim -\Delta s \langle V, D_T T \rangle.$$

This suggests a definition of the correction term W (remembering $\langle T, T \rangle = -1$),

$$W = \Delta s T \langle V, D_T T \rangle.$$

Thus we get the definition of **Fermi Transport**: A vector $V(0)$ at point p ($s = 0$) which is orthogonal to $T(0)$ is Fermi transported to q ($s = \Delta s$) to become

$$V(s + \Delta s) = V_\|(s + \Delta s) + \Delta s T \langle V, D_T T \rangle.$$

Or, bringing the right-hand side to the left and dividing by Δs, the Fermi transported vector field along the curve satisfies

$$D_T^F V \equiv D_T V - T \langle V, D_T T \rangle = 0. \tag{12.5}$$

This definition is restricted to vectors which are orthogonal to the curve but we can extend the definition to any vector field V and say that it is Fermi transported along a curve if $D_T V - T \langle V, D_T T \rangle = 0$.

We can check that a Fermi transported vector V remains orthogonal to the curve

$$\frac{d}{ds} \langle V(s), T(s) \rangle = D_T \langle V(s), T(s) \rangle = \langle D_T V, T \rangle + \langle V, D_T T \rangle = 0$$

because $D_T V - T \langle V, D_T T \rangle = 0$.

Moreover, two vectors orthogonal to the curve preserve their innner product.

$$\begin{aligned}
D_T \langle V(s), W(s) \rangle &= \langle D_T V, W \rangle + \langle V, D_T W \rangle \\
&= \langle T \langle V, D_T T \rangle, W \rangle + \langle V, T \langle W, D_T T \rangle \rangle \\
&= 0
\end{aligned}$$

because $\langle V, T \rangle = 0$ and $\langle W, T \rangle = 0$ as well.

An orthogonal set of vectors $X(0), Y(0), Z(0)$ chosen normal to $T(0)$ can be taken by Fermi transport along the curve to give an orthonormal set of vectors $X(s), Y(s), Z(s)$ normal to $T(s)$ for all s.

12.5   Fermi-Walker Transport

The Fermi transport has the following shortcomings.

Vectors which are not orthogonal to T when Fermi transported, do not preserve inner products. In particular, they change their norm or length. Moreover, the vector T itself is not Fermi transported $D^F{}_T T = D_T T \neq 0$ unless the curve is a geodesic.

This is corrected by the **Fermi-Walker transport** which is defined as the condition

$$D_T^{FW} V \equiv D_T V - T\langle V, D_T T\rangle + D_T T\langle V, T\rangle = 0. \tag{12.6}$$

We call $D_T^{FW} V$ the **Fermi-Walker derivative** of V with respect to T along the curve.

With this definition the following properties are satisfied:

1. If the curve is a geodesic, then the Fermi-Walker transport is identical to parallel transport: if $D_T T = 0$, then $D_T^{FW} V = D_T V$.

2. $D_T^{FW} T = 0$, that is, the tangent to the curve is always Fermi-Walker transported.

3. If $V(s), W(s)$ are Fermi-Walker transported vectors, then their inner product remains constant along the curve

$$\frac{d}{ds}\langle V(s), W(s)\rangle = \langle D_T V, W\rangle + \langle V, D_T W\rangle.$$

Use $D_T^{FW} V = 0 = D_T V - T\langle V, D_T T\rangle + D_T T\langle V, T\rangle$ and the similar equation for W to get

$$
\begin{aligned}
\frac{d}{ds}\langle V(s), W(s)\rangle &= \langle T\langle V, D_T T\rangle - D_T T\langle V, T\rangle, W\rangle \\
&\quad + \langle V, T\langle W, D_T T\rangle - D_T T\langle W, T\rangle\rangle \\
&= \langle T, W\rangle\langle V, D_T T\rangle - \langle D_T T, W\rangle\langle V, T\rangle \\
&\quad + \langle V, T\rangle\langle W, D_T T\rangle - \langle V, D_T T\rangle\langle W, T\rangle \\
&= 0.
\end{aligned}
$$

4. If there is a space-like vector V orthogonal to the curve at all points (that is, $\langle V(s), T(s)\rangle = 0$ for all s), then its Fermi-Walker derivative along the curve is the same as taking the "horizontal part" of the covariant derivative. The horizontal part $h(X)$ of a vector field X with respect to the unit tangent vector T to a time-like curve is obtained by subtracting from X its projection along T:

$$h(X) \equiv X + T\langle T, X\rangle$$

(the reason we have a plus sign is because $\langle T, T\rangle = -1$). The horizontal part has the property that $\langle h(X), T\rangle = 0$.

We know that if V is orthogonal to T along the curve, then

$$\frac{d}{ds}\langle V, T\rangle = \langle D_T V, T\rangle + \langle V, D_T T\rangle = 0.$$

Therefore,

$$
\begin{aligned}
D_T^{FW} V &= D_T V - T\langle V, D_T T\rangle \\
&= D_T V + T\langle D_T V, T\rangle \\
&= h(D_T V).
\end{aligned}
$$

The physical meaning of this last property is this: V is already orthogonal to T all along. So any *horizontal* change in it along the curve can only come from rotation of the vector in a plane perpendicular to T. The condition of Fermi-Walker transport $D_T^{FW} V = 0$ means that the *vector is transported without any rotation*. This is essential to find the behaviour of gyroscopes when carried by accelerated observers.

12.6 Penrose Diagrams

The nature of the physical spacetime is determined by the structure of its light cones and the time-like and null geodesics. Let there be two different metrics on the same manifold, (or on two different manifolds with a mapping $\phi : p \to q$ defined between them). One is the physical spacetime ds^2 and the other, a 'fictitious' metric $\widetilde{ds}^{\,2}$ "conformally" related to it. That is,

$$
ds^2(p) = (\Omega(q))^{-2}\, \widetilde{ds}^{\,2} (q)
$$

where Ω is a smooth function (therefore $\Omega^2 > 0$). The time-like, space-like and null tangent vectors at any point p are mapped then into similar vectors in the other metric at q. This is specially useful for an analysis of the asymptotic regions of the physical spacetime if such regions can be mapped onto finite regions (that is, a set of points with finite values of the coordinates) of the other spacetime. This will happen when $\Omega(q) \to 0$ when we make p approach an asymptotic region of the given spacetime.

Penrose diagrams are pictures of the second metric with finite 'boundary points' representing 'infinity' of the physical spacetime.

The idea is best illustrated by the Penrose diagram for the Minkowski space. The metric of Minkowski space is (changing to polar coordinates in spatial coordinates to bring out spherical symmetry)

$$
ds^2 = -dt^2 + dr^2 + r^2(d\theta^2 + \sin^2\theta\, d\phi^2) \equiv -dt^2 + dr^2 + r^2 d\omega^2
$$

where we have denoted the angular part $d\theta^2 + \sin^2\theta\, d\phi^2$ by $d\omega^2$ and write t for ct.

Define the null coordinates for outgoing and ingoing light rays,

$$
v = t + r, \qquad u = t - r
$$

which brings the metric to

$$ds^2 = -dudv + \frac{1}{4}(v-u)^2 d\omega^2.$$

In order to map an infinite region onto a finite region we use the simplest trigono-
metric functions $u = \tan U$ and $v = \tan V$ which map infinite intervals $(-\infty, \infty)$
of u and v to $-\pi/2 < U, V < \pi/2$ in a one-one onto manner.

$$ds^2 = \frac{1}{4\cos^2 U \cos^2 V} \left[-4dUdV + \sin^2(V-U)d\omega^2\right].$$

Defining R and T as

$$V = (T+R)/2, \qquad U = (T-R)/2$$

we get the desired form

$$ds^2 = \Omega^{-2}[-dT^2 + dR^2 + \sin^2 R d\omega^2] = \Omega^{-2}\, \widetilde{ds}^2$$

with $\Omega = 2\cos U \cos V$.

The following diagrams show the mapping of the region in these coordinates.

The θ and ϕ coordinates are to be associated with each point with fixed values
of T and R. In other words, each point of the triangular region in R, T coordinates
is associated with a 2-sphere S^2 of directions θ, ϕ. It is traditional (and convenient)
to double the set of points by joining another triangle and completing a diamond
shape region with each point representing only *half a 2-sphere*. This can be done by
dividing S^2 into two halves by a plane and assigning a point T, R with directions
in one half and the point $-R, T$ the opposite directions in the other half. The time
axis common to both parts still has the full sphere associated to its points and the
end points i_0 corresponding to spatial infinity $R = \pm\pi$ have to be identified.

The spatial part in the metric $\widetilde{ds}^2$,

$$dR^2 + \sin^2 R d\omega^2,$$

is precisely the metric of the three-dimensional maximally symmetric space with
positive curvature. This is just the space S^3 as a subset of R^4 with the induced
metric.

We can see the following features of the Penrose diagram:

1. All time-like geodesics begin at the past time-like infinity i^- and end at
   future time-like infinity i^+.

2. Null geodesics are straight lines at $\pm 45°$. They start at $\mathcal{I}^-$ and end at $\mathcal{I}^+$.

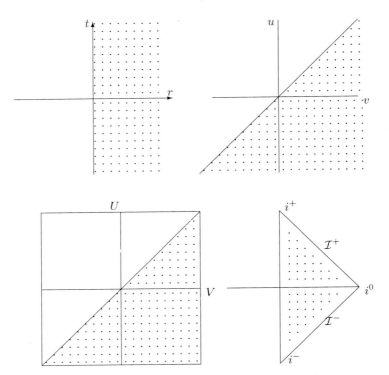

Fig. 12.1: Mapping the region $r \in (0, \infty), t \in (-\infty, \infty)$ into the finite region $R \in (0, \pi), T \in (-\pi, \pi)$ through the transformation given in the text. The points i^+, i^- correspond to $T = \pi, T = -\pi$ and are called *future time-like infinity* and *past time-like infinity* respectively. The point i^0 which corresponds to $R = \pi$ (or $r = \infty$) is called *spatial infinity*. The segments $\mathcal{I}^+$ and $\mathcal{I}^-$ are called *future null infinity* and *past null infinity* respectively.

Relation with the Einstein Static Universe

The underlying manifold of the Einstein static universe is $\mathcal{R} \times S^3$ where the real line $\mathcal{R}$ represents time and S^3 is the three-sphere.

The time coordinate T of the Einstein universe runs from $-\infty$ to $+\infty$ and R from 0 to π. Each point of this 'strip' is associated with the other two coordinates, θ and ϕ which constitute a 2-sphere S^2. Just as we did for the Penrose diagram above, we can take a copy of the strip and let R run from 0 to $-\pi$. By joining the two strips along $R = 0$ and $R = \pm\pi$ lines we get a *cylindrical* geometry for the Einstein static universe $R \times S^3$ with metric $\overset{\sim}{ds}{}^2$.

Then it is obvious that the diamond-shaped finite Penrose diagram which represents the *entire* Minkowski space can be regarded as conformally equivalent to a portion of the Einstein universe, namely the part of the cylinder with appropriate range of coordinates. This makes it possible to do analysis at the future, past and null infinity in a Minkowski spacetime as these regions correspond to finite points and line segments in the Einstein universe.

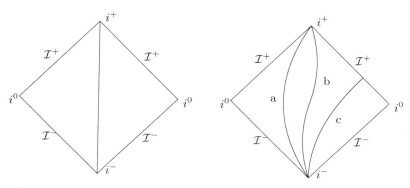

Fig. 12.2: The Penrose diagram with each point representing half a 2-sphere except the time axis (the middle vertical line in the first diagram) whose points represent the full 2-sphere. The diagram on the right shows the trajectory of (a) a particle sitting at a fixed point, (b) a particle travelling from the (spatial) infinity at constant velocity, passing through the origin and going off to infinity (c) a stationary particle in remote past getting acclerated in one direction and asymptotically moving with velocity of light.

Of course for the Minkowski space this elaborate exercise may not be needed, but the same idea can be applied to more complicated spacetimes. The usefulness of Penrose diagrams arises from the fact that the light cones always open at right angles (null geodesics run along 45° lines).

We shall construct Penrose diagrams for illustrating the causal structure of Schwarzschild spacetime in chapter 14.

12.7 Solutions of Einstein Equations

It can be proved that on any differentiable manifold there exist an infinite number of possible Riemanninan metrics. Do any of these represent physical situations?

In a trivial sense, solving Einstein equations is very easy: just take any four-dimensional Riemannian space with signature two, calculate the Einstein tensor $G_{\mu\nu}$ and declare that the chosen metric is the exact solution to the Einstein equation for the matter distribution whose stress tensor is

$$T_{\mu\nu} = \frac{c^4}{8\pi G} G_{\mu\nu}.$$

But how do we ensure that this stress tensor represents any *physically reasonable* mass-energy distribution? In other words, if we are given a symmetric tensor $T^{\mu\nu}$ such that $T^{\mu\nu}{}_{;\nu} = 0$, what conditions must it satisfy to qualify as a possible tensor for some gravitating energy distribution?

The stress-energy tensor of a perfect fluid has the standard form $T^{\mu\nu} = pg^{\mu\nu} + (\rho + p/c^2)U^\mu U^\nu$ where U^μ is the four-velocity of the fluid. Therefore $\rho c^4 = T_{\mu\nu}U^\mu U^\nu \geq 0$. The velocity vector of a matter distribution is a time-like vector. Therefore we expect our stress-energy tensor to satisfy the condition $T_{\mu\nu}U^\mu U^\nu \geq 0$ for every time-like vector U^μ. This inequality is called the **weak energy condition**. (Since the set of time-like vectors has light-like vectors on its boundary, continuity requires that the energy condition also hold for light-like vectors U as well.)

When we discuss the Raychaudhuri equation in the last chapter of the book we will encounter another reasonable expectation from the tensor called the **strong energy condition** which amounts to requiring

$$\left(T_{\mu\nu} - \frac{1}{2}g_{\mu\nu}T\right)U^\mu U^\nu \geq 0 \tag{12.7}$$

for every time-like vector. Energy conditions on a stress-energy tensor are crucial in proving the singularity theorems . Roughly speaking these theorems state that under very general and reasonable assumptions on the stress tensor, the resulting spacetime may contain points where trajectories of particles or light rays converge and focus. Such points correspond to infinite values of curvature and are called singularities. A completely spherical collapse of dust will lead to such a singularity. These theorems tell us that singularities are not a feature of a highly symmetric gravitational collapse but are generic in nature.

12.8   Tutorial

Exercise 62. A connection on two-dimensional space has a component $\Gamma^1_{11} = 1$ and other components zero in some coordinates x^1, x^2. Choose Riemann normal coordinates about point $(0,0)$ and show that the space is in fact flat Euclidean space.

Answer 62. $\omega^1{}_1 = dx^1$, others zero. $\Omega^i{}_j = 0$ so the space is flat. The geodesic equations are

$$\frac{d^2x^1}{dt^2} + \left(\frac{dx^1}{dt}\right)^2 = 0, \qquad \frac{d^2x^2}{dt^2} = 0.$$

The geodesics which pass through $(0,0)$ and have tangent components (u^1, u^2) are

$$x^1 = \ln(u^1 t + 1), \qquad x^2 = u^2 t.$$

At $t = 1$ these equations give the Riemann normal coordinates u^1, u^2 in terms of x^1, x^2:

$$u^1 = e^{x^1} - 1, \qquad u^2 = x^2.$$

Project on Vaidya-Like Spacetime

The Schwarzschild metric

$$(ds)^2 = -a(r)c^2(dt)^2 + (1/a(r))(dr)^2 + r^2(d\theta)^2 + r^2\sin^2\theta(d\phi)^2$$

where $a(r) = 1 - 2GM/rc^2$ takes a useful form; if we define a new coordinate v in place of time cdt by

$$dv = cdt + \frac{dr}{a(r)}, \tag{12.8}$$

then the metric looks like

$$(ds)^2 = -a(r)(dv)^2 + 2(dv)(dr) + r^2(d\theta)^2 + r^2\sin^2\theta(d\phi)^2 \tag{12.9}$$

in these outgoing **Eddington-Finkelstein coordinates**. This project requires calculating the curvature and Einstein tensors for a metric of this type where function a has dependence on both r and v.

Exercise 63. Given a metric

$$(ds)^2 = -a(r,v)(dv)^2 + 2(dv)(dr) + r^2(d\theta)^2 + r^2\sin^2\theta(d\phi)^2 \tag{12.10}$$

calculate the connection, Riemann curvature and Einstein tensor for this spacetme.

Answer 63. Main steps
In coordinates $x^i, i = 0, 1, 2, 3$ with $x^0 = v, x^1 = r, x^2 = \theta, x^3 = \phi$. The metric components are (those not written are zero)

$$\left\langle \frac{\partial}{\partial x^i}, \frac{\partial}{\partial x^j} \right\rangle = g_{ij} = \begin{pmatrix} -a & 1 & & \\ 1 & 0 & & \\ & & r^2 & \\ & & & r^2\sin^2\theta \end{pmatrix},$$

$$\langle dx^i, dx^j \rangle = g^{ij} = \begin{pmatrix} 0 & 1 & & \\ 1 & a & & \\ & & 1/r^2 & \\ & & & 1/r^2\sin^2\theta \end{pmatrix}.$$

Let $b = \sqrt{a}$, define an orthonormal basis of 1-forms $\alpha^a, a = 0, 1, 2, 3$. (Hopefully the indices $a, b = 0, \ldots, 3$ would not be confused with the functions $a(r,v), b(r,v)$ or c with speed of light!)

$$\begin{aligned} \alpha^0 &= -bdv + (1/b)dr \\ \alpha^1 &= (1/b)dr \\ \alpha^2 &= rd\theta \\ \alpha^3 &= r\sin\theta d\phi \end{aligned}$$

$$\langle \alpha^a, \alpha^b \rangle = \eta^{ab}$$

where η^{ab} has diagonal terms $-1, 1, 1, 1$. Others are zero. The inverse relations are

$$
\begin{aligned}
dv &= (\alpha^1 - \alpha^0)/b \\
dr &= b\alpha^1 \\
d\theta &= \alpha^2/r \\
d\phi &= \alpha^3/r\sin\theta.
\end{aligned}
$$

The Cartan equation identify the non-zero coefficients of f_{abc} as

$$
\begin{aligned}
f_{001} &= & B & = & -f_{010} \\
f_{101} &= & C & = & -f_{110} \\
f_{212} &= & b/r & = & -f_{221} \\
f_{313} &= & b/r & = & -f_{331} \\
f_{323} &= & \cot\theta/r & = & -f_{332}
\end{aligned}
$$

where

$$
B = -b_{,r} + \frac{b_{,v}}{b^2},
$$

$$
C = \frac{b_{,v}}{b^2}.
$$

The matrix of connection 1-forms calculated with the help of Ricci rotation coefficients is

$$
\omega^a{}_b = \begin{pmatrix}
0 & B\alpha^0 + C\alpha^1 & 0 & 0 \\
B\alpha^0 + C\alpha^1 & 0 & -b\alpha^2/r & -b\alpha^3/r \\
0 & b\alpha^2/r & 0 & -\cot\theta\alpha^3/r \\
0 & b\alpha^3/r & \cot\theta\alpha^3/r & 0
\end{pmatrix}.
$$

The curvature matrix

$$
\Omega^a{}_b = d\omega^a{}_b + \omega^a{}_c \wedge \omega^c{}_b
$$

can be calculated easily and the non-zero components of

$$
\Omega_{ab} \equiv \eta_{ac}\Omega^c{}_b
$$

are

$$
\begin{aligned}
\Omega_{01} &= D(\alpha^0 \wedge \alpha^1) \\
\Omega_{02} &= bB(\alpha^0 \wedge \alpha^2)/r + bC(\alpha^1 \wedge \alpha^2)/r \\
\Omega_{03} &= bB(\alpha^0 \wedge \alpha^3)/r + bC(\alpha^1 \wedge \alpha^3)/r \\
\Omega_{12} &= \frac{1}{b}\left(\frac{b}{r}\right)_{,v}(\alpha^0 \wedge \alpha^2) - E(\alpha^1 \wedge \alpha^2) \\
\Omega_{13} &= \frac{1}{b}\left(\frac{b}{r}\right)_{,v}(\alpha^0 \wedge \alpha^3) - E(\alpha^1 \wedge \alpha^3) \\
\Omega_{23} &= \left(\frac{1 - b^2}{r^2}\right)(\alpha^2 \wedge \alpha^3)
\end{aligned}
$$

where

$$D = bb_{,rr} + (b_{,r})^2$$
$$E = \frac{bb_{,r}}{r} + \frac{b_{,v}}{br}.$$

To convert curvature components to the original coordinates v, r, θ, ϕ we use the matrix which connects α^a to $dx^i = dv, dr, d\theta, d\phi$,

$$\alpha^a = U^a{}_i dx^i,$$

and use expressions of $\alpha^a \wedge \alpha^b$ in terms of $dv \wedge dr, dv \wedge d\theta, dr \wedge d\phi$ etc. The curvature matrix of two-forms $\underline{\Omega}^i{}_j$ in these coordinates is then given by

$$\underline{\Omega}^i{}_j = U^{-1}{}^i{}_a \Omega^a{}_b U^b{}_j.$$

The Riemann curvature components in v, r, θ, ϕ coordinates then can be read off from

$$\underline{\Omega}^i{}_j = \frac{1}{2} \underline{R}^i{}_{jkl} dx^k \wedge dx^l.$$

The non-zero components are in terms of original function $b = \sqrt{a}$:

$$\underline{R}^0{}_{001} = -[bb_{,rr} + (b_{,r})^2]$$
$$\underline{R}^0{}_{202} = -brb_{,r}$$
$$\underline{R}^0{}_{303} = -brb_{,r} \sin^2 \theta$$
$$\underline{R}^1{}_{001} = -b^2[bb_{,rr} + (b_{,r})^2]$$
$$\underline{R}^1{}_{101} = bb_{,rr} + (b_{,r})^2$$
$$\underline{R}^1{}_{202} = -brb_{,v}$$
$$\underline{R}^1{}_{212} = -brb_{,r}$$
$$\underline{R}^1{}_{303} = -brb_{,v} \sin^2 \theta$$
$$\underline{R}^1{}_{313} = -brb_{,r} \sin^2 \theta$$
$$\underline{R}^2{}_{002} = -\frac{b^3}{r}\left(b_{,r} - \frac{b_{,v}}{b^2}\right)$$
$$\underline{R}^2{}_{012} = bb_{,r}/r$$
$$\underline{R}^2{}_{102} = bb_{,r}/r$$
$$\underline{R}^2{}_{323} = (1 - b^2) \sin^2 \theta$$
$$\underline{R}^3{}_{003} = -\frac{b^3}{r}\left(b_{,r} - \frac{b_{,v}}{b^2}\right)$$
$$\underline{R}^3{}_{013} = bb_{,r}/r$$
$$\underline{R}^3{}_{103} = bb_{,r}/r$$
$$\underline{R}^3{}_{223} = b^2 - 1.$$

The Ricci tensor has just four non-vanishing components

$$\underline{R}_{00} = b^2[bb_{,rr} + (b_{,r})^2 + 2bb_{,r}/r] - 2bb_{,v}/r$$
$$\underline{R}_{01} = -[bb_{,rr} + (b_{,r})^2 + 2bb_{,r}/r]$$
$$\underline{R}_{22} = 1 - b^2 - 2brb_{,r}$$
$$\underline{R}_{33} = [1 - b^2 - 2brb_{,r}] \sin^2 \theta.$$

The Ricci scalar is

$$\underline{R} = g^{ij}\underline{R}_{ij} = -2bb_{,rr} - 2(b_{,r})^2 - 8bb_{,r}/r + 2(1 - b^2)/r^2$$

and the Einstein tensor $\underline{G}_{ij} = \underline{R}_{ij} - (1/2)g_{ij}\underline{R}$ is

$$
\begin{aligned}
\underline{G}_{00} &= b^2(1 - b^2)/r^2 - 2b^3b_{,r}/r - 2bb_{,v}/r \\
\underline{G}_{01} &= 2bb_{,r}/r - (1 - b^2)/r^2 \\
\underline{G}_{22} &= br^2b_{,rr} + 2brb_{,r} + r^2(b_{,r})^2 \\
\underline{G}_{33} &= \sin^2\theta\underline{G}_{22}.
\end{aligned}
$$

If there is no dependence on the time-related coordinate v, then the vacuum equations give the Schwarzschild solution

$$a(r) = b^2 = 1 - \frac{\text{constant}}{r}.$$

If that constant is replaced by a function of v, then we get the metric discovered by P.C.Vaidya in 1953,

$$a(r, v) = 1 - \frac{2GM(v)}{rc^2}.$$

In this case only $\underline{G}_{00}$ is non-zero. The Vaidya spacetime corresponds to an exact solution of the Einstein equation,

$$\underline{G}_{ij} = \frac{8\pi G}{c^4}T_{ij},$$

with a stress-energy tensor whose v-v component alone is non-zero in these coordinates:

$$T_{00} = \frac{c^4}{8\pi G}G_{00} = \frac{c^2}{4\pi r^2}\frac{dM}{dv}.$$

Such matter is sometimes called **null dust**.

Exercise 64. OPTICAL METRIC
Let **g** be a static metric of the form

$$g_{\mu\nu} = \begin{pmatrix} g_{00} & 0 \\ 0 & g_{ij} \end{pmatrix}$$

with $t = x^0$ the time veriable and $x^i, i = 1, 2, 3$ space coordinates. (Remember that $g_{\mu\nu}$ are not dependent on t for static spacetime and that in our choice of notation g_{00} is a negative quantity.)

Let

$$\gamma_{ij} = -\frac{1}{g_{00}}g_{ij}$$

be a metric in three-dimensional space. Show that the three-dimensional geodesics of γ_{ij} are the spatial projections of *null* geodesics of the four-dimensional metric $g_{\mu\nu}$.

Answer 64. We have the following relations for the Christoffel symbols:

$$\Gamma^0_{00} = 0$$

$$\Gamma^0_{0i} = \frac{1}{2}g^{00}(g_{00,i}) = (\ln\sqrt{-g_{00}})_{,i} \equiv A_{,i}$$

$$\Gamma^0_{ij} = 0$$

$$\Gamma^i_{00} = \gamma^{im}A_{,m}, \qquad \gamma^{ij} = -g_{00}g^{ij}$$

$$\Gamma^i_{0j} = 0$$

$$\Gamma^i_{jk} = \tilde{\Gamma}^i_{jk} + \delta^i_j A_{,k} + \delta^i_k A_{,j} - \gamma_{jk}\gamma^{im}A_{,m},$$

where $\tilde{\Gamma}^i_{jk}$ are Christoffel symbols for the metric γ_{ij}. A null geodesic in the four-dimensional metric satisfies (for x^i coordinates)

$$
\begin{aligned}
0 &= \frac{d^2x^i}{d\lambda^2} + \Gamma^i_{00}\left(\frac{dx^0}{d\lambda}\right)^2 + \Gamma^i_{jk}\frac{dx^j}{d\lambda}\frac{dx^k}{d\lambda} \\
&= \frac{d^2x^i}{d\lambda^2} + \tilde{\Gamma}^i_{jk}\frac{dx^j}{d\lambda}\frac{dx^k}{d\lambda} + \gamma^{im}A_{,m}\left[\left(\frac{dx^0}{d\lambda}\right)^2 - \gamma_{jk}\frac{dx^j}{d\lambda}\frac{dx^k}{d\lambda}\right] + 2A_{,\lambda}\frac{dx^i}{d\lambda} \\
&= \frac{d^2x^i}{d\lambda^2} + \tilde{\Gamma}^i_{jk}\frac{dx^j}{d\lambda}\frac{dx^k}{d\lambda} + 2A_{,\lambda}\frac{dx^i}{d\lambda},
\end{aligned}
$$

the quantity in square brackets being zero because $dx^\mu/d\lambda$ is a null vector. The parameter λ is an affine parameter for the null geodesic in four-dimensional space. To change to an affine parameter for three-dimensional space we choose the length ds which is related to dx^0 by

$$ds^2 = \gamma_{jk}dx^j dx^k = -\frac{1}{g_{00}}g_{jk}dx^j dx^k = (dx^0)^2,$$

therefore we can choose $x^0 = t = s$ as the affine parameter. Changing

$$\frac{d}{d\lambda} = \frac{dx^0}{d\lambda}\frac{\partial}{\partial t}$$

we get

$$\frac{d^2x^i}{d\lambda^2} = \frac{d}{d\lambda}\left(\frac{dx^0}{d\lambda}\frac{dx^i}{dt}\right) = \frac{d^2x^0}{d\lambda^2}\frac{dx^i}{dt} + \left(\frac{dx^0}{d\lambda}\right)^2\frac{d^2x^i}{dt^2},$$

therefore the null-geodesic equation becomes

$$\left(\frac{dx^0}{d\lambda}\right)^2\left[\frac{d^2x^i}{dt^2} + \tilde{\Gamma}^i_{jk}\frac{dx^j}{dt}\frac{dx^k}{dt}\right] + \frac{d^2x^0}{d\lambda^2}\frac{dx^i}{dt} + 2A_{,\lambda}\frac{dx^i}{d\lambda} = 0.$$

The last two terms add up to zero because of the x^0 component of the null geodesic equation

$$
\begin{aligned}
\frac{d^2x^0}{d\lambda^2}\frac{dx^i}{dt} + 2A_{,\lambda}\frac{dx^i}{d\lambda} &= \frac{d^2x^0}{d\lambda^2}\frac{dx^i}{dt} + 2A_{,\lambda}\frac{dx^i}{dt}\frac{dx^0}{d\lambda} \\
&= \frac{dx^i}{dt}\left[\frac{d^2x^0}{d\lambda^2} + 2A_{,m}\frac{dx^m}{d\lambda}\frac{dx^0}{d\lambda}\right] \\
&= \frac{dx^i}{dt}\left[\frac{d^2x^0}{d\lambda^2} + 2\Gamma^0_{0m}\frac{dx^m}{d\lambda}\frac{dx^0}{d\lambda}\right] \\
&= 0,
\end{aligned}
$$

thus

$$
\frac{d^2x^i}{dt^2} + \tilde{\Gamma}^i_{jk}\frac{dx^j}{dt}\frac{dx^k}{dt} = 0.
$$

[The metric γ_{ij} has been called the optical metric by G.W. Gibbons and M.J. Perry in Proc. R. Soc. Lond., A358, 467 (1978)].

Chapter 13

Weak Gravitational Fields

In all ordinary situations Einstein's theory of gravity differs from the Newtonian gravity by very small corrections. But conceptually it gives many predictions which are startling. For example a spherical mass rotating slowly about an axis will "drag inertial frames" so that the spin of a gyroscope far from the mass will precess. Another prediction of the general theory of relativity is that perturbations of the metric $g_{\mu\nu}$ propagate as waves.

13.1 Einstein Tensor for Weak Fields

We assume that coordinates can be chosen such that the metric tensor is given by

$$g_{\mu\nu} = \eta_{\mu\nu} + h_{\mu\nu} \tag{13.1}$$

where the spacetime dependence is only in $h_{\mu\nu}$ which are assumed to be much smaller than the 'background' Minkowski coordinates with metric $\eta_{\mu\nu}$. We will neglect quantities which are second or higher power in $h_{\mu\nu}$'s.

The contravariant form of a metric tensor is

$$g^{\mu\nu} = \eta^{\mu\nu} + k^{\mu\nu}$$

where $k^{\mu\nu}$ are small. From $g^{\mu\nu} g_{\nu\sigma} = \delta^\mu_\sigma$ we see that up to terms of first-order

$$k^{\mu\nu} = -h^{\mu\nu}, \qquad g^{\mu\nu} = \eta^{\mu\nu} - h^{\mu\nu}, \qquad h^{\mu\nu} = \eta^{\mu\sigma}\eta^{\nu\tau}h_{\sigma\tau}.$$

We shall use the metric $\eta_{\mu\nu}$ and $\eta^{\mu\nu}$ to raise and lower indices in $h_{\mu\nu}$, for example $h^\mu_\nu \equiv \eta^{\mu\sigma}h_{\sigma\nu}$. As η's are constants the derivative indices can also be raised or lowered, for example $h^{\mu\nu,\sigma} = \eta^{\sigma\tau}(h^{\mu\nu}),_\tau$ etc.

The Christoffel coefficients, Γ's, are first-order quantities because η are constants

$$\Gamma^\mu_{\nu\sigma} = \frac{1}{2}\eta^{\mu\lambda}(h_{\lambda\nu,\sigma} + h_{\lambda\sigma,\nu} - h_{\nu\sigma,\lambda})$$

and the curvature and Ricci tensors

$$R^\mu{}_{\nu\sigma\lambda} = -\Gamma^\mu{}_{\nu\sigma,\lambda} + \Gamma^\mu{}_{\nu\lambda,\sigma},$$
$$R_{\nu\sigma} = -\Gamma^\mu{}_{\nu\mu,\sigma} + \Gamma^\mu{}_{\nu\sigma,\mu},$$

because the last two terms in the curvature tensor are quadratic in Γ's. Substituting the expression for Γ's we get

$$2R_{\nu\tau} = \eta^{\mu\sigma}[-h_{\mu\sigma,\nu\tau} + h_{\mu\nu,\sigma\tau} + h_{\mu\tau,\nu\sigma} - h_{\nu\tau,\mu\sigma}]$$

and further

$$R = (h^{\mu\nu}){}_{,\mu\nu} - \partial^2 h$$

where $h \equiv h^\mu_\mu = \eta^{\mu\nu} h_{\mu\nu}$ and

$$\partial^2 \equiv \eta^{\mu\nu} \partial_\mu \partial_\nu = -\left(\frac{\partial}{\partial x^0}\right)^2 + \nabla^2$$

is the d'Alembertian operator in Minkowski space.

The Einstein tensor for the weak field case is, therefore

$$R_{\nu\tau} - \frac{1}{2}\eta_{\nu\tau}R = \frac{1}{2}\left[(h^\mu_\nu){}_{,\mu\tau} + (h^\mu_\tau){}_{,\mu\nu} - h_{,\nu\tau} - \partial^2 h_{\nu\tau}\right]$$
$$+ \frac{1}{2}\eta_{\nu\tau}\left[-(h^{\mu\sigma}){}_{,\mu\sigma} + \partial^2 h\right] \tag{13.2}$$

and the Einstein equation is

$$-\frac{16\pi G}{c^4}T_{\nu\tau} = \partial^2 h_{,\nu\tau} + (h){}_{,\nu\tau} - [(h^\mu_\nu){}_{,\mu\tau} + (h^\mu_\tau){}_{,\mu\nu}]$$
$$- \eta_{\nu\tau}\left[-(h^{\mu\sigma}){}_{,\mu\sigma} + \partial^2 h\right]. \tag{13.3}$$

This can be further simplified by changing to the "reversed trace" perturbation defined by

$$\bar{h}_{\mu\nu} \equiv h_{\mu\nu} - \frac{1}{2}\eta_{\mu\nu}h. \tag{13.4}$$

It is called reversed trace because $\bar{h} = \eta^{\mu\nu}\bar{h}_{\mu\nu} = h - 2h = -h$. We can write the original quantities in terms of the $\bar{h}$ quantities as

$$h_{\mu\nu} = \bar{h}_{\mu\nu} - \frac{1}{2}\eta_{\mu\nu}\bar{h}, \qquad h^\mu_\nu = \bar{h}^\mu_\nu - \frac{1}{2}\delta^\mu_\nu\bar{h}.$$

In terms of these the Einstein equation simplifies to

$$\partial^2 \bar{h}_{\nu\tau} - [(\bar{h}^\mu_\nu){}_{,\mu\tau} + (\bar{h}^\mu_\tau){}_{,\mu\nu}] + \eta_{\nu\tau}(\bar{h}^{\mu\sigma}){}_{,\mu\sigma} = -\frac{16\pi G}{c^4}T_{\nu\tau}. \tag{13.5}$$

13.2 'Fixing a Gauge'

Recall that we can make use of diffeomorphic invariance to simplify formalism.

Any metric $\mathbf{g}$ is as good as the pullback $\Phi^*(\mathbf{g})$ for a diffeomorphism Φ of the spacetime. We choose a one-parameter group of diffeomorphisms Φ_t generated by a vector field $X = \xi^\mu \partial_\mu$. Under this, the metric tensor changes, for small t by

$$(\Phi_t^* g)_{\mu\nu} = g_{\mu\nu} + t\delta g_{\mu\nu}, \qquad \delta g_{\mu\nu} = \xi_{\mu;\nu} + \xi_{\nu;\mu}.$$

We shall absorb the infinitesimal t into ξ itself and regard ξ as a small quantity of first order like h. Thus $\xi_\mu = g_{\mu\nu}\xi^\nu = \eta_{\mu\nu}\xi^\nu$ and we can replace the covariant derivative in $\xi_{\mu;\nu}$ by ordinary derivatives because the Christoffel symbols are small too.

Choosing a suitable diffeomorphism so that Einstein equations in the new (pulled-back) metric are obtained in some desired form is called **fixing a gauge** in analogy with choosing the gauge in electrodynamics or in gauge theories of elementary particle physics.

Let us write the pullback

$$(\Phi_t^* g)_{\mu\nu} = \eta_{\mu\nu} + H_{\mu\nu}, \qquad H_{\mu\nu} = h_{\mu\nu} + \xi_{\mu,\nu} + \xi_{\nu,\mu}$$

and construct quantities $\overline{H}_{\mu\nu} \equiv H_{\mu\nu} - \frac{1}{2}\eta_{\mu\nu}H$ as before. Then the 'gauge' (that is choice of ξ^μ) that simplifies the equation is the one for which

$$(\overline{H}^{\mu\nu})_{,\nu} = 0, \qquad \text{"Lorentz Gauge"}. \tag{13.6}$$

That this can actually be done is straightforward to see. We have

$$
\begin{aligned}
H &= h + 2\eta^{\mu\nu}\xi_{\mu,\nu}, \\
\overline{H}^{\mu\nu} &= H^{\mu\nu} - \frac{1}{2}\eta^{\mu\nu}H \\
&= \overline{h}^{\mu\nu} + \xi^{\mu,\nu} + \xi^{\nu,\mu} - \eta^{\mu\nu}\eta^{\sigma\tau}\xi_{\sigma,\tau},
\end{aligned}
$$

and

$$(\overline{H}^{\mu\nu})_{,\nu} = (\overline{h}^{\mu\nu})_{,\nu} + \partial^2\xi^\mu.$$

So $(\overline{H}^{\mu\nu})_{,\nu} = 0$ if ξ is chosen such that

$$\partial^2\xi^\mu = -(\overline{h}^{\mu\nu})_{,\nu}.$$

This is always possible. In fact there is a **residual freedom** still in the choice of ξ^μ. We can add to the chosen ξ^μ any ξ_1^μ which satisfies $\partial^2\xi_1^\mu = 0$.

13.3   The Solution

Having chosen the gauge we can *call our H quantities again by the original name h* with the understanding that now it satisfies the additional gauge condition

$$(\overline{h}^{\mu\nu})_{,\nu} = 0. \tag{13.7}$$

With the gauge having been fixed the Einstein equation is

$$\partial^2 \overline{h}^{\mu\nu} = -\frac{16\pi G}{c^4} T_{\mu\nu} \tag{13.8}$$

whose solution is well known from classical Maxwell electrodynamics,

$$\overline{h}^{\mu\nu} = \frac{4G}{c^4} \int \frac{T^{\mu\nu}(x^0 - |\mathbf{x} - \mathbf{x}'|/c, \mathbf{x}')}{|\mathbf{x} - \mathbf{x}'|} d^3 x'. \tag{13.9}$$

13.4   Static Mass Distribution

We first take the simpler case of a static mass distribution near the origin. The stress-energy tensor has only one non-zero component $T^{00} = \rho c^2$:

$$T^{\mu\nu} = \begin{pmatrix} \rho c^2 & 0 & 0 & 0 \\ 0 & 0 & 0 & 0 \\ 0 & 0 & 0 & 0 \\ 0 & 0 & 0 & 0 \end{pmatrix}. \tag{13.10}$$

Therefore the gravitational field at $\mathbf{x}$ (away from the mass distribution which is concentrated near the origin $\mathbf{x}' = \mathbf{0}$) is obtained by integrating with $r = |\mathbf{x} - \mathbf{x}'|$ pulled out of the integral. The only non-zero component of $\overline{h}^{\mu\nu}$ is

$$\overline{h}^{00} = \frac{4G}{c^4} \int \frac{\rho c^2}{r} dv = \frac{4GM}{rc^2}$$

where

$$M = \int \rho c^2\, dv.$$

The covariant tensor

$$\overline{h}_{00} = (\eta_{00})^2 \overline{h}^{00} = \overline{h}^{00}.$$

The trace is $\overline{h} = \eta^{\mu\nu} \overline{h}_{\mu\nu} = -4GM/rc^2$. Therefore from $h_{\mu\nu} = \overline{h}_{\mu\nu} - \eta_{\mu\nu}\overline{h}/2$ we get

$$h_{00} = h_{11} = h_{22} = h_{33} = \frac{2GM}{rc^2} \qquad \text{(Static mass distribution)}.$$

Thus, our metric for the weak static field is

$$ds^2 = -\left(1 - \frac{2GM}{c^2 r}\right) c^2 (dt)^2$$
$$+ \left(1 + \frac{2GM}{c^2 r}\right) [(dx^1)^2 + (dx^2)^2 + (dx^3)^2]. \qquad (13.11)$$

As simple applications of the metric in this approximation we rederive the formula for light deflection and another effect called the Shapiro time delay.

13.4.1 Light Ray in a Weak Field

Let $K(\lambda)$ be the tangent vectors of a null geodesic, $\langle K, K \rangle = 0$ and $D_K K = 0$ with affine parameter λ.

$$K^\mu(\lambda) = \frac{dx^\mu}{d\lambda}.$$

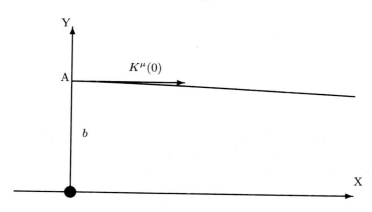

Fig. 13.1: Deflection of light.
Light ray starts at A parallel to X axis. It bends and develops a small
component in the negative Y direction due to mass at the origin.

We assume the light ray to start $(t = 0, \lambda = 0)$ from a point A (see Figure 13.1) $(x^\mu(A) = (0, 0, b, 0))$ and with initial direction along the x^1 axis. Thus the tangent vector to the geodesic is initially

$$K^\mu(0) = (1, 1, 0, 0) + O(h)$$

where we generally denote small quantities by the symbol $O(h) \sim h_{00} = h_{11}$ etc. The geodesic will stay in the x^1-x^2 plane under the assumption that all mass is concentrated near the origin.

The equation for parallel transport of K along the geodesic is

$$\frac{dK^\mu}{d\lambda} = -\Gamma^\mu_{\nu\sigma} K^\nu K^\sigma.$$

We are interested in finding the first-order change in the direction of the geodesic far away ($x^1 \to \infty$). This change comes from $O(h)$ changes in K^2. To calculate to lowest order we can replace $K^\nu K^\sigma$ on the right-hand side by their initial values $K^0 \approx K^1 \approx 1$. Thus the following equation holds,

$$\begin{aligned}
\frac{dK^2}{d\lambda} &= -\Gamma^2_{\nu\sigma} K^\nu K^\sigma \\
&= -(\Gamma^2_{00} + 2\Gamma^2_{01} + \Gamma^2_{11}) \\
&= h_{00,2} \\
&= -\frac{2GM}{c^2} \frac{x^2}{r^3},
\end{aligned}$$

using $h_{00} = h_{11} = 2GM/(rc^2)$. To lowest order x^2 stays close to b, so that

$$\frac{x^2}{r^3} \approx \frac{b}{(b^2 + (x^1)^2)^{3/2}}$$

and we can write

$$d\lambda = \frac{d\lambda}{dx^1} dx^1 = \frac{dx^1}{K^1} \approx dx^1.$$

Therefore

$$dK^2 = -\frac{2GMb}{c^2} \frac{dx^1}{(b^2 + (x^1)^2)^{3/2}}.$$

Integrating on x^1 from 0 to ∞ we get

$$K^2(\infty) = -\frac{2GM}{bc^2}.$$

The ratio of $K^2(\infty)/K^1(\infty) \approx K^2(\infty)$ gives the angle made by asymptotic direction of the light ray with the x^1 axis .

$$\delta = -\frac{2GM}{bc^2}. \qquad (13.12)$$

For a light ray coming from infinity, going past the source at a distance b and then observed again at infinity, the geometry of the curve is symmetric and the two deflections add up. Therefore the deflection is twice this value, that is $4GM/bc^2$. This is what we calculated as the light deflection in a Schwarzschild field.

13.4.2 Time Delay

As another example we estimate the background coordinate time that light will take to go from the point $A(0, b, 0)$ in the figure to point $B(X, b, 0)$. In the absence of gravitating body the time will be just X/c. In presence of a big mass at the origin the time taken is slightly more than this. The excess is called the Shapiro time delay.

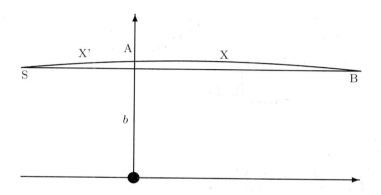

Fig. 13.2: Time delay.
The time taken by light to travel from A to B according to flat space clocks is an amount $[2GM/c^3]\ln(2X/b)$ greater than X/c which it would take in the absence of the gravitating body.

We make assumptions similar to the last section. Neglecting the vertical change in the path of the ray,

$$(ds)^2 = -c^2(dt)^2(1 - H) + (1 + H)(dx)^2 = 0$$

where

$$H \equiv \frac{2GM}{rc^2}$$

implies

$$dt \approx \frac{1}{c}(1 + H)dx.$$

Therefore

$$\int dt = \frac{1}{c}\int_0^X \left[1 + \frac{2GM}{c^2\sqrt{x^2 + b^2}}dx\right]$$

$$= \frac{X}{c} + \frac{2GM}{c^3}\int_0^X \frac{1}{\sqrt{x^2 + b^2}}dx.$$

Using the standard integral

$$\int \frac{ds}{\sqrt{s^2 + 1}} = \ln[s + \sqrt{s^2 + 1}]$$

the delay is to leading order

$$\Delta t = \frac{2GM}{c^3} \ln\left(\frac{2X}{b}\right).$$

If light comes from some source S at a large distance X' and after passing the gravitating body at perpendicular distance b is received by a detector B at a distance X, the time delay compared to the time when gravitating body is absent is therefore the sum

$$\Delta t_{SA} + \Delta t_{AB} = \frac{2GM}{c^3} \ln\left(\frac{4XX'}{b^2}\right). \tag{13.13}$$

If the source is kept at B and a signal is sent to S which reflects it back, then the round-trip delay is twice this value.

We have estimated the delay under simplified assumptions. A more careful calculation gives

$$\Delta t_{\text{round trip}} = \frac{4GM}{c^3}\left[1 + \ln\left(\frac{4RR'}{r_0^2}\right)\right] \tag{13.14}$$

where R, R' are the distances of B and S from the gravitating body and r_0 is the distance of closest approach of light to the body.

13.5 Slowly Rotating Mass Distribution

Let us calculate the field of a mass distribution slowly rotating with a constant angular velocity about the z-axis of a cartesian coordinate system. We assume the mass to be concentrated around the origin and symmetric about the z-axis. We calculate the field at a point far away from the body.

Matter in the non-relativistic regime will have negligible pressure (the kinetic theory of gases gives pressure proportional to average velocity square times mass density ρ which is much smaller than ρc^2). Therefore the stress-energy tensor is of the form $T^{\mu\nu} = \rho U^\mu U^\nu$. If the mass distribution is rotating slowly with constant angular velocity ω about the z-axis, then

$$U^\mu \approx (c, v^1 = -\omega x^2, v^2 = \omega x^1, 0)$$

which is, neglecting terms of $O((v^i)^2/c^2)$,

$$T^{\mu\nu} \approx \rho U^\mu U^\nu \approx \rho \begin{pmatrix} c^2 & -c\omega x^2 & c\omega x^1 & 0 \\ -c\omega x^2 & 0 & 0 & 0 \\ c\omega x^1 & 0 & 0 & 0 \\ 0 & 0 & 0 & 0 \end{pmatrix}. \tag{13.15}$$

From our formula for weak fields (written for a contravariant metric tensor) the non-zero $\overline{h}^{\mu\nu} = \overline{h}^{\nu\mu}$ are for $\mu, \nu = 00, 01, 02$. We calculate these.

First of all there is no time dependence in $T^{\mu\nu}$. The masses are moving, but velocities are constant in time. This is the stationary situation analogous to stationary currents in magnetostatics.

$$T^{\mu\nu}(x^0 - |\mathbf{x} - \mathbf{x}'|/c, \mathbf{x}') = T^{\mu\nu}(\mathbf{x}').$$

Therefore using the fact that $r \equiv |\mathbf{x}| >> |\mathbf{x}'|$,

$$
\begin{aligned}
\frac{1}{|\mathbf{x} - \mathbf{x}'|} &= (|\mathbf{x}|^2 + |\mathbf{x}'|^2 - 2\mathbf{x}.\mathbf{x}')^{-1/2} \\
&\approx \frac{1}{r} + \frac{x^1 x^{1'} + x^2 x^{2'} + x^3 x^{3'}}{r^3},
\end{aligned}
$$

we can write the leading order contributions to $\overline{h}^{\mu\nu}$,

$$
\begin{aligned}
\overline{h}^{00} &= \frac{4G}{c^4} \int \frac{\rho(\mathbf{x}')c^2}{|\mathbf{x} - \mathbf{x}'|} d^3 x' \\
&= \frac{4G}{c^2 r} \int \rho(\mathbf{x}') d^3 x' \\
&\equiv \frac{4GM}{c^2 r}, \\
\overline{h}^{01} &= -\frac{4\omega G}{c^3 r^3} \int \rho(\mathbf{x}') x^{2'}(x^1 x^{1'} + x^2 x^{2'} + x^3 x^{3'}) d^3 x' \\
&= -\frac{4\omega G x^2}{c^3 r^3} \int \rho(\mathbf{x}') (x^{2'})^2 d^3 x' \\
&= -\frac{4\omega G x^2}{c^3 r^3} I_{22},
\end{aligned}
$$

where (i) M is the total mass, (ii) the integration over $x^{2'} x^{1'}$ and $x^{2'} x^{3'}$ vanish because of symmetry of mass distribution and (iii) I_{22} is the moment-of-inertia component. Similarly,

$$\overline{h}^{02} = \frac{4\omega G x^1}{c^3 r^3} I_{11}.$$

The mass distribution is symmetric about the z-axis. Therefore we can use the perpendicular axis theorem of moments of inertia,

$$I_{11} = I_{22} = (I_{11} + I_{22})/2 = I_{33}/2.$$

The angular momentum is equal to

$$J = \omega I_{33}. \tag{13.16}$$

To the order of approximation used, the components $\overline{h}^{\mu\nu}$ are

$$\overline{h}^{00} = \frac{4GM}{c^2 r}, \qquad \overline{h}^{01} = -\frac{2JGx^2}{c^3 r^3}, \qquad \overline{h}^{02} = \frac{2JGx^1}{c^3 r^3}. \qquad (13.17)$$

The trace $\overline{h}$ is $\overline{h} = \overline{h}^0_0 = -\overline{h}^{00}$ and therefore $h_{\mu\nu} = \overline{h}_{\mu\nu} - \eta_{\mu\nu}\overline{h}/2$ are

$$
\begin{aligned}
h^{00} &= \frac{2GM}{c^2 r} = h_{00}, \\
h^{11} &= h^{22} = h^{33} = \frac{2GM}{c^2 r} = h_{11} = h_{22} = h_{33}, \\
h^{01} &= -\frac{2JGx^2}{c^3 r^3} = -h_{01}, \\
h^{02} &= \frac{2JGx^1}{c^3 r^3} = -h_{02}.
\end{aligned}
$$

The metric for this case is thus $g_{\mu\nu} = \eta_{\mu\nu} + h_{\mu\nu}$,

$$
\begin{aligned}
(ds)^2 &= -\left(1 - \frac{2GM}{c^2 r}\right)(dx^0)^2 \\
&\quad + \left(1 + \frac{2GM}{c^2 r}\right)[(dx^1)^2 + (dx^2)^2 + (dx^3)^2] \\
&\quad - \frac{4JG}{c^3 r^2}\left(\frac{x^1(dx^2) - x^2(dx^1)}{r}\right)(dx^0) \qquad (13.18)
\end{aligned}
$$

where, as is physicists' custom, we write g_{01} and g_{10} terms together.

   Actually, the form of this metric is more general. Even if the source is not the Newtonian, slowly moving matter but fully relativistic, the fields at a large distance will have this form.

13.6   Gravi-Magnetic Effects

13.6.1   Freely Falling Gyroscope

Newtonian gravity depends only on instantaneous mass distributions or density, whereas Einstein's theory involves all components of the stress-energy tensor. These include mass and energy flow. We have seen in the last section how gravitational fields are modified by stationary mass-energy currents. These gravitational fields produce effects proportional to v/c much like the magnetostatic effects of stationary currents in electromagnetism. If a gyroscope falls along the axis of a rotating mass distribution such that its spin points in a direction in the plane perpendicular to the axis, then this spin direction will precess in the plane in the same sense as the rotating mass as the gyroscope falls. These effects are called **gravi-magnetic** effects.

An early discussion of such effects was by H.Thirring and J.Lense in 1918 who considered effects at the center of a slowly rotating massive spherical shell. Einstein had come to the same conclusion in 1913 even before the general theory of relativity was formulated. See discussion on Mach's principle in Misner, Thorne and Wheeler (section 21.12) and Weinberg (section 9.7).

A gyroscope is a small heavy symmetric body rotating with its axis supported frictionlessly on frames which are again mounted so that they are free to move in all possible directions. The gyroscope can be transported by moving the outermost frame. The design of frames ensures that to a very good approximation no torques can be applied to the rotating body. If the gyroscope is in free fall, its spin will be transported without change, that is, it will be parallel transported. If the gyroscope is carried by an accelerated observer, then the gyroscope's spin will be Fermi-Walker transported. We discuss the free fall in this section and motion along the accelerated path in the next.

We can represent the spin by a space-like vector S orthogonal to the time-like velocity vector U to the gyroscope's trajectory in spacetime. If the gyroscope is falling freely (that is following a geodesic) then S will remain orthogonal to U all along the trajectory because U is also parallel transported along the geodesic.

We now see how the spin of a gyroscope behaves for the mass distribution of the last section. In order to simplify our calculation we assume a gyroscope falling freely along the z-axis with its spin pointing in the x-y plane. As the fall is along the z-axis, $x^1 = x = 0$ and $x^2 = y = 0$ along the trajectory. Therefore tangent vector $U(\tau)$ of the geodesic has components only along time and z-direction $U^\mu = (U^0, 0, 0, U^3)$. The fact that $x^1(s) = 0, x^2(s) = 0$ makes many Christoffel symbols equal to zero along the trajectory. The parallel transport equation

$$\frac{dS^\mu}{d\tau} + \Gamma^\mu_{\nu\sigma} U^\nu S^\sigma = 0 \tag{13.19}$$

gives to leading order,

$$\frac{dS^1}{d\tau} = -\frac{2GJ}{c^2 z^3} S^2, \qquad \frac{dS^2}{d\tau} = \frac{2GJ}{c^2 z^3} S^1. \tag{13.20}$$

More details are given in the tutorial to this chapter. If we are in the frame of reference of the falling gyroscope (an inertial frame) we would not see the precession because the spin is being parallel transported. But it is the background Minkowski coordinate system (distant stars) that seems to precess with respect to the gyroscope. The Lense-Thirring precession frequency at 'height' z is

$$\Omega_{LT} = \frac{2GJ}{c^2 z^3} \tag{13.21}$$

in the same sense as the rotating mass. We express this phenomenon by saying that the freely falling inertial frame (in which the gyroscope spin is a constant vector) is **dragged along** by the rotating mass distribution.

13.6.2   Gyroscope at Rest

We can generalise the metric of a slowly rotating mass distribution whose angular momentum is in an arbitrary direction instead of being along the z-axis:

$$
\begin{aligned}
(ds)^2 \;=\; & -\left(1 - \frac{2GM}{c^2 r}\right)(dx^0)^2 \\
& +\left(1 + \frac{2GM}{c^2 r}\right)[(dx^1)^2 + (dx^2)^2 + (dx^3)^2] \\
& -\frac{4G\epsilon_{ijk}J^j x^k}{c^3 r^3}(dx^i dx^0),
\end{aligned}
\tag{13.22}
$$

where $\mathbf{J}$ is the angular momentum vector and the ϵ_{ijk} is the antisymmetric symbol equal to 1 if ijk is an even permutation of 123, -1 if odd, and zero if any two indices in ijk happen to be the same.

If a gyroscope is 'nailed' to a point $\mathbf{x}$ in the coordinate system with the help of forces applied to its centre of mass (so that its spin is not disturbed) then the gyroscope follows the non-geodesic trajectory

$$
x^\mu(\tau) = (x^0 = c\tau, \mathbf{x}).
\tag{13.23}
$$

The unit tangent vector T, $\langle T, T \rangle = -1$ of this path is given by

$$
T = \left(1 - \frac{2GM}{c^2 r}\right)^{-\frac{1}{2}} \frac{\partial}{\partial x^0}.
\tag{13.24}
$$

The gyroscope's spin is Fermi-Walker transported along the trajectory. Its components $S = (0, S^i)$ with respect to the global background Minkowski coordinate system ("fixed stars") change as follows:

$$
\frac{dS^i}{d(c\tau)} + \Omega_{ij}S^j = 0
\tag{13.25}
$$

where the precession frequency vector

$$
\mathbf{\Omega} \equiv c(\Omega_{23}, \Omega_{31}, \Omega_{12})
$$

is given by

$$
\mathbf{\Omega} = \frac{G}{r^3 c^2}\left[-\mathbf{J} + 3\frac{\mathbf{x}(\mathbf{x} \cdot \mathbf{J})}{r^2}\right].
\tag{13.26}
$$

This is worked out in the tutorial. Note that if the $\mathbf{J}$ is along the z-axis and the gyroscope is at rest at a point along the z-axis with its spin $\mathbf{S}$ in the x-y plane, then the precession frequncy vector $\mathbf{\Omega}$ points in the z-direction with magnitude $2JG/z^3 c^2$. To lowest order the two precession frequencies (of the freely falling and the fixed on z-axis gyroscopes) match because the correction due to time dilation factor between coordinate time and the proper time is of one higher order.

13.7 Energy and Momentum

The Einstein equation relates a geometric quantity, the Einstein tensor $G_{\mu\nu}$, to the stress energy tensor $T_{\mu\nu}$ of matter. This tensor $T_{\mu\nu}$ is the total stress-energy tensor of *all* matter and radiation fields *except gravity*. Gravitational field is characterized by a conspicuous absence of a separate stress-energy tensor. This is due to the non-linear nature of the Einstein equations.

The Bianchi identities $G^{\mu\nu}_{;\nu} = 0$ imply $T^{\mu\nu}_{;\nu} = 0$. But this is *not a conservation law for energy and momentum of matter*. By a conservation of some quantity we understand that the change in the amount of that quantity inside a closed three-dimensional volume can be accounted for by the flowing out of that quantity from the surface of that volume.

We have discussed Gauss' divergence theorem in section 10.11. It can be summarised by saying that if there is a vector field X such that $\text{div} X = X^{\nu}_{;\nu} = 0$, then over a four-dimensional domain D,

$$\int_D (\text{div} X)\Omega = \int_{\partial D} \langle \mathbf{n}, \mathbf{n}\rangle(\langle X, \mathbf{n}\rangle)(\Omega)_{\partial D} = 0.$$

Equivalently as

$$\begin{aligned} X^{\nu}_{;\nu} &= X^{\nu}_{,\nu} + \Gamma^{\nu}_{\mu\nu}X^{\mu} \\ &= X^{\nu}_{,\nu} + \frac{1}{\sqrt{-g}}(\sqrt{-g})_{,\mu}X^{\mu} \\ &= \frac{1}{\sqrt{-g}}[\sqrt{-g}X^{\nu}]_{,\nu}, \end{aligned}$$

therefore $X^{\nu}_{;\nu} = 0$ gives a continuity equation

$$(\sqrt{-g}X^{\nu})_{,\nu} = 0$$

which can be converted into a surface integral.

It is often stated (somewhat carelessly) that $T^{\mu\nu}_{;\nu} = 0$ gives the law of conservation of energy-momentum for matter. This is not correct. However, if there is a **Killing vector field** K, then there we can construct a conserved current: define

$$P^{\nu} \equiv T^{\mu\nu}K_{\mu}, \qquad K_{\mu} \equiv g_{\mu\sigma}K^{\sigma},$$

then

$$(P^{\nu})_{;\nu} = (T^{\mu\nu}K_{\mu})_{;\nu} = (T^{\mu\nu})_{;\nu}K_{\mu} + T^{\mu\nu}K_{\mu;\nu} = 0$$

because both the terms are zero: the first from the Einstein equation and the second because K is a Killing vector field so $K_{\mu;\nu} = -K_{\nu;\mu}$. Thus $(P^{\nu})_{;\nu} = 0$, and we can interpret P^{μ} as the conserved quantity.

If there is a Killing vector field K, then $P^\nu \equiv T^{\mu\nu} K_\mu$ gives a conservation law: If D is a four-dimensional region with boundary hypersurface ∂D then

$$\int_{\partial D} \langle \mathbf{n}, \mathbf{n} \rangle \langle P, \mathbf{n} \rangle (\Omega)_{\partial D} = 0 \tag{13.27}$$

where $\mathbf{n}$ is the unit normal vector on ∂D and $(\Omega)_{\partial D}$ is the volume on the hypersurface.

We have seen (in Chapter 10) that there cannot be more than 10 independent Killing fields in four-dimensional spacetime. That happens in the maximally symmetric space like the Minkowski space. The Killing field corresponding to time translations gives rise to energy conservation, those of space translations give linear momentum, of rotations to angular momentum and so on. In a general spacetime there will be no Killing fields, and there will be no conserved quantities either. But if it is possible to choose vector fields which approximate Killing fields in a small region of spacetime, then there can be *local* conservation of energy momentum etc. Thus it is appropriate to say that $T^{\mu\nu}{}_{;\nu} = 0$ represents the **local conservation of energy momentum**. After all, the conservation laws of energy and momentum in non-gravitational physics are of this nature. They depend on the near flatness of the spacetime region in which they are tested.

Conserved Quantities Along a Geodesic

We have discussed the conservation of energy and angular momentum of a freely falling particle in a Schwarzschild field in Chapter 4. There these quantities were seen to be conserved due to the existence of Killing fields $\partial/\partial t$ and $\partial/\partial \phi$ of the spacetime. In general, if K is the Killing field and U the four-velocity of the freely falling particle (that is tangent vector to a geodesic) then $\langle K, U \rangle$ remains constant along the path. This law represents the conservation of quantities of the **test particle**, in a symmetric gravitational field and has nothing to do with the conservation of energy-momentum of the *source* of gravitation.

Of course, one can think of the combined source plus test particle stress-energy tensor producing the gravitational field by the Einstein equation. Then the fact that test particles move along geodesics of the spacetime must also follow from the Einstein equation. This is indeed so but these calculations are not easy and can be done only approximately. They were initiated by Einstein and Infeld in the 1940s.

13.8   Energy Psuedo-Tensor

The equation $T^{\mu\nu}{}_{;\nu} = 0$ does not give a conservation law in general but rather gives the exchange of energy momentum between matter and a gravitational field. This equation can be written as

$$T^{\mu\nu}{}_{,\nu} = -\Gamma^\mu_{\sigma\nu} T^{\sigma\nu} - \Gamma^\nu_{\sigma\nu} T^{\mu\sigma} \tag{13.28}$$

where the right-hand side measures the failure of matter energy momentum to conserve due to exchange of energy and momentum with the gravitational field.

At a given fixed point, we can always choose local coordinates such that all Γ's vanish at the given point. Therefore the right-hand side can be made zero *at* a *single point*, but in order to define momentum and energy of a field system we have to integrate corresponding densities over a region and their values at a single point are of no use.

Does this mean that "gravitational field energy" has no meaning? The answer is that gravitational field energy has a meaning in those situations where we can clearly separate the matter and gravitational degrees of freedom. This may require choosing a special coordinate system.

In the following we discuss the separation of matter and gravitational energy, momentum etc. for mass distribution located in a finite volume so that we can choose a coordinate system which is **asymptotically Minkowskian**. This means we can write $g_{\mu\nu} = \eta_{\mu\nu} + h_{\mu\nu}$ where the $h_{\mu\nu}$ tend to zero as $r = (x^2 + y^2 + z^2)^{1/2} \to \infty$.

We have already separated the first-order quantities. Let us write the exact Einstein tensor as a part of the first-order in $h_{\mu\nu}$ plus the "rest",

$$G^{\mu\nu} = G^{\mu\nu}_{(1)} + G^{\mu\nu}_{\text{rest}}$$

where $G^{(1)}_{\mu\nu}$ is constructed from $\overline{h}_{\mu\nu} = h_{\mu\nu} - \eta_{\mu\nu}h/2$ as before (section 13.1). We prefer to use the contravariant version of Einstein's equations for notational convenience.

$$
\begin{aligned}
-2G^{\mu\nu}_{(1)} &= \partial_\alpha\partial_\beta[\overline{h}^{\mu\nu}\eta^{\alpha\beta} + \overline{h}^{\alpha\beta}\eta^{\mu\nu} - \overline{h}^{\mu\alpha}\eta^{\nu\beta} - \overline{h}^{\nu\beta}\eta^{\mu\alpha}] \\
&\equiv \partial_\alpha\partial_\beta H^{\nu\alpha\mu\beta} \qquad\qquad\qquad (13.29)
\end{aligned}
$$

where the second line defines the quantity $H^{\nu\alpha\mu\beta}$. It has interesting symmetry properties. It is , like the Riemann tensor, antisymmetric in ν, α and in μ, β, as well as symmetric under the interchange of the pair of joint indices $\nu\alpha$ with μ, β ; $H^{\nu\alpha\mu\beta} = H^{\mu\beta\nu\alpha}$.

If we operate by ∂_ν on $G^{\mu\nu}_{(1)}$ and sum over ν, we get zero because the symmetric operator $\partial_\nu\partial_\alpha$ acts on an expression antisymmetric in ν, α and both indices are summed. Thus $G^{\mu\nu}_{(1)}$ has vanishing ordinary, rather than covariant divergence:

$$(G^{\mu\nu}_{(1)})_{,\nu} = 0.$$

Rearrange the exact Einstein equation as

$$
\begin{aligned}
G^{\mu\nu}_{(1)} &= \frac{8\pi G}{c^4}T^{\mu\nu} - G^{\mu\nu}_{\text{rest}} \\
&= \frac{8\pi G}{c^4}(T^{\mu\nu} + t^{\mu\nu})
\end{aligned}
$$

where

$$t^{\mu\nu} \equiv -\frac{c^4}{8\pi G}G^{\mu\nu}_{\text{rest}}. \qquad\qquad\qquad (13.30)$$

Because $(G^{\mu\nu}_{(1)})_{,\nu} = 0$ we also have the ordinary divergence of $T^{\mu\nu} + t^{\mu\nu}$ zero:

$$(T^{\mu\nu} + t^{\mu\nu})_{,\nu} = 0. \qquad\qquad\qquad (13.31)$$

This can be interpreted as the conservation of 'total' energy-momentum, the first term $T^{\mu\nu}$ giving the matter and the second $t^{\mu\nu}$ the gravitational energy-momentum. Except for this warning: *the above decomposition is based on the existence of a coordinate system which is asymptotically Minkowskian where $g_{\mu\nu}$ can be written as the sum $\eta_{\mu\nu} + h_{\mu\nu}$.* The 'gravitational' energy-momentum tensor is called energy-momentum 'psuedo-tensor' because its form is dependent on the coordinate system.

13.9   Energy-Momentum for an Isolated System

In electrostatics the electric charge inside a volume can be calculated by summing the flux of electric field over the surrounding surface. This is just Gauss' theorem. Similarly the Newtonian potential which satisfies $\nabla^2\Phi = 4\pi\rho$ allows the calculation of total mass inside a volume by integrating over the gravitational field (or acceleration) ($\mathbf{g} = -\nabla\Phi$)

$$M = -\frac{1}{4\pi}\int \mathbf{g} \cdot \mathbf{n} dS.$$

The idea is that since charge or mass produce the fields, the field carries the information of the source which causes the field.

In the previous section we have identified the "psuedo tensor" $t^{\mu\nu}$ as an expression for gravitational energy-momentum. It is of second or higher order in h. If we integrate $T^{0\nu} + t^{0\nu}$ over a large volume, we expect to get the total energy and momentum contained in it. Call it P^ν,

$$\begin{aligned}
P^\nu &= \frac{c^4}{8\pi G}\int_V d^3x\, G_1^{0\nu} \\
&= -\frac{c^4}{16\pi G}\int_V d^3x\, \partial_\alpha\partial_\beta H^{\nu\alpha 0\beta} \\
&= -\frac{c^4}{16\pi G}\int_V d^3x\, \partial_\alpha\partial_j H^{\nu\alpha 0j} \\
&= -\frac{c^4}{16\pi G}\int_S dS\, n^j (H^{\nu\alpha 0j})_{,\alpha}
\end{aligned} \qquad (13.32)$$

where the antisymmetry property reduces the integrand to a spatial divergence $\partial_i(...)^i$ which can be converted into a two-dimensional surface integral with surface normal $\mathbf{n} = (n^1, n^2, n^3)$.

It is important to realise that this expression for the total mass is independent of any gauge choice. Moreover, under global Lorentz transformation of Minkowski coordinates $x^\mu \to \Lambda^\mu_\nu x^\nu$ the quantity P^ν transforms like a four-vector.

It is also important to see that the surface integral is to be calculated far away from the source. Thus even if the source is strong, the asymptotic field at the point of evaluation will give the right expression for total energy and momentum. As an example one can evaluate the total energy for the Schwarzschild field. A short calculation (see tutorial for this chapter) yields the total energy equal to $P^0 = Mc^2$. Thus constant M appearing in the Schwarzschild solution *is the total energy of the static mass distribution plus the gravitational energy.*

The formulas for angular momentum can similarly be written. The angular momentum with respect to the origin of coordinates is

$$M^{kl} = \int_V d^3x [x^k(T^{0l} + t^{0l}) - x^l(T^{0k} + t^{0k})] \tag{13.33}$$

which can be written similar to the case for total energy momentum

$$M^{kl} = -\frac{c^4}{16\pi G} \int_S dS \left(x^k n^j (H^{l\alpha 0j})_{,\alpha} - x^l n^j (H^{k\alpha 0k})_{,\alpha}\right). \tag{13.34}$$

13.10 $t^{\mu\nu}$ up to Second-Order

The gravitational stress-energy tensor $t^{\mu\nu}$ is determined by the Einstein tensor from which its linear terms have been subtracted, $t^{\mu\nu} = -c^4 G^{\mu\nu}_{\text{rest}}/8\pi G$. For weak fields it may be sufficient to calculate $G^{\mu\nu}_{\text{rest}}$ only up to second-order,

$$G^{\mu\nu}_{\text{rest}} \approx G^{\mu\nu}_{(2)} = [R^{\mu\nu} - \frac{1}{2}g^{\mu\nu}R]_{(2)}.$$

We start with $g_{\mu\nu} = \eta_{\mu\nu} + h_{\mu\nu}$ and following the procedure in section 13.1 of this chapter get the following expressions which help us calculate the psuedo tensor $t^{\mu\nu}$ to second-order.

$$R_{\mu\nu} \quad = \quad R^{(1)}_{\mu\nu} + R^{(2)}_{\mu\nu} + \cdots$$

where $R^{(1)}_{\mu\nu}$ was calculated in section 13.1,

$$2R^{(1)}_{\nu\tau} \quad = \quad \eta^{\mu\sigma}[-h_{\mu\sigma,\nu\tau} + h_{\mu\nu,\sigma\tau} + h_{\mu\tau,\nu\sigma} - h_{\nu\tau,\mu\sigma}].$$

The contribution to $R^{(2)}_{\mu\nu}$ comes from various sources. First of all Γ's are of structure $g^{(\cdot\cdot)}[g_{(\cdot,\cdot,)} + ..] = (\eta - h)[h + ..]$. They gives second-order contribution to the $\Gamma^\mu_{\nu\sigma,\tau}$ terms in the Riemann tensor, in addition to the Γ terms. After a straightforward calculation we get

$$
\begin{aligned}
R^{(2)}_{\mu\nu} \quad = \quad & +\frac{1}{2}h^{\alpha\beta}[h_{\mu\nu,\alpha\beta} + h_{\alpha\beta,\mu\nu} - h_{\mu\alpha,\nu\beta} - h_{\nu\beta,\mu\alpha}] \\
& -\frac{1}{2}\left[(h^{\alpha\beta})_{,\beta} - \frac{1}{2}\eta^{\alpha\beta}(h)_{,\beta}\right][h_{\alpha\mu,\nu} + h_{\alpha\nu,\mu} - h_{\mu\nu,\alpha}] \\
& +\frac{1}{4}h^{\alpha\beta}_{,\mu}h_{\alpha\beta,\nu} \\
& +\frac{1}{4}\eta^{\sigma\tau}\eta^{\alpha\beta}(h_{\sigma\mu,\alpha} - h_{\alpha\mu,\sigma})(h_{\tau\nu,\beta} - h_{\beta\nu,\tau}).
\end{aligned} \tag{13.35}
$$

Similarly $R = g^{\mu\nu} R_{\mu\nu}$ can be written

$$
\begin{aligned}
R &= R^{(1)} + R^{(2)} + \cdots, \\
R^{(1)} &= \eta^{\mu\nu} R^{(1)}_{\mu\nu} \\
&= (h^{\mu\nu})_{,\mu\nu} - \partial^2 h, \\
R^{(2)} &= \eta^{\mu\nu} R^{(2)}_{\mu\nu} + h^{\mu\nu} R^{(1)}_{\mu\nu},
\end{aligned}
$$

and so

$$
g_{\mu\nu} R = \eta_{\mu\nu} R^{(1)} + h_{\mu\nu} R^{(1)} + \eta_{\mu\nu} R^{(2)}, \tag{13.36}
$$

which gives the desired expression for the Einstein tensor,

$$
G^{(2)}_{\mu\nu} = -\frac{1}{2}(\eta_{\mu\nu} h^{\alpha\beta} + h_{\mu\nu} \eta^{\alpha\beta}) R^{(1)}_{\alpha\beta} + \left(R^{(2)}_{\mu\nu} - \frac{1}{2}\eta_{\mu\nu}\eta^{\alpha\beta} R^{(2)}_{\alpha\beta}\right). \tag{13.37}
$$

Note that if we need second-order $G^{\mu\nu} = g^{\mu\alpha} g^{\nu\beta} G_{\alpha\beta}$ we would need to expand again and obtain

$$
G^{\mu\nu}_{(2)} = \eta^{\mu\alpha}\eta^{\nu\beta} G^{(2)}_{\alpha\beta} - \eta^{\mu\alpha} h^{\nu\beta} G^{(1)}_{\alpha\beta} - h^{\mu\alpha}\eta^{\nu\beta} G^{(1)}_{\alpha\beta}. \tag{13.38}
$$

13.11   Gravitational Waves

13.11.1   Isolating the Radiative Part

So far we have considered the solutions of $\partial^2 \overline{h}^{\mu\nu} = -16\pi G T^{\mu\nu}/c^4$ for stationary sources $T_{\mu\nu}$. For a time-varying mass distribution the solutions will be in the form of a gravitational wave. $\overline{h}^{\mu\nu}$ are ten quantities which can be subjected to four 'gauge conditions' (for example $(\overline{h}^{\mu\nu})_{,\nu} = 0$).

Using the convenient background of Minkowski coordinates x^{μ} and the smallness of the deviation $h_{\mu\nu}$ of the metric, we can decompose these into components which remain unaffected by diffeomorphic invariance. There are six independent degrees of freedom which are immune to diffeomorphic invariance. They are called "gauge invariant" in the following sense. Under an infinitesimal diffeomorphism where a point with coordinates x^{μ} is mapped into the point with coordinates $x^{\mu} + \xi^{\mu}$, the metric tensor is pulled back with components

$$
h_{\mu\nu} + \xi_{\mu;\nu} + \xi_{\nu;\mu} \approx h_{\mu\nu} + \xi_{\mu,\nu} + \xi_{\nu,\mu}
$$

where we can replace the covariant derivatives by ordinary derivatives because the ξ and Γ's are both small.

We use the fact that both the original and the transformed metric have weak fields decomposition $(\eta_{\mu\nu} + h_{\mu\nu})$ which allows identifying components with respect to the background flat geometry. Thus the h_{00} component is the weak field riding

$\eta_{00} = -1$ and so on. We can find suitable combinations of $h_{\mu\nu}$ and their derivatives such that they remain unchanged under infinitesimal diffeomorphism.

As an example, note that h_{00} involves change through ξ_0 (it changes by $2\partial_0\xi_0$) and so does h_{0i}. But the change in h_{0i} will also involve ξ^i just as changes in h_{ij} would. Now it turns out that there is a combination of h_{00}, h_{0i} and h_{ij} which will cancel the ξ_0 dependence from h_{00} with that from h_{0i} while the ξ_i dependence (that comes in with h_{0i}) cancels with a certain combination from h_{ij}. This is achieved by looking at the transformation properties of various components under three-dimensional rotations of the background Minkowski coordinates. Thus h_{00} is a scalar, h_{0i} a three-dimensional vector and so on. A three-dimensional vector can be decomposed into a gradient part and a divergence-free or transverse part. A three-dimensional symmetric tensor has a more elaborate decomposition

$$h_{ij} = \frac{1}{3}\delta_{ij}H + \left(\partial_i\partial_j - \frac{1}{3}\partial^2\right)C + (\partial_iD_j + \partial_jD_i) + h_{ij}^{TT}$$

where the vector D_i is transverse, $\partial_iD_i = 0$, and h_{ij}^{TT} is the transverse traceless (TT) part satisfying four conditions,

$$\sum_i h_{ii}^{TT} = 0, \qquad \sum_i \partial_i h_{ij}^{TT} = 0.$$

There are six quantities h_{ij} on the left, six on the right (two scalars H, C two D's and two $(6-4)$ h^{TT}'s).

We can separate them as follows. H is simply the trace $H = \sum h_{ii}$ and C and D_i are related to the gradient and transverse parts of $\nabla^{-2}\sum\partial_ih_{ij}$,

$$\sum\partial_ih_{ij} = \nabla^2\left(\frac{2}{3}\partial_jC + D_j\right).$$

Under diffeomorphisms the TT-part remains unchanged. It is 'gauge invariant' in the commonly used language. As expected, there are six gauge invariant quantities because the diffeomorphism introduces four arbitrary functions and out of ten $h_{\mu\nu}$, six combinations can manage to remain unaffected.

Gauge invariant quantities by definition have the same values whether the gauge is fixed or not. For weak fields the curvature tensor (and thus the Einstein tensor) remains 'gauge invariant'. This is easily shown in a tutorial exercise.

We can decompose the stress tensor $T_{\mu\nu}$ in a way analogous to the decomposition of $h_{\mu\nu}$ and write down Einstein equation $\partial^2\bar{h}_{\mu\nu} = -16\pi GT_{\mu\nu}/c^4$ in terms of gauge invariant quantities by comparing the decomposition on both sides. The $h_{\mu\nu}$ contain both the wave-like ("radiative") and non-wavelike solutions. It follows that out of six gauge invariant quantities only the 'TT-part' satisfies a wave equation whereas others satisfy Poisson-like equations. Thus the radiative or *wave-like nature of the metric perturbation resides in the transverse-traceless part.* We refer the student to the introductory article by Eanna Flanagan and Scott Hughes in arxiv: gr-qc/0501041 for details.

13.11.2   Wave Nature

In a region away from the sources the wave equation reads

$$(-\partial_0^2 + \nabla^2)h_{ij}^{TT} = 0. \tag{13.39}$$

We can determine h_{ij} (or rather $\overline{h}_{ij}$) from the formula for the solution given above. The question we face now is how to separate the TT-part from the $\overline{h}_{ij}$.

This is seen easily through an example. Suppose we find a simple 'plane wave' (that is constant amplitude and phase on a plane), in a small region of spacetime away from sources

$$h_{ij} = a_{ij}f(ct - z) = a_{ij}f(x^0 - x^3)$$

with a_{ij} a symmetric numerical matrix. The transverse condition requires $h_{ij,j} = 0, i = 1, 2, 3$ which implies $a_{i3} = a_{3i} = 0$. That leaves three quantities $h_{11}, h_{22}, h_{12} = h_{21}$. The traceless condition requires $h_{11} + h_{22} = 0$. As we expect there are only two degrees of freedom. We can write these two polarization amplitude $a = h_{11} = -h_{22}$ and $b = h_{12} = h_{21}$. Therefore the TT-part of h_{ij} is obtained by projecting the matrix a_{ij} into the subspace perpendicular to the direction of propagation of the wave, and then enforcing the tracelessness condition.

$$(a_{ij}) = \begin{pmatrix} a_{11} & a_{12} & a_{13} \\ a_{21} & a_{22} & a_{23} \\ a_{31} & a_{32} & a_{33} \end{pmatrix} \rightarrow (a_{ij}^{TT}) = \begin{pmatrix} a & b & 0 \\ b & -a & 0 \\ 0 & 0 & 0 \end{pmatrix}. \tag{13.40}$$

This suggests how to separate the TT-parts from the general expression. If we know the direction of propagation of the wave, say, along the unit vector $\mathbf{n}, n_i n_i = 1$, then we use a projection operator perpendicular to $\mathbf{n}$. The projection matrix P_{ij} is a symmetric matrix with the following properties:

$$P_{ij} = \delta_{ij} - n_i n_j, \qquad P_{ij}P_{jk} = P_{ik}, \qquad P_{ii} = 2. \tag{13.41}$$

We first obtain the transverse ('T-part') $\overline{h}_{ij}^T$,

$$\overline{h}_{ij}^T = P_{ik}P_{jl}\overline{h}_{kl}, \tag{13.42}$$

then subtract its trace from it to get the TT-part:

$$\begin{aligned} \overline{h}_{ij}^{TT} &= \overline{h}_{ij}^T - \frac{1}{2}P_{ij}\overline{h}_{kk}^T \\ &= P_{ik}P_{jl}\overline{h}_{kl} - \frac{1}{2}P_{ij}(\overline{h}_{kl}P_{kl}). \end{aligned} \tag{13.43}$$

Note that we have calculated the TT-part of the 'reverse trace' perturbation $\overline{h}_{ij}$ rather than h_{ij} because our wave equation is written for $\overline{h}_{ij}$. Happily, because of the tracelessness the two are the same:

$$\overline{h}_{ij}^{TT} = h_{ij}^{TT}. \tag{13.44}$$

Remark on Spin of Gravitational Radiation

In a quantum theory of gravity, the gravitational waves will be quantized into particles, 'gravitons'. The wave equation above corresponds to a relativistic wave equation for a particle of zero mass. What is the spin? The spin of a relativistic particle can be inferred from the transformation properties of the classical field of the particle under global Lorentz transformations. A second-rank symmetric tensor will in general have spin states of spin 0, 1, 2. In quantum theory of massless particles the spin is called helicity and there are always two helicity states with full angular momentum $\pm J$ corresponding to the forward and backward alignment of spin with the direction of momentum. This is in contrast to the $(2J+1)$ states for a particle with finite mass.

The TT-conditions reduce the degrees of freedom only to the two helicity or polarization degrees of freedom. That the helicity for gravitons will be two is seen most easily by applying a global rotation about the z-axis by an angle θ to the plane wave along the z-axis given above $h_{ij}^{TT} = a_{ij}^{TT} f(ct - z)$. Under the rotation

$$a_{ij}^{TT} \rightarrow R_{ik} R_{jl} a_{kl}^{TT}$$

where the transformation on $h^{\mu\nu}$ is by the Lorentz matrix for rotation $h^{\mu\nu} \rightarrow \Lambda_\sigma^\mu \Lambda_\tau^\nu h^{\nu\sigma}$ with $\Lambda_0^0 = 1, \Lambda_i^0 = \Lambda_0^i = 0, \Lambda_j^i = R_{ij}$. For

$$R_{ij} = \begin{pmatrix} \cos\theta & \sin\theta & 0 \\ -\sin\theta & \cos\theta & 0 \\ 0 & 0 & 1 \end{pmatrix}$$

the two polarizations of the gravitational wave change as

$$a_{ij}^{TT} = \begin{pmatrix} a & b & 0 \\ b & -a & 0 \\ 0 & 0 & 0 \end{pmatrix} \rightarrow \begin{pmatrix} a\cos 2\theta + b\sin 2\theta & b\cos 2\theta - a\sin 2\theta & 0 \\ b\cos 2\theta - a\sin 2\theta & -a\cos 2\theta - b\sin 2\theta & 0 \\ 0 & 0 & 0 \end{pmatrix}.$$

In quantum theory the two polarizarion amplitudes a and b will become annihilation operators of the gravitons and the operator for the circular polarizations $a \pm ib$ will transform as $(a \pm ib) \rightarrow U(a + ib)U^\dagger$ with $U = \exp(iJ_3\theta/\hbar)$. Thus

$$(a \pm ib) \rightarrow e^{\mp 2i\theta}(a \pm ib)$$

showing that the helicity or spin of gravitons is $2\hbar$.

13.11.3 'Quadrupole Formula'

After showing that the radiative degrees of freedom of weak gravitational fields reside in h_{ij}^{TT} we go back to the solution for the field in the Lorentz gauge $h_{,\nu}^{\mu\nu} = 0$,

$$\overline{h}^{\mu\nu} = \frac{4G}{c^4} \int \frac{T^{\mu\nu}(x^0 - |\mathbf{x} - \mathbf{x}'|, \mathbf{x}')}{|\mathbf{x} - \mathbf{x}'|} d^3x'. \tag{13.45}$$

At very large distance $r \equiv |\mathbf{x}|$ from the source the multipole terms obtained from the expansion of $|\mathbf{x} - \mathbf{x}'|$ can be neglected and we get simply for the spatial part

$$\overline{h}^{ij} = \frac{4G}{c^4 r} \int T^{ij}(x^0 - r, \mathbf{x}') d^3x'. \tag{13.46}$$

There is a standard way to relate the integral on the right-hand side with the second time derivative of the moment of inertia tensor. We know that the local energy-momentum conservation looks like $T^{\mu\nu}_{,\nu} = 0$ for weak fields. Separating the space and time components, we have two equations $(T^{00})_{,0} + (T^{0i})_{,i} = 0$ and $(T^{i0})_{,0} + (T^{ij})_{,j} = 0$. Therefore

$$(T^{00})_{,00} = -((T^{0i})_{,0})_{,i} = (T^{ij})_{,ij}.$$

Now we derive a useful identity. Calculate

$$(T^{ij}x^k x^l)_{,i} = (T^{ij})_{,i}x^k x^l + T^{kj}x^l + T^{lj}x^k$$

and then

$$
\begin{aligned}
(T^{ij}x^k x^l)_{,ij} &= (T^{ij})_{,ij}x^k x^l + (T^{ik})_{,i}x^l + (T^{il})_{,i}x^k \\
&\quad + (T^{kj})_{,j}x^l + T^{kl} + (T^{lj})_{,j}x^k + T^{kl} \\
&= (T^{00})_{,00}x^k x^l - 2T^{kl} + 2[T^{ik}x^l + T^{il}(x^k)]_{,i}.
\end{aligned}
$$

Both $(T^{ij}x^k x^l)_{,ij}$ and $[T^{ik}x^l + T^{il}(x^k)]_{,i}$ are divergences. Integrating these terms over a volume enclosing all matter we get zero on the surface. Therefore using the above identity to express T^{kl} in terms of $(T^{00})_{,00}x^k x^l$ we get

$$
\begin{aligned}
\bar{h}^{kl} &= \frac{4G}{c^4 r} \int T^{kl}(x^0 - r, \mathbf{x}')\, d^3 x' \\
&= \frac{2G}{c^4 r} \frac{d^2}{d(x^0)^2} \int T^{00} x^{k'} x^{l'}\, d^3 x' \\
&= \frac{2G}{c^2 r} \frac{d^2}{d(x^0)^2} \int \rho x^{k'} x^{l'}\, d^3 x' \\
&= \frac{2G}{c^4 r} \ddot{I}_{kl} \quad\quad\quad\quad\quad\quad\quad\quad (13.47)
\end{aligned}
$$

where the dot denotes differentiation with respect to time t, $(x^0 = ct)$ and I_{kl} are the moment of inertia tensor components.

13.11.4 Separation of the TT-Part

Our solution $\bar{h}^{kl}$ is equal to a numerical matrix times $1/r$: $h^{kl} \sim A_{kl}/r$. Therefore the transverse condition $(A_{kl}/r)_{,l} = -A_{kl}x^l/r^3 = 0$ requires that we project A_{kl} to the subspace perpendicular to the unit vector $\mathbf{n} \equiv \mathbf{x}/r$. Define the TT-part as

$$\bar{h}^{TT}_{ij} = \frac{2G}{c^4 r} \ddot{I}^{TT}_{ij}$$

where

$$\ddot{I}_{ij}^{TT} = P_{ik}P_{jl}\ddot{I}_{kl} - \frac{1}{2}P_{ij}(\ddot{I}_{kl}P_{kl}).$$

It is more convenient to use the 'quadrupole inertia tensor'

$$Q_{kl} = I_{kl} - \frac{1}{3}\delta_{kl}I, \qquad I \equiv I_{ii}$$

in place of I_{kl}. The quadrupole tensor is zero for a spherically symmetric body. The TT-part of both the tensors are the same:

$$
\begin{aligned}
Q_{ij}^{TT} &= P_{ik}P_{jl}Q_{kl} - \frac{1}{2}P_{ij}(Q_{kl}P_{kl}) \\
&= P_{ik}P_{jl}(I_{kl} - \delta_{kl}I/3) - \frac{1}{2}P_{ij}((I_{kl} - \delta_{kl}I/3)P_{kl}) \\
&= P_{ik}P_{jl}I_{kl} - P_{ij}I/3 - \frac{1}{2}P_{ij}(I_{kl}P_{kl}) + P_{ij}I/3, \qquad (P_{kk} = 2) \\
&= I_{ij}^{TT}.
\end{aligned}
$$

Therefore the gravitational spherical wave amplitude at $\mathbf{x}$ at time t far away from the gravitating masses is given by

$$\bar{h}_{ij}^{TT} = h_{ij}^{TT} = \frac{2G}{c^4 r}\ddot{Q}_{kl}^{TT} \tag{13.48}$$

where the dependence on the direction $\mathbf{n} = \mathbf{x}/r$ is hidden in the projection operator $P_{ij} = \delta_{ij} - n_i n_j$ occuring in the expression for Q_{kl}^{TT}.

13.11.5   Energy Radiated Away as Gravitational Waves

The flux, that is, energy carried by a gravitational wave per unit area per unit time (per unit x^0 coordinate actually) in the i-th direction is contained in the gravitational energy pseudo-tensor component t^{0i}. As the amplitude of the wave $h_{\mu\nu}$ undulates with time we have to average t^{0i} suitably just as we do for any wave.

We can calculate $t^{\mu\nu}$ from its second-order Einstein tensor formula

$$t^{\mu\nu} = -\frac{c^4}{8\pi G}G_{(2)}^{\mu\nu}.$$

It is easiest to calculate it for a plane wave along the z-direction in TT-gauge: $h_{11} = -h_{22} = af(x^0 - x^3), h_{12} = h_{21} = bf(x^0 - x^3)$, the rest of the components of $h_{\mu\nu}$ are zero. It is a very short calculation. Of t^{0i} only the 03-component is non-zero because the wave is in that direction. The only dependence on coordinates is on x^0 and x^3 and that only through the combination $x^0 - x^3$ which allows a derivative with respect to one to be converted into the other. We get

$$G_{03}^{(2)} = R_{03}^{(2)} = -(a^2 + b^2)\left(f'' + \frac{1}{2}(f')^2\right)$$

where f', f'' are the first and second derivatives of $f = f(x^0 - x^3)$ with respect to its argument. Upon averaging the fluctuating term f'' will vanish and we get

$$
\begin{aligned}
\langle t_{03} \rangle &= \frac{c^4}{16\pi G} \langle (a^2 + b^2)(f')^2 \rangle \\
&= \frac{c^5}{32\pi G} \langle \dot{h}_{ij}^{TT} \dot{h}_{ij}^{TT} \rangle
\end{aligned}
$$

where we have written it back in terms of the gravitational field and replaced the derivatives with respect to $(x^0 - x^3)$ by derivatives with respect to time ($ct = x^0$). In this form the formula is applicable to not just the z-direction but any direction.

By substituting our formula for h_{ij}^{TT} and integrating over a spherical surface (which involves integrating over all directions in the flux formula) we get the total flux from the gravitational source. It is given by the famous 'quadrupole formula' (derived first by Einstein) for energy loss per unit time from a radiating source,

$$
P = \frac{G}{5c^5} \langle \dddot{Q}_{ij} \dddot{Q}_{ij} \rangle. \tag{13.49}
$$

The details are left for the tutorial.

13.12   Detection of Gravitational Waves

A gravitational wave changes the metric locally. A massive matter distribution with non-zero quadrupole tensor may start changing rapidly starting at some fixed time. The effects on the metric at a point far away will be felt when waves reach there. The metric at the point which was Minkowskian will get a part $h_{\mu\nu}^{TT}$ added to it. If only the gravitational waves are present, we can choose a local coordinate system so that the metric has the form $\eta_{\mu\nu} + h_{\mu\nu}^{TT}$ with the wave propagating along the z-direction. This means $h_{\mu\nu}^{TT}$ is of the form

$$
h_{\mu\nu}^{TT} = A_{\mu\nu} f(x^0 - x^3)
$$

with non-zero components $A_{11} = -A_{22} = a, A_{12} = A_{21} = b$.

A particle which is at rest before the arrival of the wave , that is with its velocity four-vector $U^\mu = (c, 0, 0, 0)$, and otherwise not influenced by any other force, remains at rest after the arrival of the wave because the particle follows the geodesic of free fall and its acceleration is given by

$$
\frac{d^2 x^i}{d\tau^2} + \Gamma^i_{00} c^2 = 0.
$$

But $\Gamma^i_{00} = 0$ for the TT-metric. Thus the velocity remains zero, the coordinate position of the particle does not change!

This does not mean that there is no effect of the wave. Indeed, if there are two neighboring particles lying in the x-y plane with coordinates (x, y, z) and

(x', y', z) their numerical coordinates may not change, but the *physical distance* between them changes. Let us take the two particles a coordinate distance L apart connected by a light spring with unextended (or uncompressed) length L. The second particle kept at an angle θ with respect to the first, that is, $x' = x + L\cos\theta, y' = y + L\sin\theta$. The physical distance L_{phy} between them when the wave arrives is

$$
\begin{aligned}
L_{\text{phy}} &= [(\eta_{\mu\nu} + h_{\mu\nu})(x' - x)^\mu (x' - x)^\nu]^{1/2} \\
&= [\{(1+a)L^2\cos^2\theta + 2bL^2\sin\theta\cos\theta + (1-a)L^2\sin^2\theta\}f]^{1/2} \\
&\approx L + \frac{1}{2}L(a\cos 2\theta + b\sin 2\theta)f.
\end{aligned}
$$

Thus the passing gravitational wave produces a strain

$$
\frac{\delta L_{\text{phy}}}{L_{\text{phy}}} = \frac{1}{2}(a\cos 2\theta + b\sin 2\theta)f.
$$

It is interesting to note the 2θ appearing in the formulas again due to the spin of the field being 2.

One way to see what it means is to imagine first a wave 'polarised' so that $a \neq 0, b = 0$. When two mass points in the x-y plane kept along the x-axis ($\theta = 0$) at coordinates $x = 0$ and $x = L$ feel a positive strain (will be pushed out) then, at the same time mass points kept along the y-axis $\theta = \pi/2, \cos 2\theta = -1$ will feel the negative strain (pushed in). Mass points in the 45° direction will remain unaffected. This is the horizontal-vertical polarization denoted by the plus symbol '+'.

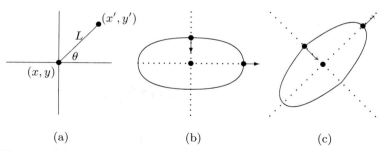

(a)   (b)   (c)

Fig. 13.3: (a) A pair of particles in a plane perpendicular to the direction of gravitational wave. (b) The plus (+) polarization (c) The cross (×) polarization.

On the other hand if $a = 0, b \neq 0$, then similar maximum strains are produced at $\pm 45°$ lines and no strain in horizontal-vertical directions. This is the polarization corresponding to h_{12} denoted popularly by '×'.

Detection of gravitational waves depends on the above principle. However, no gravitational waves have been detected yet because of the smallness of the effect. A big international program for measuring gravitational wave signals is underway.

13.13 Tutorial

Exercise 65. TIDAL FORCES OF WEAK FIELDS
Calculate the Riemann tensor for a static weak field

$$ds^2 = -\left(1 + \frac{2\Phi}{c^2}\right)c^2 dt^2 + \left(1 - \frac{2\Phi}{c^2}\right)[(dx^1)^2 + (dx^2)^2(dx^2)^2 + (dx^3)^2]$$

where Φ is the Newtonian potential. Note that the Riemann tensor components are just the tidal forces of Chapter 1.

Answer 65.

$$R^0{}_{0\sigma\tau} = 0, \qquad R^0{}_{i0j} = -\frac{2}{c^2}\frac{\partial^2 \Phi}{\partial x^i \partial x^j} = R^i{}_{00j}.$$

Exercise 66. GEODETIC PRECESSION
A gyroscope falls freely in a stable circular orbit of radius R in Schwarzschild spacetime. Show that the angle by which the spin of the gyroscope changes after completing one revolution is equal (to leading order)

$$\Delta\phi\text{one revo.} = \frac{3\pi GM}{Rc^2}$$

in the same direction (same sense) as the orbit.

Answer 66. The gyroscope spin is a space-like four-vector S orthogonal to the velocity vector U of the orbit. We have discussed the circular orbit in exercise 28 in Chapter 4. In Schwarzschild coordinates $x^\mu = (ct, r, \theta, \phi)$ the $\theta = \pi/2$ circular orbit has constant angular frequancies

$$\Omega \equiv \frac{d\phi}{dt} = \left(\frac{GM}{R^3}\right)^{1/2},$$

$$\omega \equiv \frac{d\phi}{d\tau} = \left(\frac{GM}{R^3}\right)^{1/2}\left(1 - \frac{3GM}{Rc^2}\right)^{-1/2},$$

where τ is the proper time along the trajectory.

The tangent (or velocity) vector of the trajectory $U^\mu = dx^\mu/d\tau$ also has constant components (i.e., independent of τ).

$$U^\mu = \left(c\frac{dt}{d\tau}, 0, 0, \frac{d\phi}{d\tau}\right) = \omega(c/\Omega, 0, 0, 1)$$

$$= \left(1 - \frac{3GM}{Rc^2}\right)^{-1/2}\left(c, 0, 0, \left(\frac{GM}{R^3}\right)^{1/2}\right).$$

Let $S(\tau)$ be the spin four-vector at proper time τ. The spin vector S will be parallel transported along the orbit by the equation

$$\frac{dS^\mu}{d\tau} + \Gamma^\mu_{\nu\sigma}U^\nu S^\sigma = 0.$$

A look at the connection components of the Schwarzschild metric (section 4.1) shows that S^2 remains constant in time. We can take it to be zero without any loss of generality. We need also not worry about S^0 because $\langle S, S \rangle$ remains constant under parallel transport and we can calculate S^0 from this relation.

The transport equations for S^1 and S^3 are

$$\frac{dS^1}{d\tau} - R\left(1 - \frac{3GM}{Rc^2}\right)U^3 S^3 = 0,$$

$$\frac{dS^3}{d\tau} + \frac{1}{R}U^3 S^1 = 0.$$

Both equations thus imply

$$\frac{d^2 S^1}{d\tau^2} + \Omega^2 S^1 = 0,$$

$$\frac{d^2 S^3}{d\tau^2} + \Omega^2 S^3 = 0.$$

In fact S^0 also satisfies the same equation. One can verify it directly.

We can choose the initial condition so that at $\tau = 0$ the gyroscope is at $r = R, \theta = \pi/2, \phi = 0$ and the spin vector points in the radial direction $S^\mu(0) = (s^0, s^1, 0, 0)$. The solution for these initial conditions is

$$S^1(\tau) = s^1 \cos(\Omega \tau), \qquad S^3(\tau) = -\frac{\omega}{R\Omega} s^1 \sin(\Omega \tau).$$

At the end of one revolution $\phi = 2\pi$ when $\tau = T = 2\pi/\omega$. To leading order (that is, neglecting quantities of order $(GM/Rc^2)^2$) S^1 and S^0 do not change. Only S^3 which was zero at $\tau = 0$ becomes equal to $s^1 3GM/R^2 c^2$ at $\tau = T$. The spin vector thus gets rotated by an infinitesimal angle

$$S^1(T) \approx s^1, \qquad S^3(T) \approx \frac{3GM s^1}{R^2 c^2} = \frac{1}{R}\Delta\phi_{\text{one revol}}, \qquad S^0(T) \approx s^0.$$

Exercise 67. GRAVI-MAGNETIC EFFECTS
Complete the derivation of precession for a freely falling gyroscope along the z-axis in the field of a mass distribution rotating slowly about the z-axis.

Answer 67. A freely falling gyroscope will have its spin parallel transported along the geodesic of fall. Let U be the time-like velocity vector (tangent to the geodesic).

$$U = \left(U^0 = c\frac{dt}{d\tau}, 0, 0, T^3 = \frac{dx^3}{d\tau}\right).$$

The spin vector has $S^3 = 0$ because the vector is orthogonal to the direction of the geodesic. The equation for parallel transport for S^1 is

$$\frac{\partial U^1}{\partial \tau} + \Gamma^1_{\mu\nu} U^\mu S^\nu = 0$$

with a similar equation for S^2 involving $\Gamma^2_{\mu\nu}$.

For our weak field

$$h_{00} = h_{11} = h_{22} = h_{33} = 2GM/c^2 r, \qquad h_{01} = 2GJx^2/c^3 r^3 = h_{10}$$

$$h_{02} = -2GJx^1/c^3 r^3 = h_{20},$$

the $\Gamma^1_{\mu\nu}$'s and $\Gamma^2_{\mu\nu}$'s which are non-zero on the path of free-fall $(x^1 = 0, x^2 = 0)$ are

$$
\begin{aligned}
\Gamma^1_{02} &= \Gamma^1_{20} = 2GJ/c^3 z^3, & \Gamma^1_{13} = \Gamma^1_{31} = -GM/c^2 z^2, \\
\Gamma^2_{01} &= \Gamma^2_{10} = -2GJ/c^3 z^3, & \Gamma^2_{23} = \Gamma^2_{32} = -GM/c^2 z^2.
\end{aligned}
$$

We can take $U^0 \approx c$ in the second term of the transport equation because the correction to it would be one order of v/c smaller where v is the velocity of the falling gyroscope. Thus, the parallel transport equation for the gyroscope spin components are

$$
\frac{\partial S^1}{\partial \tau} + \Gamma^1_{02} c S^2 = 0,
$$

$$
\frac{\partial S^2}{\partial \tau} + \Gamma^2_{01} c S^2 = 0.
$$

This gives the Lense-Thirring precession frequency

$$
\Omega_{LT} = \frac{2GJ}{c^2 z^3}
$$

given in the text.

Exercise 68. PRECESSION OF A GYROSCOPE AT REST
Show that a gyroscope at rest at $\mathbf{x}$ precesses according to the equation

$$
\frac{dS^i}{d(c\tau)} + \Omega_{ij} S^j = 0
$$

with angular frequency $\mathbf{\Omega} = c(\Omega_{23}, \Omega_{31}, \Omega_{12})$ where

$$
\mathbf{\Omega} = \frac{G}{r^3 c^2}\left[-\mathbf{J} + 3\frac{\mathbf{x}(\mathbf{x} \cdot \mathbf{J})}{r^2} \right] \tag{13.50}
$$

with respect to fixed stars in the field of rotating mass distribution with angular momentum $\mathbf{J}$.

Answer 68. The trajectory is defined by $\mathbf{x} =$ constant. Therefore the unit tangent vector T is given by $T^\mu = [(1 - 2GM/c^2 r)^{-1/2}, 0, 0, 0]$. The Fermi-Walker equation

$$
D_T S - T\langle S, D_T T\rangle + D_T T\langle S, T\rangle = 0
$$

determines the transport of the spin components by

$$
\frac{dS^i}{d(c\tau)} + \Gamma^i_{0j} T^0 S^j + D_T T\langle S, T\rangle = 0
$$

where the middle term drops out because $T^i = 0$. Now

$$
\begin{aligned}
(D_T T)^i &= T^\nu T^i_{,\nu} = \Gamma^i_{00} T^0 T^0 = \left(1 - \frac{2GM}{c^2 r}\right)^{-1} \frac{1}{2}(-h_{00,i}) \\
&\approx \frac{GMx^i}{c^2 r^3}
\end{aligned}
$$

is small and so is $\langle S, T \rangle = S^1 T^0 g_{i0}$. Therefore we can drop the last term as well. This gives up to leading order,

$$\frac{dS^i}{d(c\tau)} + \Gamma^i_{0j} S^j = 0.$$

The Christoffel symbol effectively determines the precesion frequency vector:

$$\Gamma^i_{0j} = \frac{1}{2}(g_{i0,j} - g_{0j,i})$$

$$= \frac{G}{c^3}\left[\frac{\partial}{\partial x^j}\left(\frac{\epsilon_{ilm} x^l J^m}{r^3}\right) - (i \leftrightarrow j)\right]$$

which directly gives the expression for Ω_{ij}.

Exercise 69. ENERGY AND MOMENTUM
Calculate the total energy-momentum four-vector for the Schwarzschild spacetime.

Answer 69. We calculate the energy $E = P^0$.
  Express the Schwarzschild metric

$$ds^2 = -\left(1 - \frac{2GM}{rc^2}\right)(dx^0)^2 + \left(1 - \frac{2GM}{rc^2}\right)^{-1}(dr)^2 + r^2(d\theta)^2 + r^2\sin^2\theta(d\phi)^2$$

in terms of cartesian coordinates $x^3 = r\cos\theta, x^2 = r\sin\theta\sin\phi, x^1 = r\sin\theta\cos\phi$ (and here $rdr = x^i dx^i$ is useful) we obtain

$$h_{00} = \frac{2GM}{rc^2}, \qquad h_{0i} = 0, \qquad h_{ij} = \frac{2GM}{rc^2}\frac{x^i x^j}{r^2}.$$

This allows us to evaluate

$$(H(\overline{h}, \eta)^{0\alpha 0j})_{,\alpha} = -\frac{4GM x^j}{c^2 r^3}$$

which gives P^0 when evaluated over a spherical surface at large distance $(n^j = x^j/r)$

$$E = -\frac{c^4}{16\pi G}\int_S n^j (H(\overline{h}, \eta)^{0\alpha 0j})_{,\alpha}$$

$$= Mc^2.$$

Exercise 70. GAUGE INVARIANCE OF CURVATURE TENSORS FOR WEAK FIELDS
Show that $R^\mu{}_{\nu\sigma\tau}$ (and therefore $R_{\mu\nu}$, R and $G_{\mu\nu}$) for weak fields do not change under an infinitesimal diffeomorphism.

Answer 70. We know that $h_{\mu\nu}$ changes as

$$h_{\mu\nu} \rightarrow h_{\mu\nu} + \xi_{\mu,\nu} + \xi_{\nu,\mu}$$

so that

$$\Gamma^\mu_{\nu\sigma} \rightarrow \Gamma^\mu_{\nu\sigma} + (\xi^\mu)_{,\nu\sigma}$$

and this gives the invariance of $R^\mu{}_{\nu\sigma\tau}$. (The $\Gamma\Gamma$ terms do not contribute.)

Exercise 71. Work out the details of the quadrupole radiation formula.

Answer 71. OUTLINE OF SOLUTION

$$Q_{ij}^{TT} = (P_{ik}P_{jl} - \frac{1}{2}P_{ij}P_{kl})Q_{kl}.$$

Substituting $P_{ij} = \delta_{ij} - n_i n_j$ and simplifying

$$Q_{ij}^{TT} = Q_{ij} - (nQ)_i n_j - (nQ)_j n_i + \frac{1}{2}(nQn)(\delta_{ij} + n_i n_j)$$

where $(nQ)_i = n_j Q_{ji}$ and $(nQn) = n_i Q_{ij} n_j$. Next we calculate the sum $Q_{ij}^{TT}Q_{ij}^{TT}$ which similarly simplifies to

$$Q_{ij}^{TT}Q_{ij}^{TT} = Q_{ij}Q_{ij} - 2(nQ)_i(nQ)_i + \frac{1}{2}(nQn)(nQn).$$

These can now be integrated using the integrals

$$\int n_i n_j d\Omega = \frac{4\pi}{3}\delta_{ij},$$

$$\int n_i n_j n_k n_l d\Omega = \frac{4\pi}{15}[\delta_{ij}\delta_{kl} + \delta_{ik}\delta_{jl} + \delta_{il}\delta_{jk}],$$

where the integration is over $d\Omega = \sin\theta d\theta d\phi$ and the unit vector $\mathbf{n}$ can be expressed in terms of angles θ, ϕ. The direct calculation of these integrals is not required. We know that the first integral is going to be a constant second-rank tensor. The only such tensor is δ_{ij}. therefore $\int n_i n_j d\Omega$ is a constant times δ_{ij}. The constant is determined by contracting the indices (that is taking trace of both sides).

Similarly the second integral is a fourth-rank tensor symmetric under exchange of any two indices. From δ_{ij} it can only be made as the given combination. Again, the constant in front can be determined by contracting a pair of indices.

Chapter 14

Schwarzschild and Kerr Solutions

For the purpose of simplifying formulas, we use a unit of time so that t stands for ct and unit of mass so that $2M$ stands for $2GM/c^2$ in this chapter except for section 14.4 where c and G appear as usual.

We have already discussed in Chapter 4 the motion of bodies (like planets) or light moving in the Schwarzschild field. The Schwarzschild solution holds the same importance in relativistic theory of gravitation as the Coulomb field of a point charge (or spherically symmetric charge distribution) holds for electrostatics. What is important about this solution is its uniqueness: any spherically symmetric mass distribution confined within a finite radius will produce a gravitational field outside that mass distribution given by the Schwarzschild metric. The spacetime is static where there is no matter. Even if the mass distribution changes with time in any arbitrary way, so long as it keeps spherical symmetry, the spacetime is static outside the distribution. This is the content of Birkhoff's theorem.

A rigorous proof of the Birkhoff theorem cannot be given in this elementary book. The argument can be summarised as follows. By a spherically symmetric spacetime one means a spacetime on which every element of the rotation group $SO(3)$ acts as an isometry, that is, a diffeomorphism which preserves the metric or inner product. If we start with any point in this spacetime and keep applying all possible group diffeomorphisms, we get an "orbit" of $SO(3)$ which is a surface. The definition of spherical symmetry further requires that these orbits should be 2-dimensional space-like surfaces. It is then shown that coordinates can be chosen so that the metric can be brought to the form

$$(ds)^2 = -A(t,r)(dt)^2 + B(t,r)(dr)^2 + C(t,r)[(d\theta)^2 + \sin^2\theta(d\phi)^2].$$

The substitution of these metric components in Einstein equations (for matter with perfect fluid tensor) then gives a general form of the Schwarzschild metric

in those regions where there is no matter. The proof can be found for example in
Appendix B of Hawking and Ellis, *Large scale structure of spacetime*.

We discuss the geometry of the Schwarzschild spacetime when the mass M
is contained in a small region near $r = 0$.

14.1   The Schwarzschild Solution

The coordinate t in the Schwarzschild solution is the time shown by a clock sitting
with fixed coordinates in the asymptotic region $(r \to \infty)$.

We know that near the gravitating body the clock will run slow. We may
expect that the clock will run slower and slower for small r and stop running
altogether at $r = 0$. But that is not true. The clocks stop running much before
that.

The infinitesimal proper time shown by a clock nailed to the coordinate
system (that is for r, θ, ϕ fixed) is

$$ d\tau = \left(1 - \frac{2M}{r} \right)^{1/2} (dt) $$

where dt is the coordinate time shown by the clock at infinity. The interval $d\tau$ goes
to zero at the Schwarzschild radius $r = 2M$. Two light signals sent from a source
on this fixed clock near $r = 2M$ with an interval of one second between them
(according to this clock) may be received by a stationary observer near $r \to \infty$
with a gap of many years between them.

We can see that light cones begin to shrink (as seen in these coordinates)
near the surface $r = 2M$. Their opening angles become smaller and smaller as
$r \to 2M$ from the $r > 2M$ side. For light we must have $ds = 0$ which implies that
radial light velocity in Schwarzschild coordinates, $\pm dr/dt = (1 - 2M/r)$ goes to
zero as one gets closer and closer to the surface. It is remarkable that the light
velocity becomes smaller in both directions: radially out as well as radially in!

At $r = 2M$ light stays put at $r = 2M$ and the surface contains light rays
running along the time axis t. This means the tangent vectors $\partial/\partial t$ are not time-
like but become null. This is obvious from the expression for g_{00}. On the inside of
the surface the tangent vector $\partial/\partial t$ becomes space-like because of the change in
sign of g_{00}, while $\partial/\partial r$ becomes the time-like vector. A light signal starting from a
point inside, $r < 2M$, can only move towards the 'future' which is $r = 0$. It cannot
come out. For this reason a spherical mass distribution which is entirely contained
within its Schwarzschild radius is called a **Schwarzschild black hole**.

The fact that light (or material bodies) falling towards the gravitating body
(which is assumed to be so compact that it is contained inside an $r = 2M$ surface)
take an infinite amount of Schwarzschild coordinate time t does not mean that
they never reach $r = 0$. A clock carried by a falling observer who begins to fall
from $r = R$ sees the clock ticking normally along its trajectory and reaching

$r = 0$ in finite proper time $R^{3/2}\pi/\sqrt{8M}$. (See Tutorial at the end of the chapter.) This just means that t is not an appropriate coordinate to use on and inside $r = 2M$. Although g_{11} and g^{00} become singular at $r = 2M$ there is no singularity in any physically relevant quantity, for example, scalar quantities (whose values are independent of coordinates) constructed out of curvature tensors: $R = g^{\mu\nu} R_{\mu\nu}$ or $R_{\mu\nu\sigma\tau} R^{\mu\nu\sigma\tau}$. But these quantities do become infinity at $r = 0$. That happens to be a real singularity.

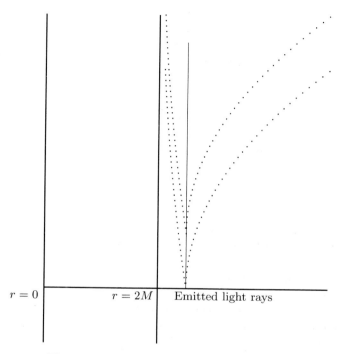

Fig. 14.1: Light rays originating near $r = 2M$.

The surface $r = 2M$ is a three-dimensional surface in four-dimensional space. Its normal vector $\mathbf{n} = (dr)^\sharp$ is null $\langle \mathbf{n}, \mathbf{n} \rangle = \langle dr, dr \rangle = g^{11} = 0$. Any light ray starting from a point on the surface stays on the surface (or wraps around the surface if the direction is not radial). It doesn't fall in or get out. The surface is generated by such null geodesics and is called a **null surface**.

In order to explore this surface we must use coordinates which are not singular on it. The surface $r = 2M$ is also an **event horizon** because it is like a one-way membrane, objects can go in from outside ($r > 2M$) to inside ($r < 2M$) but not the other way round.

We now explore the nature of the coordinate problem or coordinate singularity at $r = 2M$ by first studying a simpler case.

The mathematically minded student may wonder why there should be any coordinate singularity at all. After all, aren't we supposed to **define** our coordinate charts first before we study other properties of the manifold? Why did we admit a problematical coordinate in the first place?

What we actually do in practice is that we assume a certain coordinate system on a manifold and a certain form of the metric and solve the Einstein equation. In the Schwarzschild case we did precisely that and obtained an exact solution. The solution happens to hold for all values of r except that it has metric components $g_{\mu\nu}$ or $g^{\mu\nu}$ being undefined for $r = 2M$ and $r = 0$. The spacetime for $r > 2M$ which we can observe and which has been experimentally tested should be matched (or 'continued' or 'extended') with the spacetime for $r < 2M$ if possible. In doing this we find the nature of the singularities. The process is akin to extending the domain of analyticity of an analytic function of a complex variable.

14.2 Kruskal-Szekeres Coordinates

14.2.1 Rindler Wedge as an Example

We cannot discuss the rigorous mathematical definition of extendibility in this elementary book. Roughly speaking, a spacetime is called extendible if there is a larger spacetime of which it can be considered a subset and the metric of the larger space matches smoothly with the metric of the given space.

As an example consider the two-dimensional space called the **Rindler wedge**. It is a space with coordinates $(t, x); -\infty < t < +\infty, 0 < x < +\infty$ and line element

$$ds^2 = -x^2 \, dt^2 + dx^2.$$

So, to begin with, the space is defined only on the half (x, t) plane as shown.

In order to explore the nature of this spacetime, we look for the light rays, or null geodesics along which $ds = 0$. In two dimensions there is the advantage that the null geodesics provide a system of coordinate mesh just as the coordinates $u = t - x, v = t + x$ do for the Minkowski space.

The null geodesics require $ds = 0$ which means $dx/dt = \pm x$ or $t = \pm \ln x +$ const. The path of an incoming light ray, which corresponds to a negative sign, is shown in the diagram.

Along the path of light rays (or null geodesics), u and v, defined by

$$u \equiv t - \ln x, \qquad v \equiv t + \ln x$$

remain constant along outgoing and incoming rays respectively and characterise a particular geodesic.

Using u, v as coordinates with ranges $-\infty < u, v < +\infty$ we find

$$ds^2 = -e^{v-u} dudv.$$

We can easily absorb the expoentials and redefine

$$U = -e^{-u}, \qquad \in (-\infty, 0),$$

$$V = e^{v}, \qquad \in (0, \infty).$$

Since U is a function of u only and V of v, U, V also remain constant along the null geodesics. The incoming light ray shown in the (x, t) diagram is now a straight line parallel to the U-axis.

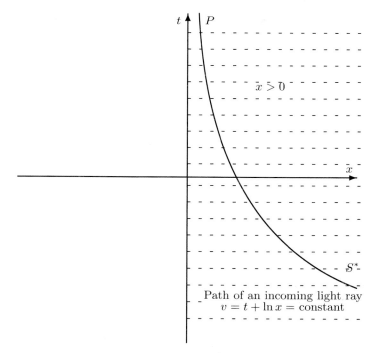

Fig. 14.2: Incoming light in $ds^2 = -x^2 \, dt^2 + dx^2$.

In these coordinates the metric becomes the *Minkowski space metric* $ds^2 = -dU \, dV$! It can be made more explicit by choosing $U = T - X$ and $V = T + X$ to give

$$ds^2 = -dT^2 + dX^2.$$

But note the range of coordinates $U : (-\infty, 0), V : (0, \infty)$. The original spacetime with coordinates $(t, x), -\infty < t < +\infty, 0 < x < +\infty$ (the shaded half plane in Figure 14.2) is now mapped to one quarter wedge shaped region of the Minkowski space in Figure 14.3.

The question naturally arises: what happens to the light ray that was coming in? Should it stop at $U = 0$ or should it keep going and enter the region which corresponds to negative values of x?

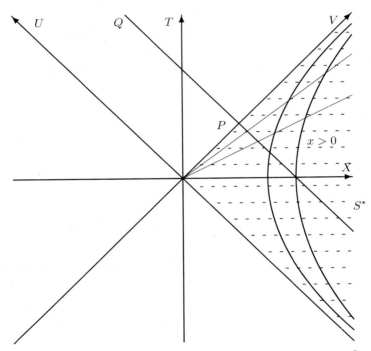

Fig. 14.3: The original (x, t) coordinates are related to these as $x^2 = -UV$ and $\exp(2t) = -V/U$. Lines of constant t are straight lines with $t = -\infty$ corresponding to the $-U$ half-axis and $t = +\infty$ corresponding to the V half-axis. Lines of constant x are rectangular hyperbolas. The line $x = 0$ becomes the two half-axes. The incoming light signal travels along the straight line $V =$ constant. The path of light ray S^*P of Figure 14.2 continues in this extended spacetime along S^*PQ.

It is difficult to see what happens in the original diagram (Figure 14.2) because the light signal reaches $x = 0$ only asymptotically for $t \to \infty$. But it is clear here. There is no reason to stop trajectories of light or material particles abruptly. One should extend the spacetime by joining portions which naturally allow such continuation. This is the essence of 'extendibility' of spacetime.

So, if we allow U, V to run over all possible values, we recover the full Minkowski space which cannot be extended any further. The original half-plane is **maximally extended**.

The original coordinate patch with x between zero and infinity could not be extended to negative values of x because $g^{00} = -1/x^2$ becomes singular at

$x = 0$. Usually if any of the $g_{\mu\nu}$'s or $g^{\mu\nu}$'s become infinity at some point, then it may mean that the coordinate system has a problem which can be cured by choosing a proper set of coordinates and extending the spacetime if necessary. Such a situation is called a **coordinate singularity**. At a coordinate singularity the $g_{\mu\nu}$ or $g^{\mu\nu}$ may become singular but scalars constructed from the curvature tensor $R, R_{\mu\nu\sigma\tau}R^{\mu\nu\sigma\tau}$ etc. are finite. If these scalars also become infinity at a point then we have a genuine singularity. We shall see next that the singularity at $r = 0$ of the Schwarzschild spacetime is a genuine singularity whereas the singularity at $r = 2M$ is a coordinate singularity.

14.2.2 Kruskal-Szekeres Coordinates

The Schwarzschild spacetime is spherically symmetric. Therefore we choose fixed values for θ, ϕ and concentrate on the t, r plane.

The metric can be written with the simplified notation and dropping the θ, ϕ part simply as

$$ds^2 \;\; = \;\; -\left(1 - \frac{2M}{r}\right) dt^2 + \left(1 - \frac{2M}{r}\right)^{-1} dr^2.$$

This metric clearly has a problem at $r = 2M$. We must determine the nature of the singularity using the techniques we learnt in the last section for the Rindler wedge. We shall notice the similarity of the region $r > 2M$ and the region $x > 0$ of the Rindler wedge.

We treat the ranges $0 < r < 2M$ and $r > 2M$ separately.

14.2.3 Region I: $r > 2M$

The null geodesics are given by $ds = 0$ which implies

$$\frac{dr}{dt} = \pm\left(1 - \frac{2M}{r}\right)$$

or

$$dr_I^* \equiv \frac{dr}{(1 - 2M/r)} = \pm dt.$$

This is easily integrated to

$$t = \pm r_I^* + \text{const.}$$

where

$$r_I^* = r + 2M \ln\left(\frac{r}{2M} - 1\right).$$

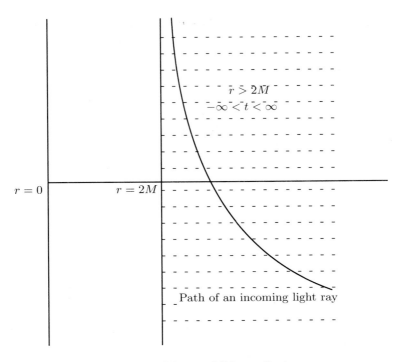

Fig. 14.4: Schwarzschild coordinates.

An incoming light ray has $v_I \equiv t + r_I^* = $ constant, where $r_I^* = r + 2M \ln(r/2M - 1)$. Notice the remarkable similarity of this diagram with Figure 14.2.

As r ranges from $2M$ to ∞, r_I^* ranges from $-\infty$ to ∞. It is important to notice that there is a one-to-one mapping between r and r_I^* even if we are unable to write a closed formula for r in terms of r_I^*. The constants along the null geodesics are

$$u_I = t - r_I^*, \qquad v_I = t + r_I^*,$$

u_I, v_I take all values $-\infty < u_I, v_I < \infty$. With these

$$
\begin{aligned}
ds^2 &= -\left(1 - \frac{2M}{r}\right)(dt)^2 + \left(1 - \frac{2M}{r}\right)^{-1}(dr)^2 \\
&= -\left(1 - \frac{2M}{r}\right)[(dt)^2 - (dr_I^*)^2] \\
&= -\left(1 - \frac{2M}{r}\right)(dt - dr_I^*)(dt + dr_I^*) \\
&= -\frac{2M}{r}e^{-r/2M}e^{(v_I - u_I)/4M}du_I dv_I
\end{aligned}
$$

where r appearing in the formulas is supposed to have been expressed in terms of $r_I^* = (v_I - u_I)/2$. Looking at this expression we are prompted to use

$$U_I = -e^{-u_I/4M}, \qquad V_I = e^{v_I/4M}$$

with ranges $U_I : (-\infty, 0); V_I : (0, \infty)$ so that

$$ds^2 = -\frac{32M^3}{r} e^{-r/2M} dU_I dV_I.$$

Since r is in the range of $(2M, \infty)$ the factor in front of $dU_I dV_I$ is finite. There is no singularity at $r = 2M$. We can go a step further by defining

$$U_I = T - X, \ V_I = T + X$$

and see that the region $r > 2M$ is mapped into the region I in the figure which corresponds to the Rindler wedge region in the Minkowski diagram. Except for the overall r-dependent factor the metric is like the Minkowski metric. In particular, all light rays move along the $45°$ lines as in two-dimensional Minkowski space.
    To see where our original coordinates t, r are located, calculate

$$U_I V_I = -e^{(v_I - u_I)/4M} = -e^{r_I^*/2M} = -e^{r/2M} \left(\frac{r}{2M} - 1 \right).$$

This shows that curves of constant r are the hyperbolas $T^2 - X^2 < 0$. Points for which $r = 2M$ are the two half axes: the positive half axis for V_I and negative half for U_I. Curves of constant t are curves of constant $u_I + v_I$ or curves of constant T/X,

$$\frac{T}{X} = \frac{V_I + U_I}{V_I - U_I} = \frac{1 - e^{-t/2M}}{1 + e^{-t/2M}}.$$

This shows that in the T-X plane straight lines in region I passing through the origin and making angles between $\pm 45°$ with the X-axis are lines of constant t: $t = 0$ the X-axis itself, $t = -\infty$ the $-45°$ line and $t = \infty$ the $+45°$ line.

14.2.4   Region II: $r < 2M$

Now, take the region $r < 2M$. The null geodesic equation which requires $dr/dt = \pm(1 - 2M/r)$ can be integrated as

$$t = \pm r_{II}^* + \text{const.}$$

where

$$r_{II}^* = r + 2M \ln \left(1 - \frac{r}{2M} \right).$$

The difference from the previous case is that r_{II}^* is always negative. The region $r = (0, 2M)$ is mapped to $r_{II}^* = (0, -\infty)$. As before we define

$$u_{II} = t - r_{II}^*, \qquad v_{II} = t + r_{II}^*;$$

u_{II}, v_{II} take all values $-\infty < u_I, v_I < \infty$. This makes

$$ds^2 = \frac{2M}{r} e^{-r/2M} e^{(v_{II} - u_{II})/4M} du_{II} dv_{II}$$

where r is understood to have been expressed in terms of $r_{II}^* = (v_{II} - u_{II})/2$. This time we choose

$$U_{II} = e^{-u_{II}/4M}, \qquad V_{II} = e^{v_{II}/4M}$$

with ranges $U_{II} : (0, \infty); V_{II} : (0, \infty)$ so that

$$ds^2 = -\frac{32M^3}{r} e^{-r/2M} dU_{II} dV_{II}.$$

Now if we define

$$U_{II} = T - X, \ V_{II} = T + X$$

we see that the region $t, 0 < r < 2M$ is mapped into the region-II of the X-T plane. Constant r points lie on

$$T^2 - X^2 = U_{II} V_{II} = e^{-r/2M} \left(1 - \frac{r}{2M}\right) > 0$$

which are hyperbolas in region-II lying between the coordinate axes which correspond to $r = 2M$ and the hyperbola $T^2 - X^2 = 1$ *which is the singularity* $r = 0$.

We cannot extend beyond this $r = 0$ hyperbola because the metric becomes genuinely singular due to the $32M^3/r$ factor in the metric. One can check that it is not just a coordinate singularity but a genuine singularity by calculating curvature scalars.

Lines of constant t (along which r steadily decreases to zero) are **time-like** straight lines passing through the origin and lying in region-II. These lines end at the $r = 0$ hyperbola

$$\frac{T}{X} = \frac{V_{II} + U_{II}}{V_{II} - U_{II}} = \frac{1 + e^{-t/2M}}{1 - e^{-t/2M}} > 1.$$

The Killing vectors $\partial/\partial t$ which are tangent to hyperbolas $r = $ constant are space-like in this region.

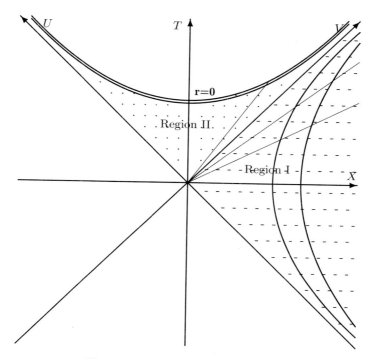

Fig. 14.5: Kruskal-Szekeres coordinates.
Lines of constant t are straight lines, with $t = -\infty$ corresponding to the $-U$ half-axis and $t = +\infty$ corresponding to the V half-axis. Lines of constant $r > 2M$ are rectangular hyperbolas in Region I with $r = 2M$ corresponding to the two half-axes. In Region II constant r lines are hyperbolas but spacetime cannot be extended beyond the singularity $r = 0$ shown by a double-lined hyperbola.

The coordinates of Region I and Region II join together smoothly and can be denoted by the same symbol:

$$U = U_I = U_{II}, \qquad V = V_I = V_{II}.$$

14.2.5 Horizon and the Black Hole

The extension of spacetime in the last section reveals that in the Region II corresponding to $r < 2M$ all future directed time-like or light-like curves end up at the singularity $r = 0$. If somehow all mass is compressed in a spherically symmetric way inside the Schwarzschild radius $r = 2M$, then not only would nothing come out of the surface, all matter inside would eventually collapse into the singularity $r = 0$ because that is the fate of all future-directed time- or light-like curves. The singularity is called the Schwarzschild black hole.

The surface $r = 2M$ consists of all those light rays which neither fall in or can come out. In fact the surface is generated by such null geodesics as the construction above shows.

The surface $r = 2M$ is an **event horizon** which means that events beyond the surface (that is those in the region II corresponding to $r < 2M$) cannot influence or communicate with events in region I. The surface of any future half of a light cone in the flat Minskowski space is an example of a horizon because the events inside the cone can only influence events inside it. What makes the horizon $r = 2M$ non-trivial is that it is a "trapped surface" of finite extension and area. The singularity at $r = 0$ is hidden or covered by the surface, the singularity is not a 'naked singularity'.

Because the event horizon is a null surface, the induced metric on it is degenerate. But there is no problem in defining the area of the $t =$constant subset which has positive definite space-like metric. For $r=$constant and $t =$ constant the metric of the surface is $ds^2 = r^2(d\theta^2 + \sin^2\theta d\phi^2)$ which gives $4\pi r^2$ as the area of such surfaces.

A remarkable feature of the singularity is its space-like nature. From the naive picture of the Schwarzschild coordinates one is tempted to think of the singularity $r = 0$ as a one-dimensional time-like straight line. But the Kruskal-Szekeres coordinates show that it is actually a space-like subset.

14.3   Extension of Schwarzschild Spacetime

It is natural to ask the question, what corresponds to the remaining half-plane corresponding to negative values of V?

The radial lines of constant t (from $-\infty$ to $+\infty$) move in Region I from negative U-axis to positive V-axis when t is a time-like coordinate. The coordinate t becomes space-like from here. The time coordinate continues as the decreasing variable r from $(t = \infty) \leftrightarrow (r = 2M)$ up to $r = 0$. The radial lines of constant t pick up the role of the space-like variable r which has been decreasing from ∞ to $r = 2M$ in Region I and now t runs from $+\infty$ to $-\infty$ when it reaches the positive U-axis.

The radial lines can be extended backwards in Regions III and IV for the same constant values. Similarly the hyperbolas of constant r have their mirror images in Region III and Region IV. The mirror singularity in Region IV is such that light coming from it can come to the physical Region I but no light or particle can go from Region I to Region IV. This hypothetical singularity is called a **white hole**.

Region III corresponds to a space-like coordinate r running from $2M$ to ∞ but that region is inaccessible to us in Region I. No time-like curve can start in Region I and move forward into Regions III or IV.

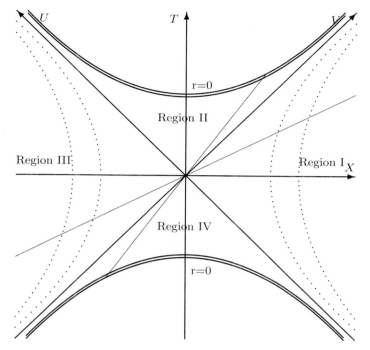

Fig. 14.6: Maximal Extension of Schwarzschild Spacetime.
The Regions III and IV are mathematically natural extensions of the
Schwarzschild spacetime. All geodesics (time-like, light-like or space-
like) can either be extended indefinitely on both sides or they end up
at the singularities (the **black hole** or the **white hole** in Regions II and
IV respectively).

One can wonder why there has to be a white-hole. Why can't we extend the
spacetime throughout Region IV? The reason is that there is no problem with the
differential manifold or coordinates. But the Riemannian metric has been extended
to these regions by continuation of the metric which has a factor of $1/r$. Thus to
use a continuously differentiable metric everywhere we need to have the symmetric
location of the singularity.

14.4 Spherical Mass Distribution: Interior Solution

The Schwarzschild solution represents a gravitational field outside a spherical mass
distribution. We now solve the Einstein equations in the special but unrealistic
case of static spherical matter distribution in the form of a perfect fluid. Even
this simple case is able to show how relativistic theory differs qualitatively from
Newtonian theory.

Let the metric be of the Schwarzschild form

$$ds^2 = -a(r)c^2(dt)^2 + b(r)(dr)^2 + r^2(d\theta)^2 + r^2\sin^2\theta(d\phi)^2 \tag{14.1}$$

where

$$a = a(r) = e^A, \qquad b = b(r) = e^B$$

are functions of r only.

The stress energy tensor of the fluid is

$$T^{\mu\nu} = pg^{\mu\nu} + (\rho + p/c^2)U^\mu U^\nu$$

where U^μ is the velocity field of matter flow and p and ρ are pressure and mass density respectively. p and ρ are both functions of r only. For a static equilibrium we expect U to have zero spatial components, therefore

$$U^\mu = (U^0 = c/\sqrt{a}, 0, 0, 0).$$

Notice the normalization to make U a velocity four-vector: $\langle U, U \rangle = -c^2$. This makes $T^{\mu\nu}$ "diagonal",

$$T^{00} = \rho c^2/a, \qquad T^{11} = p/b, \qquad T^{22} = p/r^2, \qquad T^{33} = p/(r^2\sin^2\theta)$$

and the covariant form of it is

$$T_{00} = \rho c^2 a, \qquad T_{11} = pb, \qquad T_{22} = pr^2, \qquad T_{33} = p(r^2\sin^2\theta).$$

The Einstein equations correspond to the four non-zero components of $G_{\mu\nu}$;

$$G_{00} = \frac{ab'}{b^2 r} + \frac{a}{r^2}\left(1 - \frac{1}{b}\right) = \frac{8\pi G}{c^4}\rho c^2 a, \tag{14.2}$$

$$G_{11} = \frac{a'}{ar} + \frac{1}{r^2}(1 - b) = \frac{8\pi G}{c^4}bp, \tag{14.3}$$

$$G_{22} = \frac{r^2}{2b}\left[2\left(\frac{a'}{a}\right)' + \left(\frac{a'}{a} + \frac{2}{r}\right)\left(\frac{a'}{a} - \frac{b'}{b}\right)\right] = \frac{8\pi G}{c^4}r^2 p. \tag{14.4}$$

We have not written the fourth equation for G_{33}. It is the same as for G_{22} except that both sides are multiplied by the same factor $\sin^2\theta$. The G_{00} equation can be written (after cancelling a on both sides) as

$$\frac{1}{r^2}\frac{d}{dr}\left[r\left(1 - \frac{1}{b}\right)\right] = \frac{8\pi G}{c^2}\rho.$$

If we define, in place of $\rho(r)$, a variable

$$m(r) = 4\pi\int_0^r \rho(r)r^2\,dr,$$

then our G_{00} equation can be solved for b,

$$r \left(1 - \frac{1}{b} \right) = \frac{2Gm}{rc^2} + \text{constant.}$$

The constant should be zero because as $r \to 0$, $m(r)$ should be zero. Thus

$$b = \left(1 - \frac{2Gm}{rc^2} \right). \tag{14.5}$$

If the mass distribution is up to radius R and after that the stress tensor is zero, the parameter m acquires a value

$$M = m(R) = 4\pi \int_0^R \rho(r) r^2 \, dr$$

and after that the metric has to be joined to the Schwarzschild metric with parameter M.

Note that M is not, as seems on first sight, the proper mass contained up to radius R as $4\pi r^2 dr$ is not the volume of the shell between r and $r + dr$. The 3-volume of the shell should include the "$\sqrt{g}$" factor and the mass contained is actually

$$M_{\text{proper}} = 4\pi \int_0^R \rho(r) r^2 [1 - 2Gm(r)/rc^2]^{-1/2} \, dr.$$

This factor is always positive and greater than 1 if the hypersurface spanned by coordinates r, θ, ϕ is purely space-like with signature $(+, +, +)$ in the static case we are considering. The interpretation of M is that it accounts for all energy, including the energy contributed by the gravitational field.

The G_{11} equation can now be solved for $A' = a'/a$ where $a = e^A$. After substituting the value of b we get

$$A' = 2G \frac{m + 4\pi r^3 p/c^2}{r(rc^2 - 2mG)}. \tag{14.6}$$

Next, instead of solving the equation for G_{22} we look at $T^{\mu\nu}{}_{;\nu} = 0$ which is actually a conseqence of the Bianchi identities in contracted form, $G^{\mu\nu}{}_{;\nu} = 0$. For $\mu = 1$ it is

$$T^{1\nu}{}_{;\nu} = T^{10}{}_{;0} + T^{11}{}_{;1} + T^{12}{}_{;2} + T^{13}{}_{;3}.$$

Note that even if $T^{10} = 0$ everywhere this does not mean $T^{10}{}_{;0}$ is zero! Actually $T^{10}{}_{;0} = T^{10}{}_{,0} + \Gamma^1_{00}T^{00} + \Gamma^1_{10}T^{11}$. Similarly for other terms. Collecting them together and using the Christoffel symbols of the Schwarzschild form of metric, we get

$$T^{1\nu}{}_{;\nu} = \frac{A'\rho c^2}{2b} + \frac{A'p}{2b} + \frac{p'}{b} = 0,$$

which gives, after substituting the expression for A', the **Oppenheimer-Volkoff equation**

$$p' = \frac{dp}{dr} = -G\frac{(\rho c^2 + p)(m + 4\pi r^3 p/c^2)}{r(rc^2 - 2mG)}. \tag{14.7}$$

We can rewrite this equation in a form which shows relativistic correction factors to the Newtonian formula $(dp = -Gm\rho dr/r^2)$ for equilibrium

$$\frac{dp}{dr} = -\frac{Gm(r)}{r^2}\rho\left[1 + \frac{p}{\rho c^2}\right]\left[1 + \frac{4\pi r^3 p}{m(r)c^2}\right]\left[1 - \frac{2m(r)G}{rc^2}\right]^{-1}.$$

This formula gives us a qualitative understanding of gravitational collapse. One can start at the centre $r = 0$ with some unknown value of pressure $p(0)$ and integrate outward up to the surface of the star, knowing the relation between density and pressure.

More realistic models of stars can be built up using the spherical symmetry. Astrophysics deals with this subject using the physical inputs about the constitution of matter, its equation of state, etc. The general picture seems to be this. Gravitation becomes very effective if the total mass of the gaseous matter which begins to collapse is very high. Gravity begins to pull matter inwards, increasing pressure and temperature, leading to nuclear processes such as conversion of hydrogen into helium with accompanying energy release. The equilibrium of thermal pressure and the gravitational pull may go on for billions of years as long as hydrogen keeps burning into helium. This is how a typical star like our Sun works.

But after the 'nuclear fuel has burnt out' (that is there are no longer nuclear processes available where by forming a heavier element the binding energy advantage can release energy as radiation), the star cools and contracts. If the 'electron degeneracy pressure' (due to the Pauli principle of electrons not liking to be too close to each other) is large enough to counter gravitational pull, it becomes a 'white dwarf', staying inactive like that. But if the mass of the star is much larger than a certain minimum it has no option but to collapse further under its own weight.

The **Chandrasekhar limit** of about one and a half solar masses gives an idea of how massive a star has to be to undergo unchecked gravitational collapse.

Thus a sufficiently massive star can go on collapsing further and violent events like supernova explosions can happen along the way. It is possible that even after a big fraction of matter is thrown off in an explosion, the remaining mass can form a 'neutron star' where practically all nuclei are touching each other and the matter density is that of nuclear matter. The matter is supported by the 'neutron degeneracy' pressure just as the electron degeneracy pressure supports a white dwarf. A neutron star of about one solar mass is a *very* compact object of about 10 kilometer size, with its Schwarzschild horizon of three kilometers not too deep inside it.

How do we know that in various violent astrophysical processes there are not mass lumps big enough so that even the neutron degeneracy pressure is unable to stop the collapse? We can conclude from known physics and from reasonable extrapolation that in such cases black holes are formed where all matter has vanished behind a horizon.

A black hole has strong gravitational fields. Matter in its neighbourhood is attracted to it. The swirling matter around a black hole in its exceptionally strong gravitational fields will produce all kinds of radiation as it falls into the hole. That is our chance of inferring a black hole. It is believed there are a large number of black holes of a few solar masses. In addition it is believed that practically every galaxy has at its center an 'active galactic nucleus' which, judging by the speeds of stars and gaseous matter around it, implies the existence of supermassive black holes of mass equal to millions of solar masses.

A Schwarzschild black hole has zero angular momentum and would be rather an exception. In general there will be rotating black holes which are described by the exact solution of the Einstein equation known as the Kerr solution.

14.5 The Kerr Solution

Realistic collapse is hardly ever spherically symmetric. Actual collapse involves rotation of collapsing mass about an axis. An axially symmetric, stationary, **exact** vacuum solution of Einstein's equations was discovered by R.P. Kerr in 1963 and is known by his name. This is also a black hole solution in the sense that it is asymptotically flat and there is a bounding surface or horizon into which matter or radiation can fall but never come out. The Kerr solution represents a "rotating black hole". For zero angular momentum the Kerr solution reduces to the Schwarzschild solution.

A truly remarkable feature of these solutions is their uniqueness. The Kerr solution depends on just two parameters, M and a related to the total mass and angular momentum of the matter contained in them. It can be shown that any asymptotically flat, stationary solution of vacuum Einstein equations which has a horizon and which is non-singular outside the horizon must be the Kerr solution for some definite values of M and a. It seems that when all collapsing matter has vanished behind the horizon and all gravitational waves radiated out, regardless of what is happening inside, the gravitational field outside is that given by the Kerr solution with fixed values of the two parameters. There are no distinguishing features of the black hole except its mass and angular momentum.

The Kerr solution is simple to look at but has a reputation of being complicated to derive. Even the verification that it is a solution is famously described as "cumbersome" by Landau and Lifshitz in their book, *Classical Theory of Fields*.

In this chapter we take the Kerr metric as given. A diligent and ambitious student can go directly to the relevant chapter of S.Chandrasekhar's book, *The Mathematical Theory of Black Holes* if she has learnt the techniques of calculating connection and curvature using Ricci coefficients as we discussed in Chapter 9. One must be patient and calculate slowly but accurately.

We first give the solution in its so-called **Kerr-Schild** form and go by steps to the other, more familiar **Boyer-Lindquist** form.

Let us consider a flat spacetime with coordinates x (they need not be Minkowski coordinates) and metric $\underline{\underline{\eta}}$ defined on it. Let $K = K^\mu \partial/\partial x^\mu$ be a null vector field, that is, null with respect to metric $\underline{\underline{\eta}}$,

$$\langle K, K \rangle_\eta = \eta_{\mu\nu} K^\mu K^\nu = 0.$$

Let $\kappa = K^\flat = \kappa_\mu dx^\mu$ be the associated covariant field

$$\kappa_\mu = \eta_{\mu\nu} K^\nu.$$

Let us define a new metric (on the same manifold)

$$\underline{\underline{g}} \equiv \underline{\underline{\eta}} + \kappa \otimes \kappa = (\eta_{\mu\nu} + \kappa_\mu \kappa_\nu) dx^\mu dx^\nu.$$

It is easy to see that the new metric is non-degenerate (with $g^{\mu\nu} = \eta^{\mu\nu} - K^\mu K^\nu$ as inverse) and K *remains a null vector field with respect to this new metric*:

$$\langle K, K \rangle_g = g_{\mu\nu} K^\mu K^\nu = 0.$$

This gives us a method to construct new metrics from a flat metric by choosing an appropriate null vector field.

An important result in general relativity is that *if the Ricci tensor $R_{\mu\nu}$ is equal to zero* (that is vacuum Einstein equations are satisfied) *then the integral curves of K form a geodesic congruence.* The proof is not difficult and can be found in Chandrasekhar's book.

The Kerr solution belongs to this special form of metric.

14.5.1   The Kerr-Schild Form

The Kerr-Schild form of metric is given in coordinates t^*, x, y, z as

$$ds^2 = -(dt^*)^2 + (dx)^2 + (dy)^2 + (dz)^2$$
$$+ \frac{2Mr^3}{r^4 + a^2 z^2} \left[\frac{r(xdx + ydy) + a(xdy - ydx)}{r^2 + a^2} + \frac{zdz}{r} - dt^* \right]^2 \quad (14.8)$$

where M and a are constants (related to total mass and angular momentum) and the variable r is implicitly defined by

$$\frac{x^2 + y^2}{r^2 + a^2} + \frac{z^2}{r^2} = 1.$$

The null covariant vector field is defined by

$$\kappa = \sqrt{\frac{2Mr^3}{r^4 + a^2 z^2}} \left[\frac{r(xdx + ydy) + a(xdy - ydx)}{r^2 + a^2} + \frac{zdz}{r} - dt^* \right]$$

and to check that it is indeed a null field, one can ignore the square-root in front. Then using the Minkowski metric, $\langle \kappa, \kappa \rangle$ is (up to the factor)

$$-1 + \frac{(xr - ay)^2}{(r^2 + a^2)^2} + \frac{(yr + ax)^2}{(r^2 + a^2)^2} + \frac{z^2}{r^2}$$

which vanishes by the definition of r.

14.5.2   Remarks on Kerr-Schild Metric

There are some obvious inferences from this form:

Asymptotic Flatness

The metric is asymptotically flat and reduces to the flat Minkowski metric for large values of x, y, z. This is so because, solving the quadratic equation defining r^2,

$$2r^2 = (x^2 + y^2 + z^2 - a^2) + \sqrt{(x^2 + y^2 + z^2 - a^2)^2 + 4a^2 z^2},$$

therefore for large $|x|, |y|, |z| >> |a|$, $r^2 \sim x^2 + y^2 + z^2$. The factor appearing in front of κ goes to zero.

$a = 0$ Corresponds to Schwarzschild

For $a = 0$ the solution reduces to the Schwarzschild case because then $r^2 = x^2 + y^2 + z^2$ and $xdx + ydy + zdz = rdr$, so

$$ds^2 = -(dt^*)^2 + dr^2 + r^2 d\theta^2 + r^2 \sin^2 \theta d\phi^2 + \frac{2M}{r}[dr - dt^*]^2.$$

Choosing $v = t^* - r$ and eliminating t^* in favour of v we get the form

$$ds^2 = -\left(1 - \frac{2M}{r}\right) dv^2 - 2dvdr + r^2 d\theta^2 + r^2 \sin^2 \theta d\phi^2.$$

This is the same as the Schwarzschild metric in the so-called ingoing **Eddington-Finkelstein** coordinates. If we convert the v coordinate into

$$\begin{aligned} v &= t - r - 2M \ln\left(\frac{r}{2M} - 1\right), & \text{For } r > 2M \\ &= t - r - 2M \ln\left(1 - \frac{r}{2M}\right), & \text{For } r < 2M \end{aligned}$$

we get back the standard Schwarzschild form.

Disc Singularity

The metric components are everywhere well defined except where the denominator $r^2 + a^2 z^2$ of the factor in κ vanishes. That obviously happens when $z = 0$ and $r = 0$. From the defining relation for r above it is obvious that surfaces of constant r are ellipsoids of revolution. This means that the surface $r =$ constant cuts a plane through the symmetry axis (the z-axis) in an ellipse. For example the plane $x = 0$ cuts the surface of constant r in

$$\frac{y^2}{r^2 + a^2} + \frac{z^2}{r^2} = 1$$

which has eccentricity $a/\sqrt{r^2 + a^2}$ and the foci at $y = \pm a$ from the z-axis. Because the position of the foci is independent of the value of r, these ellipsoids have the same common foci on the ring corresponding to $z = 0$ and $x^2 + y^2 = a^2$. They are 'confocal' ellipsoids of revolution.

When $r \to 0$ the ellipsoids become smaller and smaller and flatter and flatter in the z-direction. Their eccentricity increases. At $r = 0$ the upper surface ($z > 0$ side) and lower surface of the ellipsoid join together to form a disc $x^2 + y^2 \leq a^2$

Thus the true singularity of the Kerr solution is in the form of a disc. This is the only singularity it has.

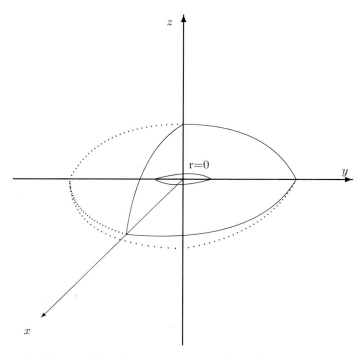

Fig. 14.7: Disc Singularity and surfaces of constant r.

14.5.3 Boyer-Lindquist Coordinates

We change the Kerr-Schild form of the metric into a physically more appealing form given by Boyer and Lindquist.

Step 1

As a first step, we notice the ellipsoidal nature of constant r surfaces

$$\frac{x^2 + y^2}{r^2 + a^2} + \frac{z^2}{r^2} = 1$$

and define new coordinates θ, ϕ_1 respecting this geometry:

$$
\begin{aligned}
x &= \sqrt{r^2 + a^2} \sin \theta \cos \phi_1, \\
y &= \sqrt{r^2 + a^2} \sin \theta \sin \phi_1, \\
z &= r \cos \theta.
\end{aligned}
$$

This gives immediately

$$
\begin{aligned}
dx^2 + dy^2 + dz^2 &= \frac{r^2 + a^2 \cos^2 \theta}{r^2 + a^2}(dr)^2 + (r^2 + a^2 \cos^2 \theta)(d\theta)^2 \\
&\quad + (r^2 + a^2) \sin^2 \theta (d\phi_1)^2,
\end{aligned}
$$

$$
\begin{aligned}
\frac{r(xdx + ydy) + a(xdy - ydx)}{r^2 + a^2} &= \frac{r^2 \sin^2 \theta}{r^2 + a^2} dr + r \sin \theta \cos \theta d\theta \\
&\quad + a \sin^2 \theta d\phi_1,
\end{aligned}
$$

$$
\frac{zdz}{r} = \cos^2 \theta dr - r \sin \theta \cos \theta d\theta.
$$

Therefore

$$
\begin{aligned}
ds^2 &= -(dt^*)^2 + \frac{r^2 + a^2 \cos^2 \theta}{r^2 + a^2}(dr)^2 + (r^2 + a^2 \cos^2 \theta)(d\theta)^2 \\
&\quad + (r^2 + a^2) \sin^2 \theta (d\phi_1)^2 \\
&\quad + \frac{2Mr}{r^2 + a^2 \cos^2 \theta}\left[-dt^* + \frac{r^2 + a^2 \cos^2 \theta}{r^2 + a^2} dr + a \sin^2 \theta d\phi_1 \right]^2.
\end{aligned}
$$

Step 2

Further simplification is achieved by defining

$$d\phi_1 = d\phi_2 + \frac{a}{r^2 + a^2} dr, \qquad \left(\text{or } \phi_1 = \phi_2 + \tan^{-1}(r/a) \right);$$

then

$$ds^2 \;=\; -(dt^*)^2 + (dr)^2 + (r^2 + a^2 \cos^2 \theta)(d\theta)^2 + (r^2 + a^2)\sin^2 \theta (d\phi_2)^2$$
$$+2a \sin^2 \theta (d\phi_2 dr) + \frac{2Mr}{r^2 + a^2 \cos^2 \theta}\left[-dt^* + dr + a \sin^2 \theta d\phi_2\right]^2 .$$

This by itself is quite a useful form and used often. However there are 'cross-terms' $drd\phi_2, drdt^*$ and $d\phi_2 dt^*$.

Step 3

To eliminate $drdt^*$ and $drd\phi_2$ we guess

$$dt^* = dt + A(r)dr, \qquad d\phi_2 = d\phi + B(r)dr$$

where A and B are functions of r, to be determined by the condition that $drdt$ and $drd\phi$ terms do not occur in the metric. Substituting and solving for A and B we get

$$A = -\frac{2Mr}{\Delta}, \qquad B = -\frac{a}{\Delta}, \qquad \Delta \equiv r^2 - 2Mr + a^2.$$

This brings the metric in the standard Boyer-Lindquist form,

$$ds^2 \;=\; -(dt)^2 + \frac{\rho^2}{\Delta}(dr)^2 + \rho^2 (d\theta)^2 + (r^2 + a^2)\sin^2 \theta (d\phi)^2$$
$$+\frac{2Mr}{\rho^2}(dt - a \sin^2 \theta \, d\phi)^2 \tag{14.9}$$

where the frequently occuring expression is abbreviated as

$$\rho^2 = r^2 + a^2 \cos^2 \theta.$$

Another form in which the metric is usually written is

$$ds^2 \;=\; -\left[1 - \frac{2Mr}{\rho^2}\right](dt)^2 + \frac{\rho^2}{\Delta}(dr)^2 + \rho^2 (d\theta)^2$$
$$+ \left[(r^2 + a^2) + \frac{2Mra^2}{\rho^2}\sin^2 \theta\right]\sin^2 \theta (d\phi)^2$$
$$-\frac{4Mar}{\rho^2}\sin^2 \theta \, d\phi dt. \tag{14.10}$$

In practice t^* of the Kerr-Schild form of metric is also written as t but we have kept a separate notation to emphasize that the two time coordinates (in the Kerr-Schild form and the Boyer-Lindquist form) are not the same.

14.5.4 Stationary and Axisymmetric Nature

None of the metric components depend on t or ϕ. This means that $T = \partial/\partial t$ and $L = \partial/\partial \phi$ are Killing vector fields. Motion of test bodies will conserve the values of $\langle T, U \rangle$ and $\langle L, U \rangle$ where U is the four-velocity of the particle.

14.5.5 Meaning of a

The Kerr solution is asymptotically flat. At large values of coordinate r the function $\rho^2 = r^2 + a^2 \approx r^2$ and the metric becomes as that of a weak field of slowly rotating mass discussed in Chapter 13. In particular the $d\phi dt$ term has coefficient $-4Ma\sin^2\theta/r$ which, when compared to $4JG(xdy - ydx)/c^3r^3$, gives us the physical meaning of parameter a. The total angular momentum J of the system described by the Kerr solution is equal to $Ma = J$ in the units of this chapter. (In ordinary units a has dimensions of length, M has dimensions of mass, then $Ma = J/c$.) In other words, a is the ratio of total angular momentum and total mass of matter causing the Kerr spacetime.

14.5.6 $g_{00} = 0$, Horizons and Ergosphere

We saw in the case of the Schwarzschild solution that g_{00} goes to zero at the surface $r = 2M$ leading to clocks fixed to the coordinate system stopping altogether. Similarly, in Kerr spacetime the proper time shown by a clock with fixed value of r, θ, ϕ stands still when r has the value given by $g_{00} = -1 + 2Mr/\rho^2 = 0$ or at

$$r = M \pm \sqrt{M^2 - a^2\cos^2\theta}.$$

In the Schwarzschild case the surface of infinitely slow clocks (the surface on which $g_{00} = 0$) coincided with the event horizon. But here, a particle or light can still escape to infinity from the surface where $g_{00} = 0$.

The event horizon, which is the null surface of finite extent on which light beams hover and neither go in nor away to infinity, is actually given by

$$\Delta = 0 \iff r_\pm = M \pm \sqrt{M^2 - a^2}.$$

The surfaces $r = r_\pm$ are event horizons. The surfaces of $g_{00} = 0$ coincide with $r_\pm$ along the z-axis, but for other values of θ there is a region between the two surfaces: the horizon $r_+ = M + \sqrt{M^2 - a^2}$ is inside the surface of infinitely slow clocks, $r = M + \sqrt{M^2 - a^2\cos^2\theta}$. The region between these surfaces from where a particle can escape to infinity is called the **ergosphere**.

The reason for that name is this. Take a particle falling from rest from some fixed value of r in the plane $\theta = \pi/2$. The particle will keep within this plane because of axial symmetry but will not follow a radial trajectory. Because of the 'rotating' nature of whatever mass distribution (or black hole) is creating the Kerr spacetime, it will drift along the direction of rotation and acquire angular velocity in the same direction as the mass distribution. But we can counter it by giving an initial condition (of sufficient opposite angular momentum) if it is outside the $g_{00} = 0$ surface so that it doesn't move in the angular direction of the Kerr spacetime.

But if a particle is in the ergosphere it is *forced* to move in the angular direction.

Even light is forced to sweep along in the direction of rotation of the Kerr spacetime.

For a light signal in the x-y plane ($\theta = \pi/2$) the condition $ds = 0$ gives

$$
\begin{aligned}
0 \;=\; & -\left[1 - \frac{2Mr}{\rho^2}\right](dt)^2 + \frac{\rho^2}{\Delta}(dr)^2 \\
& + \left[(r^2 + a^2) + \frac{2Mra^2}{\rho^2}\right](d\phi)^2 \\
& - \frac{4Mar}{\rho^2}d\phi dt.
\end{aligned}
$$

Dividing by $(dt)^2$ and evaluating on the $g_{00} = 0$ surface where $\theta = \pi/2, r = 2M, \rho^2 = 4M^2, \Delta = a^2$, we get

$$
\frac{4M^2}{a^2}\left(\frac{dr}{dt}\right)^2 + (4M^2 + 2a^2)\left(\frac{d\phi}{dt}\right)^2 - 2a\left(\frac{d\phi}{dt}\right) = 0.
$$

An immediate conclusion from this equation is that $d\phi/dt$ cannot be negative because in that case all three terms are strictly positive and the equation cannot be satisfied. Even for $d\phi/dt = 0$ we must have $dr/dt = 0$ so light doesn't move at all but stays put at the point with fixed r, ϕ and $\theta = \pi/2$. With $d\phi/dt > 0$, non-zero solutions for dr/dt are allowed if

$$
0 < (2M^2 + a^2)\left(\frac{d\phi}{dt}\right) < a.
$$

In the ergosphere all material particles as well as light have to rotate in the same sense as that of the mass distribution that is causing the gravitational field.

14.5.7 Penrose Process

The Killing vector field $T = \partial/\partial t$ is time-like in the outer region and space-like in the ergosphere.

The Killing vector field determines a conserved quantity $\langle T, U \rangle$ along the trajectory. If m is the rest mass of the particle falling along the geodesic, then $E = -m\langle T, U \rangle$ is the energy of the particle.

Since U remains a future-directed time-like vector, and T becomes space-like in the ergosphere, $\langle T, U \rangle$ being the inner product of a time-like and a space-like vector can be positive, negative or zero depending on the point and the direction.

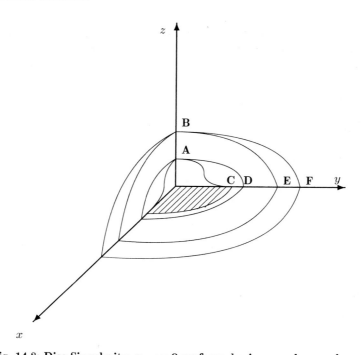

Fig. 14.8: Disc Singularity, $g_{00} = 0$ surfaces, horizons and ergosphere.

Only one actant is shown for clarity. Various surfaces are indicated by their intersection curves with the coordinate planes. The disc singularity is shown by hatched lines. The two $g_{00} = 0$ surfaces correspond to $AC(r = M - \sqrt{M^2 - a^2 \cos^2 \theta})$ and $BF(r = M + \sqrt{M^2 - a^2 \cos^2 \theta})$. Point C corresponds to $r = 0$ and touches the disc singularity. F corresponds to $r = 2M$. The two horizons are $AD(r = r_- = M - \sqrt{M^2 - a^2})$ and $BE(r = r_+ = M + \sqrt{M^2 - a^2})$.

The Penrose process is a hypothetical way of extracting energy from a rotating black hole. The idea is to throw a body with positive energy $E = -m\langle T, U \rangle > 0$ with suitably defined direction so that it falls into the Ergosphere and arrange so that it breaks into two pieces, with rest masses m_1 and m_2. The 'explosion' of the particle is to be so timed and managed that the particle with mass m_1 follows a geodesic with a starting four-velocity U_1 corresponding to negative $E_1 = -m_1\langle T, U_1 \rangle < 0$ and the other particle (rest mass m_2 with positive energy $E_2 = -m_2\langle T, U_2 \rangle > 0$ and direction U_2 so that it eventually gets out of the ergosphere. (Actually if particle 2 has to come out, particle 1 must fall into the black hole.) At the point of breaking, the four-momenta of the three particles obey $mU = m_1 U_1 + m_2 U_2$ therefore $E = E_1 + E_2$. Thus the particle 2 which returns to the outside region has energy $E_2 = E - E_1 = E + |E_1| > E$.

The particle which falls into the black hole has negative energy, and one can show that it must have negative angular momentum as well. So the total energy and the angular momentum of the black hole is reduced. The Penrose process allows extraction of energy from a rotating black hole by slowing down its spinning. This sets a limit on the total energy that can be so extracted because if the black hole stops spinning altogether the separation of the two surfaces which enclose the ergosphere will vanish too.

14.6   Tutorial

Exercise 72. Start with the flat spacetime metric

$$-dt^2 + dr^2 + r^2 d\theta^2 + r^2 \sin^2 \theta d\phi$$

for which $dr + dt$ is a null one-form field. Add $f(r)(dr + dt)^2$ to get the new metric

$$ds^2 = -dt^2 + dr^2 + r^2 d\theta^2 + r^2 \sin^2 \theta d\phi + f(r)(dr + dt)^2.$$

Find the Ricci tensor for this metric and show that the Ricci tensor is zero only for f of the form $f = (\text{constant})/r$ which gives the Schwarzschild metric. What are the Riemann tensor components for $f(r) = \exp(-r/a)$?

Answer 72. Steps:
Choose coordinates $v = r + t, r, \theta, \phi$ and an o.n. "coframe". One choice is

$$\alpha^0 = F(r)dv - \frac{1}{F(r)}, \qquad \alpha^1 = \frac{1}{F(r)}dr, \qquad \alpha^2 = rd\theta, \qquad \alpha^3 = r \sin \theta$$

where $F = \sqrt{1 - f}$. Inversely,

$$dv = \frac{1}{F}(\alpha^0 + \alpha^1), \qquad dr - F\alpha^1, \qquad d\theta = \frac{1}{r}\alpha^2, \qquad d\phi = \frac{1}{r \sin \theta}\alpha^3.$$

The Ricci rotation coefficients can be calculated as for example in

$$d\alpha^0 = F' dr \wedge dv = -F' \alpha^0 \wedge \alpha^1$$

where $F' = dF/dr$. The non-zero ones are

$$F^0{}_{01} = -F', \qquad F^2{}_{12} = \frac{F}{r} = F^3{}_{13}, \qquad F^3{}_{23} = \frac{\cot \theta}{r}.$$

These determine $f_{abc} = \eta_{ad}F^d{}_{bc}$ which in turn determine ω_{abc}. The connection matrix is (non-zero) components

$$\omega^0{}_1 = F'\alpha^0 = \omega^1{}_0, \qquad \omega^1{}_2 = -\frac{F}{r}\alpha^2 = -\omega^2{}_1,$$

$$\omega^1{}_3 = -\frac{F}{r}\alpha^3 = -\omega^3{}_1, \qquad \omega^2{}_3 = -\frac{\cot \theta}{r}\alpha^3 = -\omega^3{}_2.$$

The Riemann tensor is calculated from $\Omega = d\omega + \omega \wedge \omega$. The non-zero components are

$$R^0{}_{101} = -FF'' - (F')^2,$$

$$R^0{}_{202} = -\frac{FF'}{r} = R^0{}_{303} = R^1{}_{212} = R^1{}_{313},$$

$$R^2{}_{323} = \frac{1 - F^2}{r^2}.$$

The Ricci tensor is

$$\begin{aligned} R_{00} &= FF'' + (F')^2 + 2FF'/r = -f''/2 - f'/r = -R_{11}, \\ R_{22} &= -2FF'/r + (1 - F^2)/r^2 = f'/r + f/r^2 = R_{33}, \end{aligned}$$

and the scalar curvature is

$$R = f'' + 2\left(\frac{f}{r}\right)'.$$

Exercise 73. Find the force required (using a rocket engine for example) to keep a physical body stationary near $r = 2M$ in Schwarzschild spacetime.

Answer 73. The gravitational pull towards the centre is to be counter-balanced. The trajectory $r, \theta, \phi = $ constant is an accelerated trajectory. The velocity four-vector is

$$U = U^\mu \frac{\partial}{\partial x^\mu} = \frac{dx^\mu}{d\tau} \frac{\partial}{\partial x^\mu} = U^0 \frac{\partial}{\partial t};$$

from $\langle U, U \rangle = -1$ it follows that $U^0 = (1 - 2m/r)^{-1/2}$. The acceleration four-vector

$$A = D_U U = U^\nu \left(U^\mu{}_{,\nu} + \Gamma^\mu{}_{\nu\alpha} U^\alpha \right) \frac{\partial}{\partial x^\mu}$$

is orthogonal to U and is space-like:

$$A = (U^0)^2 \Gamma^1_{00} \frac{\partial}{\partial r} = \frac{m}{r^2} \frac{\partial}{\partial r}.$$

In order to find the force (on the unit mass body) that will have to be applied to keep the body at fixed values of r, θ, ϕ, one must express this acceleration in terms of basis with physical distances (or an orthonormal basis). For a 'coordinate distance' dr the physical distance is $ds = \sqrt{g_{11}} dr = (1 - 2m/r)^{-1/2} dr$. Therefore

$$A = \frac{m}{r^2} \left(1 - \frac{2m}{r}\right)^{-1/2} \frac{\partial}{\partial s}.$$

This shows that the force tends to infinity as $r \to 2m$.

Exercise 74. Find the equation of a radially falling dust particle in Schwarzschild space-time which begins to falls from rest from a point at $r = \infty$ and hits the singularity $r = 0$ at its proper time τ_0.

Answer 74.

$$r^{3/2} = \frac{3\sqrt{2M}}{2}(\tau_0 - \tau).$$

For radial fall $d\theta$ and $d\phi$ are zero. Denoting differentiation by τ the proper time by a dot, the metric

$$\begin{aligned} ds^2 &= -d\tau^2 = -f(dt)^2 + f^{-1}(dr)^2, \\ -1 &= -f\dot{t}^2 + f^{-1}\dot{r}^2, \qquad f = 1 - 2M/r. \end{aligned}$$

And the conserved quantity along the geodesic is the constant

$$c = g_{0\nu} \frac{dx^\nu}{d\tau} = -f\dot{t}.$$

Eliminating $\dot{t}$ and solving the equation for $\dot{r}$ gives the answer.

Exercise 75. Find the proper time τ elapsed by the watch of an observer who falls radially in a Schwarzschild field starting from rest at $r = R > 2M$ until she hits the singularity at $r = 0$.

Answer 75. Let $f(r) \equiv 1 - 2M/r$. The conserved quantity $C = -f(r)dt/d\tau$ can be related to $\dot{r} \equiv dr/d\tau$ as follows:

$$(d\tau)^2 = f(r)(dt)^2 - (f(r))^{-1}(dr)^2$$

relates $dt/d\tau$ to $\dot{r}$. Therefore

$$C = -\sqrt{f(r) + \dot{r}^2} = -\sqrt{f(R)}.$$

This gives

$$\dot{r}^2 = f(R) - f(r) = 2M\left(\frac{1}{r} - \frac{1}{R}\right)$$

which can be integrated to give

$$\tau = \frac{R^{3/2}\pi}{\sqrt{8M}}.$$

Exercise 76. Put $M = 0$ in the Boyer-Lindquist form of the Kerr metric. What can you say about the resulting metric?

Answer 76. It is flat Minkowski space.

Penrose Diagrams

Exercise 77. Construct a Penrose diagram for the Schwarzschild spacetime starting from the Kruskal-Szekeres coordinates.

Answer 77. For Regions I and II (Figure 14.5) map the V and U coordinates by the inverse tangent functions to finite range $\tan^{-1} V \in (0, \pi/2)$ and $\tan^{-1} U \in (-\pi/2, \pi/2)$. These regions then look like a part of the Minkowski space Penrose diagram (Figure 14.9, left). Note that the singularity $r = 0$ which corresponds to $UV = 1$ is mapped to the horizontal double line $\tan^{-1} U + \tan^{-1} V = \pi/2$ because

$$\tan\left(\tan^{-1} U + \tan^{-1} V\right) = \frac{U + V}{1 - UV} \to \infty.$$

If we include the maximally extended coordinates as well we get the diagram as shown in the second figure below.

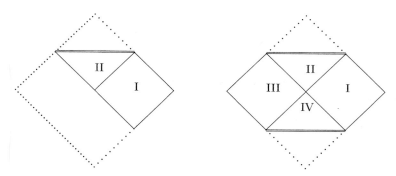

Fig. 14.9: The Penrose diagram for the Schwarzschild spacetime showing Regions I, II, III and IV of the Kruskal-Szekeres coordinates. The double line corresponds to the singularity.

Chapter 15

Cosmology

An important application of the general theory of relativity is to the universe itself. The science of the universe is called cosmology. Cosmology is a vast subject requiring a thorough knowledge of many branches of physics. We give only the barest outline of how the general theory of relativity defines the picture of our universe.

15.1 The Universe

Our universe is made up of gravitationally clustered matter in galaxies, and unclustered energy including radiation. Galaxies have been surveyed thoroughly through the optical, radio and other telescopes. They are estimated to be several billion in the observable part of the universe.

A typical galaxy, like our home galaxy Milky Way, is like a flattish disc containing about 10^{11} stars. Our Sun is just one of these stars. A typical galaxy is about several kilo parsecs in size. (A parsec is an astronomical unit of distance equal to 3.26 light years). The typical mass of a galaxy is of the order of 10^{12} solar masses. Our nearest neighbour galaxy is Andromeda about 750 kpc away.

There are two very important experimental facts about the distribution of all forms of energy and matter.

Galaxies are uniformly distributed. Typical distances between galaxies range from 50 kpc to 1 Mpc (mega parsec). These local inhomogenities smoothen out if we see the universe on a very large scale. At a scale of 1000 Mpc the universe will look like a uniform density cloud of dust, each galaxy a particle of dust.

Even though we are forced to make all our observations from our own galaxy, (in fact from our own solar system) all observations seem to suggest that there is no particular feature or mark to distinguish one place of the universe from another, nor any particular direction more preferable than the other.

The assumption that the universe looks the same from anywhere and in all directions is called the **cosmological principle** of homogeneity and isotropy of its (three-dimensional) space.

But the universe is changing along the time axis.

The galaxies are moving away from each other. If we measure distances of nearby galaxies from our galaxy, it seems that all these distances increase in time by the same proportionality factor. So if a galaxy is twice as far as compared to another, then it recedes twice as fast. The linear relationship between the speed with which a galaxy recedes away (as measured by redshift of spectral lines) and its distance, is called the **Hubble law**. We must remember that light from galaxies very far away (the farthest observed is about 3 billion light years away) takes much time to travel to us and its redshift represents the velocities of those earlier times. This has to be taken into account. Light from distant objects gives us information about how the universe was some time ago and that helps us guess how it has been changing.

The experimental validity of Hubble's law means that the galaxies have very small *peculiar velocities*, that is velocities over and above the receding velocity governed by Hubble's law.

Homogeneous and Isotropic Universe

The picture that emerges is as follows. Imagine a uniform density fluid whose particles (each a galaxy) have fixed unchanging coordinates. But the physical distances between these particles keep increasing. At any given time the physical distance is obtained by multiplying the fixed 'coordinate distance' between the particles by a positive factor $a(t)$ called the **scale factor**. The picture often given in popular books on cosmology is that of a baloon in which air is being blown in. As the balloon expands the points on its surface recede from each other. One should notice that there is no particular point on the baloon which is special: all points move away from each other without any one of them being the "center". It is a very useful and basically correct picture in two dimensions. Another picture (in three dimensions) is a cake being baked in an oven. As the cake bakes it swells in all directions, the raisins and nuts embedded in it move away from each other.

The universe changes with time, but at a fixed time it looks spatially isotropic. We further assume that it would look isotropic from all points as well. Then the universe has to be *maximally symmetric*, as far as three-dimensional space is concerned. We can choose coordinates so that *on the large scale* the metric is of the form

$$ds^2 = -c^2 dt^2 + a(t)^2 (d\sigma)^2 \tag{15.1}$$

where the spatial metric $(d\sigma)^2$ is both homogeneous and isotropic. We have discussed maximally symmetric metrics in Chapter 10. The high degree of symmetry completely determines the three-dimensional space determined by a constant k.

There are just three choices corresponding to positive $(k = 1)$, negative $(k = -1)$ or zero $(k = 0)$ curvature:

$$(d\sigma)^2 = \frac{(dr)^2}{1 - kr^2} + r^2(d\theta)^2 + r^2\sin^2\theta(d\phi)^2, \qquad k = 0, \pm 1. \qquad (15.2)$$

Our starting point therefore is the metric in coordinates $x^\mu = (x^0 = ct, r, \theta, \phi)$ with

$$g_{\mu\nu} = \begin{pmatrix} -1 & 0 & 0 & 0 \\ 0 & a^2/(1 - kr^2) & 0 & 0 \\ 0 & 0 & a^2 r^2 & 0 \\ 0 & 0 & 0 & a^2 r^2 \sin^2\theta \end{pmatrix}. \qquad (15.3)$$

This metric is called the **Friedman-Robertson-Walker** or FRW metric.

We have written the scale factor $a(t)$ as a function of time t rather than the coordinate $x^0 = ct$. This means that when we write $\dot{a}$ it means da/dt and not da/dx^0. Also, we take r, θ, ϕ to be dimensionless. This is particularly convenient as this means k is dimensionless. The scale factor $a(t)$ has dimensions of length. The time coordinate t which divides or ("foliates") the spacetime into surfaces of maximal spatial homogeneity is called the **cosmic time** coordinate. If we neglect our own peculiar velocity then the time shown by our clocks 'runs at the same rate' as the cosmic time.

Hubble Parameter

Since space is homogeneous, we can choose the origin of coordinates in our own galaxy. A galaxy has the physical distance $s(t) = a(t)|d(r, \theta, \phi)|$ from the origin where $|d(r, \theta, \phi)|$ is the coordinate distance (as measured by $d\sigma^2$). $|d(r, \theta, \phi)|$ remains constant as the universe evolves. Therefore

$$\frac{ds}{dt} = \dot{a}(t)d \equiv H(t)s(t)$$

where

$$H(t) = \frac{\dot{a}}{a}. \qquad (15.4)$$

$H(t)$ is called the **Hubble parameter**. This equation states that the speed of receding galaxies is proportional to the separation between them. The present value of H denoted by H_0 is determined by the velocities of neighbouring galaxies. Red shift of much farther galaxies gives information of H at earlier times, which is quite useful.

As a ratio of velocity and distance the Hubble parameter has dimensions of inverse time. The present value of the Hubble parameter is about $H_0 = 72$ (Km/sec)/Mpc in practical units, which corresponds to $1/H_0 = 10^6 \times 3.26 \times 3 \times 10^8 (\text{m/sec}) \times \text{year}/72\text{Km} \sim 14$ billion years.

Spatial homogeneity simplifies a lot. The distribution of matter content of the universe, that is $T_{\mu\nu}$, is only a function of t not r, θ and ϕ. If we know $T_{\mu\nu}$ at any time t, the Einstein equation can be used to find the geometrical quanties k and the scale factor function $a(t)$.

What can we say about $T_{\mu\nu}(t)$?

The universe not only has non-relativistic matter in the form of visible galaxies, but it has electromagnetic radiation in the form of background radiation permeating the whole universe, as well as 'dark matter' and 'dark energy'.

In fact, the fraction of energy contribution of visible galaxies and gas clouds is estimated to be about four to five percent in the present universe. Radiation is about 10^{-4} fraction-wise. About 25% contribution comes from dark matter whose existence is not in doubt because its gravitational effects can be seen on visible matter like stars and gas clouds. The dark matter is supposed to be clustered or clumped around galaxies but we know practically nothing about the nature of dark matter. And lastly the universe is supposed to be everywhere swathed in 'dark energy' (with negative pressure) which is about 70%. The nature of dark energy is also not understood. One can replace dark energy by assuming that there is a cosmological constant instead. But we don't know why the cosmological constant has to have a fantastically small value.

It is a strange feeling that we are floating in a fluid 95% of whose content we don't even know and the only sure knowledge we have is about the 5% portion of matter and radiation as shown by experiments in particle physics.

15.2   Friedman Equations

Assuming the distribution of matter and radiation to be that given by the stress tensor of a perfect fluid, and using the homogeneity of space, we can write

$$T_{\mu\nu} = \begin{pmatrix} \rho c^2 & 0 \\ 0 & g_{ij} p \end{pmatrix} \tag{15.5}$$

where the energy density ρc^2 and pressure p will depend on time but not on space coordinates because of homogeneity and isotropy.

The Einstein equations for this metric are called **Friedman equations**. The Einstein tensor is evaluated for this metric in the tutorials at the end of the chapter.

For the zero-zero component the Einstein equation is

$$-G_{00} = -\frac{3}{c^2}\left(\frac{\dot{a}^2}{a^2} + \frac{kc^2}{a^2}\right) = -\frac{8\pi G}{c^4}\rho c^2,$$

or

$$\frac{\dot{a}^2}{a^2} + \frac{kc^2}{a^2} = \frac{8\pi G\rho}{3}. \tag{15.6}$$

The other three equations corresponding to G_{11}, G_{22}, G_{33} all give the same equation

$$\frac{2\ddot{a}}{a} + \left(\frac{\dot{a}^2}{a^2} + \frac{kc^2}{a^2}\right) = -\frac{8\pi G p}{c^2}. \tag{15.7}$$

We can simplify this second equation by substituting the part within the parentheses from the first equation and obtaining

$$\frac{\ddot{a}}{a} = -\frac{4\pi G}{3c^2}(3p + \rho c^2). \tag{15.8}$$

Other Forms of Equations

There is a still simpler and physically appealing form of the second equation which eliminates both $\ddot{a}$ and k. Multiply the first equation by a^2 and differentiate with respect to time t. The constant k goes away upon differentiation and we get

$$2\dot{a}\ddot{a} = \frac{8\pi G}{3}\frac{d(\rho a^2)}{dt}.$$

Substitute for $\ddot{a}$ from the last equation of the previous section. The result can be written as the "first law" (of thermodynamics $pdV = -dE$);

$$-p\frac{da^3}{dt} = \frac{d(\rho c^2 a^3)}{dt}. \tag{15.9}$$

This is just the law of conservation of energy, the work done by pressure in any change of volume equals loss of energy inside: $pd(a^3) = -d(\rho c^2 a^3)$. This is not unexpected because the Einstein equations imply the local conservation law $T^{\mu\nu}{}_{;\nu} = 0$ as a consequence of the contracted Bianchi identities.

We can rewrite this equation also as

$$d(\rho c^2) = -3(\rho c^2 + p)d(\ln a). \tag{15.10}$$

Simple Consequences of Friedman Equations

One of the obvious consequences of the equation $\ddot{a}/a = -4\pi G(3p + \rho c^2)/(3c^2)$ is that (unless we can find matter with $3p + \rho c^2 = 0$) *the universe cannot be static* because a=const corresponds to both $\dot{a} = 0$ as well as $\ddot{a} = 0$! As $3p + \rho c^2 > 0$ for all familiar form of matter, $\ddot{a}(t) < 0$. We know by observation that the universe is expanding, that is $\dot{a} > 0$ today. It would have been an even larger positive number in the past because $\ddot{a}$ is negative. This means as we go back in time a keeps decreasing. Even if $\dot{a}$ were constant and retained its present value $\dot{a}_0 = a_0 H_0$ — (there is a fairly standard convention of using a subscript 0 for all present-day values of cosmological quantities) — the scale factor a will reduce to zero if we go sufficiently back in time by an amount $T = 1/H_0$:

$$a(t) = a_0 + (t - t_0)(a_0 H_0).$$

So T is a rough estimate for the age of the universe. The Hubble parameter presently has value $H_0 \approx 72$ (km/sec)/Mpc which gives the value $1/H_0 \approx 14$ billion years.

15.3   Cosmological Constant

The Hubble law of an expanding universe was established in 1929. When Einstein first applied his equation to the Universe, the conception of the Universe was that of a static distribution of galaxies. We saw in the last section that unless $3p + \rho c^2 = 0$, the second time derivative of the scale parameter $\ddot{a} \neq 0$. So the universe cannot be static; it will either expand or contract. In order to cure this 'defect' Einstein introduced an additional term in his equation:

$$R_{\mu\nu} - \frac{1}{2} g_{\mu\nu} R + \Lambda g_{\mu\nu} = \frac{8\pi G}{c^4} T_{\mu\nu}. \tag{15.11}$$

Here Λ has dimensions of inverse length squared.

Einstein is supposed to have regretted introducing Λ later on and the 'Einstein static universe' is no longer relevant today, but the 'cosmological constant' has been brought up and discarded many times in the following years for different reasons. Finally, during the last few years, data seems to indicate that it is here to stay but with an exceedingly small positive value. Why this value should be so close to zero and still not quite zero is one of the open and challenging problems of cosmology today.

We can always accomodate Λ in $T_{\mu\nu}$ by writing

$$R_{\mu\nu} - \frac{1}{2} g_{\mu\nu} R = \frac{8\pi G}{c^4} \left(T_{\mu\nu} + T_{\mu\nu}^{\text{vac.}} \right) \tag{15.12}$$

where

$$T_{\mu\nu}^{\text{vac.}} = -\frac{c^4 \Lambda}{8\pi G} g_{\mu\nu} \tag{15.13}$$

$$= \begin{pmatrix} \rho_{\text{vac.}} c^2 & 0 \\ 0 & g_{ij} p_{\text{vac.}} \end{pmatrix} \tag{15.14}$$

where

$$\rho_{\text{vac.}} c^2 = \frac{c^4 \Lambda}{8\pi G}, \qquad p_{\text{vac.}} = -\rho_{\text{vac.}} c^2.$$

When we write the cosmological constant in this way as a stress-energy tensor of a fictitious perfect fluid, we notice the strange feature that the pressure and energy density have opposite signs. If Λ is positive, then energy density of the fictitious fluid is positive and the pressure is negative.

We include the cosmological constant in this way as part of the total stress-energy tensor. This additional contribution $T_{\mu\nu}^{\text{vac.}}$ is called "vacuum" contribution because the Einstein equations with a cosmological constant can be obtained by adding a term $-2\Lambda\sqrt{-g}$ to the Lagrangian $R\sqrt{-g}$. This amounts to adding a constant term to the Hamiltonian which shifts the zero of energy by a constant amount. Traditionally in field theories such a term is called the "vacuum" energy.

15.4 Models of the Universe

All cosmology revolves round determining the function $a(t)$. The equations which govern the scale factor $a(t)$ are the **Friedman equations**:

$$\frac{\dot{a}^2}{a^2} + \frac{kc^2}{a^2} = \frac{8\pi G\rho}{3}$$

$$\frac{\ddot{a}}{a} = -\frac{4\pi G}{3c^2}(3p + \rho c^2).$$

Where convenient we can also use

$$pd(a^3) = -d(\rho c^2 a^3)$$

which follows from the other two.

It is assumed that we know the "equation of state", the relation between energy density ρc^2 and pressure p of the supposed perfect fluid filling the universe. Then we have basically two equations for two unknown functions $a(t), \rho(t)$ and an unknown constant $k = 0, \pm 1$ which determines the type of universe we live in.

In practice one starts with a   model based on some assumptions about (1) k which can be 0 ("spatially flat universe") or $k = 1$ ("spatially closed") or $k = -1$ ("open") (2) the matter content and attempts to fit the available data. The matter content could be any of the constituents: radiation, non-relativistic matter, 'vacuum' energy or cosmological constant, dark matter, dark energy or whatever is required to fit the data. It is remarkable how much can be learnt by this process of model universes. For example, it is established beyond doubt in the last few years that we live in a universe that is spatially flat, that is $k = 0$.

Flat Matter-Dominated Universe

Matter means non-relativistic 'cold' matter for which the kinetic energy and pressure can be neglected compared to energy density due to rest masses. 'Dominated' means that other forms of energy are ignored. If $k = 0$ and $p = 0$, then $d(\rho c^2 a^3) = 0$ implies

$$\rho \propto \frac{1}{a^3}.$$

From Friedman's equation for $k = 0$ it follows that

$$\dot{a}^2 \propto \frac{1}{a}$$

which gives

$$a \propto t^{\frac{2}{3}}.$$

Flat Radiation-Dominated Universe

For radiation there is no mass density ρ but the stress-energy tensor has instead the energy density u in place of ρc^2. The relation between the energy density and pressure is given by $p = u/3$ for diffuse radiation in thermal equilibrium.

If $p = u/3$ then $d(ua^3) = -uda^3/3$ implies

$$u \propto \frac{1}{a^4}$$

and then the Friedman equation implies

$$\dot{a}^2 \propto \frac{1}{a^2}$$

which gives

$$a \propto \sqrt{t}.$$

If the uniform radiation is in thermal equilibrium at temperature T, then by the Stephan-Boltzmann law $u \propto T^4$. This implies that

$$T(t) \propto \frac{1}{a(t)} \propto \frac{1}{\sqrt{t}}.$$

Flat Λ Dominated Universe

The first thing to note about the cosmological constant or 'vacuum energy' density is that it remains constant unlike matter or radiation which decrease with expansion (as a^{-3} and a^{-4} respectively)

$$\rho_{\text{vac.}} = \frac{c^2 \Lambda}{8\pi G}.$$

therefore the flat $(k = 0)$ Friedman equations imply $\dot{a}^2 = a^2 \Lambda c^2/3$ which solves to

$$a(t) = C \exp(\sqrt{c^2\Lambda/3}\, t), \qquad \text{or} \qquad a(t) = C' \exp(-\sqrt{c^2\Lambda/3}\, t).$$

We can solve exactly for $a(t)$ for single constituents as seen above. Due to the non-linear nature of Friedman equations it is not possible to deduce the time dependence of $a(t)$ when several constituents are present, none of which can be singled out as 'dominating'. This is only part of the problems of cosmology.

15.5 History of the Universe

We have seen that the Hubble parameter $H(t)$ provides an estimate of the age of the universe. If the expansion rate $\dot{a}$ were constant the age would be about 14 billion years, and the second Friedman equation implies that $\ddot{a} < 0$ implies that the expansion rate was actually greater before than now.

What was the universe like in the beginning?

Today the fraction of radiation in the universe is about 10^{-5} but as radiation density goes, at a^{-4} it would have been dominant when a was small. It would also have been hotter because density goes as T^4. At those high temperatures all particles would have been in thermal equilibrium along with their antiparticles with kinetic energies much larger than their rest masses. A model of the early universe uses the physics of elementary particles to describe what must have been going on. As the universe expanded it cooled and various light nuclei were formed. At a still later time, neutral atoms were formed when the thermal energies were not large enough to ionize atoms. The radiation became decoupled from matter at this stage and the universe entered the matter dominated era. The radiation has ever been thinning out with expansion and becoming cooler so that today its temperature is $2.73°$ K. This 'cosmic background radiation' is the incredibly accurate Planck black body spectrum corresponding to this temperature with peak wavelengths near microwaves. That is why it is called cosmic microwave background radiation or CMBR.

CMBR is a very useful tool to test theories of the universe. The CMBR is not only astonishingly true Planck spectrum it is very uniform in all directions as well. The small departures from true isotropy as seen in recent observations are a very accurate and sensitive tool which put stringent bounds on various theories of the early universe.

Research in cosmology has been going on intensely in recent years.

15.6 Tutorial on the FRW Metric

For the spatially homogeneous and isotropic universe the metric can be chosen to be in the Friedman-Robertson-Walker form,

$$ds^2 = -(dx^0)^2 + a^2 \left[\frac{(dr)^2}{1 - kr^2} + r^2(d\theta)^2 + r^2\sin^2\theta(d\phi)^2 \right], \qquad (15.15)$$

where the scale factor $a = a(t)$ depends only on the cosmological time t. The scale factor a has the dimensions of length whereas coordinates r, θ, ϕ are dimensionless.

Connection Components

Exercise 78. Find connection components for the FRW metric.

Answer 78. Straightdorward calculations. In a matrix form

$$g_{\mu\nu} = \begin{pmatrix} -1 & 0 & 0 & 0 \\ 0 & a^2/(1-kr^2) & 0 & 0 \\ 0 & 0 & a^2 r^2 & 0 \\ 0 & 0 & 0 & a^2 r^2 \sin^2\theta \end{pmatrix}. \tag{15.16}$$

The connection components for this metric are (the dot on a represents the derivative with respect to t)

$$\Gamma^0_{\mu\nu} = \begin{pmatrix} 0 & 0 & 0 & 0 \\ 0 & a\dot{a}/[c(1-kr^2)] & 0 & 0 \\ 0 & 0 & a\dot{a}r^2/c & 0 \\ 0 & 0 & 0 & a\dot{a}r^2\sin^2\theta/c \end{pmatrix},$$

$$\Gamma^1_{\mu\nu} = \begin{pmatrix} 0 & \dot{a}/[ca] & 0 & 0 \\ \dot{a}/[ca] & kr/(1-kr^2) & 0 & 0 \\ 0 & 0 & -r(1-kr^2) & 0 \\ 0 & 0 & 0 & -r(1-kr^2)\sin^2\theta \end{pmatrix},$$

$$\Gamma^2_{\mu\nu} = \begin{pmatrix} 0 & 0 & \dot{a}/[ca] & 0 \\ 0 & 0 & 1/r & 0 \\ \dot{a}/[ca] & 1/r & 0 & 0 \\ 0 & 0 & 0 & -\sin\theta\cos\theta \end{pmatrix},$$

$$\Gamma^3_{\mu\nu} = \begin{pmatrix} 0 & 0 & 0 & \dot{a}/[ca] \\ 0 & 0 & 0 & 1/r \\ 0 & 0 & 0 & \cot\theta \\ \dot{a}/[ca] & 1/r & \cot\theta & 0 \end{pmatrix}. \tag{15.17}$$

Riemann, Ricci and Einstein Tensors

Exercise 79. Find the components of the Riemann tensor.

Answer 79.

$$R^0{}_{101} = R^0{}_{110} = \frac{a\ddot{a}}{c^2(1-kr^2)}$$

$$R^0{}_{202} = -R^0{}_{220} = a\ddot{a}r^2/c^2$$

$$R^0{}_{303} = -R^0{}_{330} = a\ddot{a}r^2\sin^2\theta/c^2$$

$$R^1{}_{001} = R^2{}_{002} = R^3{}_{003} = \frac{\ddot{a}}{c^2 a}$$

$$= -R^1{}_{010} = -R^2{}_{020} = -R^3{}_{030}$$

$$R^1{}_{212} = -R^1{}_{221} = (\dot{a}^2 + kc^2)r^2/c^2$$

$$R^1{}_{313} = -R^1{}_{331} = (\dot{a}^2 + kc^2)r^2\sin^2\theta/c^2$$

$$R^2{}_{112} = -R^2{}_{121} = -\frac{\dot{a}^2 + kc^2}{c^2(1-kr^2)}$$

$$R^2{}_{323} = -R^2{}_{332} = (\dot{a}^2 + kc^2)r^2\sin^2\theta/c^2$$

$$R^3{}_{113} = -R^3{}_{131} = -\frac{\dot{a}^2 + kc^2}{c^2(1-kr^2)}$$

$$R^3{}_{223} = -R^3{}_{232} = -(\dot{a}^2 + kc^2)r^2/c^2. \tag{15.18}$$

The non-zero Ricci tensor components $R_{\mu\nu} = R^{\sigma}{}_{\mu\sigma\nu}$ are just the diagonal ones:

$$
\begin{aligned}
R_{00} &= -3\frac{\ddot{a}}{c^2 a}, \\
R_{ij} &= g_{ij}\frac{a\ddot{a} + 2\dot{a}^2 + 2kc^2}{c^2 a^2}.
\end{aligned}
\tag{15.19}
$$

The curvature scalar $R = g^{\mu\nu} R_{\mu\nu}$ is

$$
-R = -\frac{6}{c^2}\left(\frac{\ddot{a}}{a} + \frac{\dot{a}^2}{a^2} + \frac{kc^2}{a^2}\right).
\tag{15.20}
$$

Finally the Einstein tensor $G_{\mu\nu} = R_{\mu\nu} - g_{\mu\nu} R/2$ is

$$
\begin{aligned}
G_{00} &= \frac{3}{c^2}\left(\frac{\dot{a}^2}{a^2} + \frac{kc^2}{a^2}\right), \\
G_{ij} &= -\frac{g_{ij}}{c^2}\left(\frac{2\ddot{a}}{a} + \frac{\dot{a}^2}{a^2} + \frac{kc^2}{a^2}\right).
\end{aligned}
\tag{15.21}
$$

Chapter 16

Special Topics

16.1   The Gauss Equation

We have discussed Gauss' Theorema Egregium and the Gauss and Codacci equations in Chapter 2 in their original formulation of a two-dimensional surface imbedded in the three-dimensional Euclidean space. The metric and the covariant derivative on the surface was induced from the derivative of the ambient Euclidean space. The great discovery was that the metric and covariant derivative referred only to the surface, that is, *intrinsic* quantities. This result led to the Riemannian geometry we know.

The result can be generalised. If we are given an imbedded submanifold or a hypersurface in a Riemannian manifold, the metric in the larger space induces a metric on the hypersurface. Moreover, we find that there is a well-defined connection or covariant derivative on the surface. This covariant derivative is related to the induced metric in the usual manner, i.e., the Christoffel symbols of the induced connection are constructed from the induced metric components in the standard way.

This generalization can be achieved exactly as in the original formulation, that is, by defining the tangential component of the covariant derivative (of the larger space) as the covariant derivative on the surface.

For definiteness, we discuss the case of a space-like hypersurface with a time-like normal, but the result holds more generally.

Let M be a Riemannian space of dimension n with one time-like and $n-1$ space-like vectors in an orthonormal basis. We shall denote our space-like hypersurface by S. This means that at any point $p \in S \subset M$ the tangent space T_p is a direct sum of two orthogonal subspaces: a one-dimensional space generated by a time-like unit vector N, and an $(n-1)$-dimensional tangential subspace of vectors tangent to curves lying in S and passing through p. All vectors tangent to S are orthogonal to N and any vector orthogonal to N must be tangential. As point p

varies on the surface, N becomes a unit normal vector field defined on S.

$$\langle N, N \rangle = -1.$$

We shall denote vector fields defined on points of S which are tangential to S by a tilde $\widetilde{X}, \widetilde{Y}$ etc. and call them **tangential fields**; it being understood that they are tangential to S. These fields are such that

$$\langle \widetilde{X}, N \rangle = 0.$$

On the tangent space of any point p of S the inner product $\langle \widetilde{X}, \widetilde{Y} \rangle$ of two tangential vectors $\widetilde{X}$ and $\widetilde{Y}$ determines the **induced** metric $\langle \ , \ \rangle_\sim$ on S:

$$\langle \widetilde{X}, \widetilde{Y} \rangle_\sim \equiv \langle \widetilde{X}, \widetilde{Y} \rangle. \tag{16.1}$$

The induced metric is also known as the **first fundamental form**.

Let the covariant derivative of M be denoted by D. In order to calculate the covariant derivative of a vector field Y along a certain direction of field X, all we need to know is the value of vectors Y along the curve whose tangents are X. Therefore the covariant derivatives of tangential vector fields $D_{\widetilde{X}} \, \widetilde{Y}$ are well defined. Similarly, $D_{\widetilde{X}} N$ is well defined too.

Although it is intuitively plausible, for a rigorous proof of these statements see section 16.3 later in this chapter. The connection on S should be defined for all tangential vector fields and functions on S. We need to extend these fields and functions on S into M so that concepts on M can be applied. One then has to make sure that quantities defined on the surface are independent of the extensions. In the derivation which follows we use the fact that the Lie bracket $[\widetilde{X}, \widetilde{Y}]$ of two tangential fields is tangential. This particular result is also plausible in view of the Tutorial exercise (two steps forward, two steps back) in Chapter 7.

As $\langle N, N \rangle = -1$ is constant on S,

$$0 = \widetilde{X}\left(\langle N, N \rangle \right) = \langle D_{\widetilde{X}} N, N \rangle + \langle N, D_{\widetilde{X}} N \rangle.$$

Also, for tangential fields $\widetilde{X}$ and $\widetilde{Y}$, differentiating $\langle \widetilde{X}, N \rangle = 0$ along $\widetilde{Y}$ we obtain

$$0 = \widetilde{Y}\left(\langle \, \widetilde{X}, N \rangle \right) = \langle D_{\widetilde{Y}} \, \widetilde{X}, N \rangle + \langle \widetilde{X}, D_{\widetilde{Y}} N \rangle.$$

Of course, the covariant derivative of a tangential field along the direction of another tangential field need not itself be tangential. The decomposition of this derivative in tangential and normal components and defining the tangential part as the new covariant derivative for the surface was precisely the idea of Gauss.

Given any vector X in the tangent space at a point $p \in S$ we can write its tangential part as

$$h(X) = X + \langle X, N \rangle N.$$

That this is indeed the tangential part is seen by taking the inner product: $\langle h(X), N \rangle = 0$ because $\langle N, N \rangle = -1$. Therefore the splitting of a vector is

$$X = h(X) - \langle X, N \rangle N.$$

The mapping $h : T_p \to T_p$ is a linear projection operator

$$h(h(X)) = h(X).$$

Let us write the tangential-normal split of the covariant derivative as (the **Gauss'** **equation**)

$$D_{\widetilde{X}}\widetilde{Y} = \widetilde{D}_{\widetilde{X}}\widetilde{Y} + K(\widetilde{X}, \widetilde{Y})N \tag{16.2}$$

where we use new symbols for the tangential and normal parts:

$$\widetilde{D}_{\widetilde{X}}\widetilde{Y} \equiv h(D_{\widetilde{X}}\widetilde{Y}), \tag{16.3}$$

$$K(\widetilde{X}, \widetilde{Y}) \equiv -\langle D_{\widetilde{X}}\widetilde{Y}, N \rangle = \langle D_{\widetilde{X}}N, \widetilde{Y} \rangle. \tag{16.4}$$

The following result is the modern version of Theorema Egregium:

Proposition: $\widetilde{D}$ is a metric connection (or covariant derivative) on S with respect to the induced metric. Moreover, K defines a covariant symmetric tensor on S. The second-rank tensor K is called the **second fundamental form** or **extrinsic curvature**.

Proof:

To prove that $\widetilde{D}$ is a metric connection we must show that

(1) it is a connection on the surface, i.e., it satisfies the four defining conditions

$$\widetilde{D}_{\widetilde{X}+\widetilde{Z}}\widetilde{Y} = \widetilde{D}_{\widetilde{X}}\widetilde{Y} + \widetilde{D}_{\widetilde{Z}}\widetilde{Y} \dots\dots\dots\dots(i),$$

$$\widetilde{D}_{\widetilde{X}}(\widetilde{Y} + \widetilde{Z}) = \widetilde{D}_{\widetilde{X}}\widetilde{Y} + \widetilde{D}_{\widetilde{X}}\widetilde{Z} \dots\dots\dots\dots(ii),$$

$$\widetilde{D}_{f\widetilde{X}}\widetilde{Y} = f\,\widetilde{D}_{\widetilde{X}}\widetilde{Y} \dots\dots\dots\dots\dots(iii),$$

$$\widetilde{D}_{\widetilde{X}}(f\,\widetilde{Y}) = f\,\widetilde{D}_{\widetilde{X}}\widetilde{Y} + \widetilde{X}(f)\,\widetilde{Y} \dots\dots\dots(iv)$$

(where f is a smooth function on S),

(2) the torsion of $\widetilde{D}$ is zero and

(3) $\widetilde{D}$ preserves the induced metric

$$\langle \widetilde{D}_{\widetilde{X}}\widetilde{Y}, \widetilde{Z} \rangle_\sim + \langle \widetilde{Y}, \widetilde{D}_{\widetilde{X}}\widetilde{Z} \rangle_\sim = \widetilde{X}(\langle \widetilde{Y}, \widetilde{Z} \rangle_\sim).$$

(On the right-hand side, $\widetilde{X}$ acts on the function $\langle \widetilde{Y}, \widetilde{Z} \rangle_\sim$).

Connection properties (i) to (iii) are evident:

$$D_{f\tilde{X}}(\tilde{Y} + \tilde{Z}) = f D_{\tilde{X}}\tilde{Y} + f D_{\tilde{X}}\tilde{Z}.$$

Gauss' equation for the left and right-hand side of this equation can be written

$$\tilde{D}_{f\tilde{X}}(\tilde{Y} + \tilde{Z}) + K(f\tilde{X}, \tilde{Y} + \tilde{Z})N = f\tilde{D}_{\tilde{X}}\tilde{Y} + f\tilde{D}_{\tilde{X}}\tilde{Z}$$
$$+ fK(\tilde{X},\tilde{Y})N + fK(\tilde{X},\tilde{Z})N$$

and the tangential and normal components compared: therefore,

$$\tilde{D}_{f\tilde{X}}(\tilde{Y} + \tilde{Z}) = f\tilde{D}_{\tilde{X}}\tilde{Y} + f\tilde{D}_{\tilde{X}}\tilde{Z}$$

and, in addition we get the linearity of K,

$$K(f\tilde{X}, \tilde{Y} + \tilde{Z}) = fK(\tilde{X}, \tilde{Y}) + fK(\tilde{X}, \tilde{Z}).$$

For property (iv) we note that

$$D_{\tilde{X}}(f\tilde{Y}) = f D_{\tilde{X}}\tilde{Y} + \tilde{X}(f)\tilde{Y}$$

and since the second term is tangential, it only adds to the tangential part

$$\tilde{D}_{\tilde{X}}(f\tilde{Y}) = f\tilde{D}_{\tilde{X}}\tilde{Y} + \tilde{X}(f)\tilde{Y}.$$

Thus $\tilde{D}$ satisfies the conditions of a connection.

To see (2) (i.e., the connection $\tilde{D}$ is free of torsion) we just use the fact that, the original connection D has zero torsion:

$$0 = D_{\tilde{X}}\tilde{Y} - D_{\tilde{Y}}\tilde{X} - [\tilde{X}, \tilde{Y}]$$
$$= (\tilde{D}_{\tilde{X}}\tilde{Y} - \tilde{D}_{\tilde{Y}}\tilde{X} - [\tilde{X}, \tilde{Y}]) + (K(\tilde{X}, \tilde{Y}) - K(\tilde{Y}, \tilde{X}))N$$

where we combine $[\tilde{X}, \tilde{Y}]$ with the tangential part because it is tangential. Equating the tangential and normal parts separately to zero we see that $\tilde{D}$ is torsion-free,

$$\tilde{D}_{\tilde{X}}\tilde{Y} - \tilde{D}_{\tilde{Y}}\tilde{X} - [\tilde{X}, \tilde{Y}] = 0.$$

Moreover as a bonus, we get the symmetry of K,

$$K(\tilde{X}, \tilde{Y}) = K(\tilde{Y}, \tilde{X}).$$

To show (3) (that $\tilde{D}$ is metric), we calculate the action of the vector field $\tilde{X}$ on the scalar function $\langle \tilde{Y}, \tilde{Z} \rangle_\sim = \langle \tilde{Y}, \tilde{Z} \rangle$:

$$
\begin{aligned}
\tilde{X}\left(\langle \tilde{Y}, \tilde{Z} \rangle_\sim\right) &= \langle D_{\tilde{X}} \tilde{Y}, \tilde{Z} \rangle + \langle \tilde{Y}, D_{\tilde{X}} \tilde{Z} \rangle \\
&= \langle \tilde{D}_{\tilde{X}} \tilde{Y}, \tilde{Z} \rangle + \langle \tilde{Y}, \tilde{D}_{\tilde{X}} \tilde{Z} \rangle \\
&\quad + \langle K(\tilde{X}, \tilde{Y})N, \tilde{Z} \rangle + \langle \tilde{Y}, K(\tilde{X}, \tilde{Z})N \rangle \\
&= \langle \tilde{D}_{\tilde{X}} \tilde{Y}, \tilde{Z} \rangle + \langle \tilde{Y}, \tilde{D}_{\tilde{X}} \tilde{Z} \rangle \\
&\quad + K(\tilde{X}, \tilde{Y})\langle N, \tilde{Z} \rangle + \langle \tilde{Y}, N \rangle K(\tilde{X}, \tilde{Z}).
\end{aligned}
$$

The last two terms are zero bacause N is orthogonal to tangential fields $\tilde{Y}$ and $\tilde{Z}$. We can replace the metric $\langle\ ,\ \rangle$ by $\langle\ ,\ \rangle_\sim$ in the remaining terms because all fields are tangential. Thus

$$
\tilde{X} \langle \tilde{Y}, \tilde{Z} \rangle_\sim = \langle \tilde{D}_{\tilde{X}} \tilde{Y}, \tilde{Z} \rangle_\sim + \langle \tilde{Y}, \tilde{D}_{\tilde{X}} \tilde{Z} \rangle_\sim.
$$

Moreover we have shown that K is symmetric and linear in its arguments, therefore it defines a covariant second-rank tensor.

This completes the proof of the proposition.

16.2　The Gauss and Codacci Equations

The relationship between the Riemann curvature tensor of the metric in M and the Riemann curvature of the induced metric on S can be easily worked out: Write

$$
\begin{aligned}
D_{\tilde{X}}(D_{\tilde{Y}} \tilde{Z}) &= D_{\tilde{X}}(\tilde{D}_{\tilde{Y}}\tilde{Z} + NK(\tilde{Y}, \tilde{Z})) \\
&= D_{\tilde{X}}(\tilde{D}_{\tilde{Y}}\tilde{Z}) + (D_{\tilde{X}}N)K(\tilde{Y}, \tilde{Z}) \\
&\quad + \tilde{X}\,(K(\tilde{Y}, \tilde{Z}))N \\
&= [\tilde{D}_{\tilde{X}}\,(\tilde{D}_{\tilde{Y}}\tilde{Z}) + (D_{\tilde{X}}N)K(\tilde{Y}, \tilde{Z})] \\
&\quad + N[K(\tilde{X}, \tilde{D}_{\tilde{Y}}\tilde{Z}) + \tilde{X}\,(K(\tilde{Y}, \tilde{Z}))]
\end{aligned}
$$

where use has been made of the properties of the covariant derivative. We write the expression for $D_{\tilde{Y}}(D_{\tilde{X}} \tilde{Z})$ similarly, and further,

$$
D_{[\tilde{X}, \tilde{Y}]} \tilde{Z} = \tilde{D}_{[\tilde{X}, \tilde{Y}]}\tilde{Z} + NK([\tilde{X}, \tilde{Y}], \tilde{Z}).
$$

Combining these to form the Riemann curvature tensor

$$
R(\tilde{X}, \tilde{Y}) \tilde{Z} = D_{\tilde{X}}(D_{\tilde{Y}} \tilde{Z}) - D_{\tilde{Y}}(D_{\tilde{X}} \tilde{Z}) - D_{[\tilde{X}, \tilde{Y}]} \tilde{Z},
$$

we see that the expression becomes

$$R(\tilde{X},\tilde{Y})\,\tilde{Z} = \left[\tilde{R}\,(\tilde{X},\tilde{Y})\,\tilde{Z} + (D_{\tilde{X}}N)K(\tilde{Y},\tilde{Z}) - (D_{\tilde{Y}}N)K(\tilde{X},\tilde{Z})\right]$$
$$+\quad N\left[K(\tilde{X},\tilde{D}_{\tilde{Y}}\tilde{Z}) - K(\tilde{Y},\tilde{D}_{\tilde{X}}\tilde{Z})\right.$$
$$+\quad \tilde{X}\,(K(\tilde{Y},\tilde{Z})) - \tilde{Y}\,(K(\tilde{X},\tilde{Z})) - K([\tilde{X},\tilde{Y}],\tilde{Z})\Big]$$

where we have separated the tangential and normal parts in square brackets.

Take the inner product with a tangential field $\tilde{W}$ to project the tangential part (recall the definition of covariant form of Riemann curvature tensor Chapter 9)

$$R(\tilde{W},\tilde{Z};\tilde{X},\tilde{Y}) = \tilde{R}\,(\tilde{W},\tilde{Z};\tilde{X},\tilde{Y})$$
$$+\quad (D_{\tilde{X}}N,\tilde{W})K(\tilde{Y},\tilde{Z}) - (D_{\tilde{Y}}N,\tilde{W})K(\tilde{X},\tilde{Z}).$$

Using the definition of K, e.g.,

$$\langle D_{\tilde{X}}N,\tilde{W}\rangle = -\langle N, D_{\tilde{X}}\tilde{W}\rangle = K(\tilde{X},\tilde{W}),$$

we get the tangential projection, **Gauss' formula**,

$$R(\tilde{W},\tilde{Z};\tilde{X},\tilde{Y}) = \tilde{R}\,(\tilde{W},\tilde{Z};\tilde{X},\tilde{Y}) + K(\tilde{X},\tilde{W})K(\tilde{Y},\tilde{Z})$$
$$-\quad K(\tilde{Y},\tilde{W})K(\tilde{X},\tilde{Z}). \qquad (16.5)$$

By taking the inner product with the normal N, we get the orthogonal projection,

$$\langle N, R(\tilde{X},\tilde{Y})\,\tilde{Z}\rangle = -\left[K(\tilde{X},\tilde{D}_{\tilde{Y}}\tilde{Z}) - K(\tilde{Y},\tilde{D}_{\tilde{X}}\tilde{Z})\right.$$
$$+\quad \tilde{X}\,(K(\tilde{Y},\tilde{Z})) - \tilde{Y}\,(K(\tilde{X},\tilde{Z})) - K([\tilde{X},\tilde{Y}],\tilde{Z})\Big]$$

as $\langle N, N\rangle = -1$. We can write $[\tilde{X},\tilde{Y}] = \tilde{D}_{\tilde{X}}\tilde{Y} - \tilde{D}_{\tilde{Y}}\tilde{X}$ for the torsion-free connection $\tilde{D}$ and combine these with the other terms to get the **Codacci formula**

$$\left\langle N, R(\tilde{X},\tilde{Y})\,\tilde{Z}\right\rangle = -\left[(D_{\tilde{X}}K)(\tilde{Y},\tilde{Z}) - (D_{\tilde{Y}}K)(\tilde{X},\tilde{Z})\right] \qquad (16.6)$$

where the covariant derivative $D_{\tilde{X}}K$ of the second-rank covariant tensor K is, as usual, the second-rank tensor whose value on a pair of fields $\tilde{Y},\tilde{Z}$ is

$$\left(D_{\tilde{X}}K\right)(\tilde{Y},\tilde{Z}) = \tilde{X}\,(K(\tilde{Y},\tilde{Z})) - K(D_{\tilde{X}}\tilde{Y},\tilde{Z}) - K(\tilde{Y},D_{\tilde{X}}\tilde{Z})$$
$$= \tilde{X}\,(K(\tilde{Y},\tilde{Z})) - K(\tilde{D}_{\tilde{X}}\tilde{Y},\tilde{Z}) - K(\tilde{Y},\tilde{D}_{\tilde{X}}\tilde{Z})$$

(where D can be replaced by $\tilde{D}$ in the last two terms because the inner products of tangential field $\tilde{Z}$ with normal components of $D_{\tilde{X}}\tilde{Y}$ and $D_{\tilde{Y}}\tilde{X}$ are zero).

Gauss Codacci equations are important in discussions of initial value and the Hamiltonian formulation of the general theory of relativity where time development is replaced by a family of space-like hypersurfaces.

16.3 Bases on M and S

16.3.1 Basis for Tangential Fields

We now write the Gauss and Codacci formulas in explicit bases on M and S.

We can regard the surface S to be a subset of the manifold M. The surface S has one time-like vector and $n - 1$ space-like vectors in an orthonormal basis of the tangent space of its points. Let $n - 1$ coordinates $u^a, a = 1, \ldots, n - 1$ be defined on S. Then the n coordinates $x^i, i = 1, \ldots, n$ of these points (as points of M) are smooth functions of u,

$$x^i = \phi^i(u).$$

These coordinate systems define coordinate basis vector fields $\{\partial_i = \partial/\partial x^i\}$ for all vectors and $\{\partial_a = \partial/\partial u^a\}$ for tangential vector fields.

We have the following definitions and formulas which are simple to prove. See below for explanation.

$$\partial_a = \phi^i_{,a}\partial_i, \qquad (\phi^i_{,a} = \partial\phi^i/\partial u^a) \tag{16.7}$$

$$\tilde{g}_{ab} \equiv \langle \partial_a, \partial_b \rangle \tag{16.8}$$

$$\tilde{g}_{ab} = g_{ij}\phi^i_{,a}\phi^j_{,b} \tag{16.9}$$

$$h(\mathbf{v}) \equiv \mathbf{v} + N\langle N, \mathbf{v} \rangle \tag{16.10}$$

$$h_{ij} \equiv \langle h(\partial_i), h(\partial_j) \rangle \tag{16.11}$$

$$h_{ij} = g_{ij} + N_i N_j \tag{16.12}$$

$$h_{ij}N^j = 0 \tag{16.13}$$

$$h^i{}_j \equiv \delta^i_j + N^i N_j \tag{16.14}$$

$$(h(\mathbf{v}))^i = h^i{}_j v^j \tag{16.15}$$

$$h(\partial_i) \equiv \psi^a_i \partial_a \tag{16.16}$$

$$\phi^i_{,a}g_{ij} = \tilde{g}_{ab}\psi^b_j \tag{16.17}$$

$$\phi^i_{,a} = \tilde{g}_{ab}\psi^b_j g^{ij} \tag{16.18}$$

$$\psi^a_i = \tilde{g}^{ab}\phi^j_{,b}g_{ij} \tag{16.19}$$

$$h_{ij} = \psi^a_i \psi^b_j \tilde{g}_{ab} \tag{16.20}$$

$$h_{ij} = \psi^a_i \phi^k_{,a}g_{kj} \tag{16.21}$$

$$\phi^i_{,a}\psi^b_i = \delta^b_a \tag{16.22}$$

$$\phi^i_{,a}\psi^a_j = h^i_j \tag{16.23}$$

Vectors tangent to the surface are given by linear combinations of the basis vectors $\{\partial_a\}$ which span the $n - 1$ dimensional subspace tangent to S

$$\partial_a \equiv \frac{\partial}{\partial u^a} = \frac{\partial\phi^i}{\partial u^a}\frac{\partial}{\partial x^i} \equiv \phi^i_{,a}\partial_i.$$

Components of the induced metric in the basis vectors $\{\partial_a\}$ is

$$
\begin{aligned}
\tilde{g}_{ab} &= \left\langle \frac{\partial}{\partial u^a}, \frac{\partial}{\partial u^b} \right\rangle . \\
&= g_{ij} \phi^i_{,a} \phi^j_{,b}.
\end{aligned}
$$

N is the unit normal vector field defined on the surface.

$$
\left\langle N, \frac{\partial}{\partial u^a} \right\rangle = 0, \text{for } a = 1, \ldots (n-1), \qquad \langle N, N \rangle = -1.
$$

Given any tangent vector $\mathbf{v} \in T_p, p \in S$ at a point on the surface, its resolution into its normal and tangential parts is

$$
h(\mathbf{v}) = \mathbf{v} + N\langle N, \mathbf{v} \rangle.
$$

The projected vectors have inner product given by a two-form

$$
\langle h(\mathbf{v}), h(\mathbf{u}) \rangle = \langle \mathbf{v}, \mathbf{u} \rangle + \langle \mathbf{v}, N \rangle \langle N, \mathbf{u} \rangle.
$$

If the normal N is written as

$$
N = N^j \frac{\partial}{\partial x^j} = N^j \partial_j,
$$

then h_{ij} in the basis $\{\partial_i\}$ is

$$
\begin{aligned}
h_{ij} &= \langle h(\partial_i), h(\partial_j) \rangle \\
&= g_{ij} + \langle \partial_i, N \rangle \langle N, \partial_j \rangle \\
&= g_{ij} + g_{ik} g_{jl} N^k N^l \\
&= g_{ij} + N_i N_j
\end{aligned}
$$

where we have written $N_i \equiv g_{ij} N^j$. Notice that h_{ij} is purely tangential in the sense that

$$
h_{ij} N^j = 0.
$$

The projection matrix which maps components of $\mathbf{v}$ into those of $h(v)$ can be found easily, the i-th compnent

$$
\begin{aligned}
(h(\mathbf{v}))^i &= (\mathbf{v} + N\langle N, \mathbf{V} \rangle)^i \\
&= (\delta^i_j + N^i N_j) v^j \\
&\equiv h^i_j v^j.
\end{aligned}
$$

When ∂_a is expanded in the basis $\{\partial_i\}$ the components $\phi^i_{,a}$ are "tangential", that is

$$
\langle \partial_a, N \rangle = \phi^i_{,a} g_{ij} N^j = 0, \qquad a = 1, \ldots, n-1.
$$

This gives

$$
\phi^i_{,a} g_{ij} = \phi^i_{,a} h_{ij}.
$$

As $h(\partial_i)$ is tangential it can be expanded in the basis $\{\partial_a\}$. Let

$$
h(\partial_i) = \psi^a_i \partial_a,
$$

then

$$\phi^i_{,a} g_{ij} = \langle \partial_a, \partial_j \rangle = \langle \partial_a, h(\partial_j) \rangle = \widetilde{g}_{ab}\, \psi^b_j.$$

Both g_{ij} and $\widetilde{g}_{ab}$ are non-singular matrices, with inverses g^{ij} and $\widetilde{g}^{ab}$; respectively, we can express $\phi^i_{,a}$ and ψ^a_i in terms of each other as

$$\phi^i_{,a} = \widetilde{g}_{ab}\, \psi^b_j g^{ij}, \qquad \psi^a_i = \widetilde{g}^{ab}\, \phi^j_{,b} g_{ij}.$$

Other useful relations are

$$h_{ij} = \psi^a_i \psi^b_j\, \widetilde{g}_{ab} = \psi^a_i \phi^k_{,a} g_{kj}, \qquad \psi^a_i \phi^k_{,a} = \delta^k_i + g_{ij} N^j N^k.$$

16.3.2   Extension of Vector Fields on S

A tangential vector field $\widetilde{X}$ which is only defined on points of the surface S can be extended into a neighbourhood of the surface as follows. Let us expand it in the basis $\{\partial_a\}$,

$$\widetilde{X} = \widetilde{X}^a (u) \partial_a = \widetilde{X}^a (u) \phi^i_{,a}(u) \partial_i.$$

Now choose any smooth functions $X^i(x)$ in a neighbourhood of S which coincide with $X^i(x(u)) = \widetilde{X}^a (u)\phi^i_{,a}(u)$ on S. This will give us one possible extension of a tangential field $\widetilde{X}$ (which is defined only on the surface). We call such a field an extension of $\widetilde{X}$ and write X for it.

Three Lemmas About Extensions

Lemma 1 If X is an extension of the tangential field $\widetilde{X}$, and f is a smooth function on M ($f|_S = f(x(u))$ is a smooth function on S) then

$$X(f)|_S = \widetilde{X} (f|_S).$$

It follows directly that

$$X(f)|_S = X^i(x)\partial_i(f)|_S = \widetilde{X}^a (u)\phi^i_{,a}(u)\partial_i(f)|_S = \widetilde{X}^a (u)\partial_a(f(x(u))) = \widetilde{X} (f|_S).$$

Lemma 2 If X and Y are extensions of tangential fields $\widetilde{X}$ and $\widetilde{Y}$ respectively, then $[X,Y]|_S$ is independent of extensions and equal to the Lie bracket $[\widetilde{X},\widetilde{Y}]$

calculated on the surface. We see this readily using $X^i|_S = \tilde{X}^a \phi^i_{,a}$:

$$
\begin{aligned}
[X, Y]|_S &= [X^i(\partial_i Y^j) - (X \leftrightarrow Y)]\partial_j \\
&= [\tilde{X}^a (\phi^i_{,a}\partial_i Y^j) - (X \leftrightarrow Y)]\partial_j \\
&= [\tilde{X}^a (\partial_a(\tilde{Y}^b \phi^j_{,b}) - (X \leftrightarrow Y)]\partial_j \\
&= [\tilde{X}^a (\partial_a(Y^b))\partial_b + \tilde{X}^a \tilde{Y}^b \phi^j_{,ab}\partial_j - (X \leftrightarrow Y)] \\
&= [\tilde{X}^a (\partial_a(Y^b))\partial_b - (X \leftrightarrow Y)] \\
&= [\tilde{X}, \tilde{Y}]
\end{aligned}
$$

where the terms involving $\phi^j_{,ab}$ being symmetric in X and Y cancels when $X \leftrightarrow Y$ is added.

Lemma 3 If X and Y are extensions of tangential fields $\tilde{X}$ and $\tilde{Y}$, then the covariant derivative $D_X Y$ on manifold M when restricted to the surface S is independent of the extensions.

We have

$$
\begin{aligned}
D_X Y|_S &= X^i[\partial_i Y^k + \Gamma^k_{ij} Y^j]\partial_k|_S \\
&= [\tilde{X}^a \partial_a(\tilde{Y}^b \phi^k_{,b}) + \Gamma^k_{ij} \tilde{X}^a \tilde{Y}^b \phi^i_{,a}\phi^j_{,b}]\partial_k|_S.
\end{aligned}
$$

All the quantities inside the bracket are expressible in terms of u or its functions, and are explicitly independent of how $\tilde{X}$ etc. are extended. Moreover, the tangential part is

$$
h(D_X Y|_S) = [\tilde{X}^a \tilde{Y}^b_{,a} \phi^k_{,b} + \tilde{X}^a \tilde{Y}^b \phi^k_{,ab} + \Gamma^k_{ij} \tilde{X}^a \tilde{Y}^b \phi^i_{,a}\phi^j_{,b}]h(\partial_k|_S).
$$

The rightmost factor is $h(\partial_k) = \psi^c_k \partial_c$. The coefficient ψ^c_k can be taken inside the square bracket where it combines with the first term to give $\phi^k_{,b}\psi^c_k = \delta^c_b$. To the second term it simply multiplies as

$$
\psi^c_k = \tilde{g}^{cd} g_{km}\phi^m_{,d}.
$$

When combined with the third term it gives

$$
\begin{aligned}
\phi^i_{,a}\phi^j_{,b}\Gamma^k_{ij}\psi^c_k &= (1/2)\phi^i_{,a}\phi^j_{,b}[g_{mi,j} + g_{mj,i} - g_{ij,m}]g^{mk}\psi^c_k \\
&= (1/2)[g_{mi,j} + g_{mj,i} - g_{ij,m}] \tilde{g}^{cd} \phi^i_{,a}\phi^j_{,b}\phi^m_{,d}.
\end{aligned}
$$

Terms can be written

$$
\begin{aligned}
g_{mi,j}\phi^i_{,a}\phi^j_{,b}\phi^m_{,d} &= g_{mi,b}\phi^i_{,a}\phi^m_{,d} \\
&= \tilde{g}_{da,b} - g_{mi}\phi^i_{,ab}\phi^m_{,d} - g_{mi}\phi^i_{,a}\phi^m_{,db}
\end{aligned}
$$

using

$$g_{mi}\phi^i_{,a}\phi^m_{,d} = \tilde{g}_{da} .$$

Combining all the terms together we get

$$h(D_X Y|_S) =$$
$$\tilde{X}^a \left[\tilde{Y}^c_{,a} + \tilde{Y}^b \tilde{g}^{cd} \left(g_{km}\phi^m_{,d}\phi^k_{,ab} + (1/2)\left[\tilde{g}_{da,b} + \tilde{g}_{bd,a} - \tilde{g}_{ab,d} \right] \right. \right.$$
$$+ (1/2)\left[-g_{mi}\phi^i_{,ab}\phi^m_{,d} - g_{mi}\phi^i_{,a}\phi^m_{,db} - g_{mj}\phi^j_{,ba}\phi^m_{,d} - g_{mj}\phi^i_{,b}\phi^m_{,da} \right.$$
$$\left. \left. \left. + g_{ij}\phi^i_{,ad}\phi^j_{,b} + g_{ij}\phi^i_{,a}\phi^j_{,bd} \right] \right) \right] \partial_c.$$

All terms of the type $g_{xx}\phi^x_{,xx}\phi^x_{,x}$ add to zero and we get

$$\tilde{D}_{\tilde{X}}\tilde{Y} \equiv h(D_X Y|_S) = \tilde{X}^a \left[\tilde{Y}^c_{,a} + \tilde{\Gamma}^c_{ab}\tilde{Y}^b \right] \partial_c \qquad (16.24)$$

where the connection components for the surface are given by

$$\tilde{\Gamma}^c_{ab} \equiv \frac{1}{2} \tilde{g}^{cd} \left[\tilde{g}_{da,b} + \tilde{g}_{bd,a} - \tilde{g}_{ab,d} \right]. \qquad (16.25)$$

This proves that the tangential component of connection in the larger space is indeed a connection built from the induced metric in the standard way.

16.3.3 Gauss Codacci Formulas

Let us rewrite the Gauss and Codacci formulas:

$$R(\tilde{W}, \tilde{Z}; \tilde{X}, \tilde{Y}) = \tilde{R}(\tilde{W}, \tilde{Z}; \tilde{X}, \tilde{Y})$$
$$+ K(\tilde{X}, \tilde{W})K(\tilde{Y}, \tilde{Z}) - K(\tilde{Y}, \tilde{W})K(\tilde{X}, \tilde{Z}), \qquad [\text{Gauss}]$$
$$\left\langle N, R(\tilde{X}, \tilde{Y})\tilde{Z} \right\rangle = -\left[(\tilde{D}_{\tilde{X}} K)(\tilde{Y}, \tilde{Z}) - (\tilde{D}_{\tilde{Y}} K)(\tilde{X}, \tilde{Z}) \right] \qquad [\text{Codacci}]$$

in the basis for tangential fields. Choose

$$\tilde{X} = \partial_c, \qquad \tilde{Y} = \partial_d, \qquad \tilde{W} = \partial_a, \qquad \tilde{Z} = \partial_b,$$

then for the Gauss formula we put

$$R(\partial_a, \partial_b; \partial_c, \partial_d) = \phi^i_{,a}\phi^j_{,b}\phi^k_{,c}\phi^l_{,d}R_{ijkl}$$

so that

$$\phi^i_{,a}\phi^j_{,b}\phi^k_{,c}\phi^l_{,d}R_{ijkl} = \tilde{R}_{abcd} + K_{ca}K_{db} - K_{da}K_{cb} \qquad (16.26)$$

where

$$K_{ab} \equiv K(\partial_a, \partial_b).$$

The Codacci formula gives similarly

$$N_i \phi^k_{,c} \phi^l_{,d} \phi^j_{,b} R^i{}_{jkl} = K_{db\|c} - K_{cb\|d}$$

where we denote the induced covariant derivative $\widetilde{D}$ on the surface S in components-index form by a stroke $\|$ as distinguished from the semicolon ; used for the covariant derivative D:

$$K_{ab\|c} \equiv \left(\widetilde{D}_{\partial_c} K \right)(\partial_a, \partial_b).$$

The left-hand side of the Codacci equation can be written

$$N^i \phi^k_{,c} \phi^l_{,d} \phi^j_{,b} R_{ijkl} = -N^i \phi^k_{,c} \phi^l_{,d} \phi^j_{,b} R_{jikl};$$

multiply by $\widetilde{g}^{cb}$ on both sides and sum on c, b we get

$$N^i R_{il} \phi^l_{,d} = (K^b_b)_{\|d} - (K^b_d)_{\|b} \tag{16.27}$$

where

$$K^a_b = \widetilde{g}^{ac} K_{cb}.$$

Remark on Notation

The component-index Gauss Codacci formulas are written differently in the literature depending on the use. For example, the $(n-1)$-dimensional curvature tensor should have indices running from 1 to $(n-1)$ as we have indicated by $\widetilde{R}_{abcd}$. But it is often convenient to write $\widetilde{R}_{ijkl}$ with indices running 1 to n. This is done by following definitions

$$\begin{aligned}
\widetilde{R}_{ijkl} &\equiv \widetilde{R}\left(h(\partial_i), h(\partial_j); h(\partial_k) h(\partial_l)\right), \\
K_{ij} &\equiv K(h(\partial_i), h(\partial_j)), \\
K_{ij\|k} &\equiv \left(D_{h(\partial_k)} K\right)(h(\partial_i), h(\partial_j)).
\end{aligned}$$

Then Gauss and Codacci formulas become

$$R_{i'j'k'l'} h^{i'}_i h^{j'}_j h^{k'}_k h^{l'}_l = \widetilde{R}_{ijkl} + K_{ik} K_{jl} - K_{il} K_{jk} \tag{16.28}$$

and

$$N_i R^i{}_{j'k'l'} h^{j'}_j h^{k'}_k h^{l'}_l = -K_{kl\|j} + K_{jl\|k}. \tag{16.29}$$

The left-hand side of the Codacci equation is equal to

$$
\begin{aligned}
N_i R^i{}_{j'k'l'} h^{j'}_j h^{k'}_k h^{l'}_l &= N^i R_{ij'k'l'} h^{j'}_j h^{k'}_k h^{l'}_l \\
&= -N^i R_{j'ik'l'} h^{j'}_j h^{k'}_k h^{l'}_l .
\end{aligned}
$$

Multiply by g^{jk} on both sides of the Codacci equation and sum to get the left-hand side as

$$
\begin{aligned}
-N^i R_{j'ik'l'} \left(h^{j'}_j h^{k'}_k g^{jk} \right) h^{l'}_l &= -N^i R_{j'ik'l'} \left(h^{j'}_j h^{k'}_k g^{jk} \right) h^{l'}_l \\
&= -N^i (g^{j'k'} + N^{j'} N^{k'}) R_{j'ik'l'} h^{l'}_l \\
&= -N^i R_{il'} h^{l'}_l
\end{aligned}
$$

where the term containing three N factors is zero because of antisymmetry of $R_{j'ik'l'}$ in j' and i. The right-hand side becomes

$$
g^{jk}(-K_{kl\|j} + K_{jl\|k}),
$$

therefore the Codacci equation is

$$
N^i R_{il'} h^{l'}_l = g^{jk}(K_{kl\|j} - K_{jl\|k}). \tag{16.30}
$$

The student is advised to study these equations with careful attention to notation. The relevant portions in advanced books are sections 2.7 of Hawking-Ellis, 21.5 of Misner-Thorne-Wheeler and 10.2 of Wald.

16.4 The Raychaudhuri Equation

We are back to physical spacetime in this section.

A congruence of curves in a region of a manifold is a collection of curves such that only one curve from the collection passes through each point of the region.

The effects of gravitation are best seen by studying what happens to trajectories of matter particles or of photons. If we imagine particles of dust falling under gravity, their trajectories will be time-like geodesics. By observing how these trajectories behave we get an idea about the gravitational field. If we consider a fluid instead of dust, the particles of the fluid are pushed around not only by gravity but by stresses (like pressure) as well. Nevertheless, the trajectories are a congruence of time-like curves which give valuable insights into the nature of the gravitational field.

Obviously the congruence of time-like curves depends on the "initial conditions". Different initial conditions give rise to different congruences, and it is worthwhile to study them in the general context.

A congruence of time-like curves can be characterized by the unit time-like vector field N of tangent vectors along the curves.

$$
\langle N, N \rangle = -1. \tag{16.31}
$$

Conversely, given a unit time-like vector field like this we can find its integral curves to get the congruence.

Note that the normalization condition above implies that the covariant derivative of the field N with respect to any other smooth vector field X is orthogonal to N itself:

$$D_X \langle N, N \rangle = \langle D_X N, N \rangle + \langle N, D_X N \rangle = 2 \langle N, D_X N \rangle = 0.$$

In particular $D_N N$ is orthogonal to N. Of course, if the curves in the congruence are geodesics, then

$$D_N N = 0 \qquad \text{for geodesics.} \tag{16.32}$$

16.4.1   Deviation of Nearby Time-like Geodesics

Let us consider the simplest case of a congruence of time-like geodesics, which could, for example, be dust particles in a region falling freely. We want to find out how the neighbouring trajectories deviate from each other (or come closer). Let us imagine a pair of nearby geodesic trajectories and study how two points (one on each) separate when they travel the same proper time along their respective geodesics.

Let us use the proper time as the affine parameter along each of the trajectories. We take a curve $\lambda : s \to \lambda(s)$ which is nowhere tangent to any of the congruences. Those curves of the congruence which pass through points $\lambda(s)$ can be labelled by the parameter s.

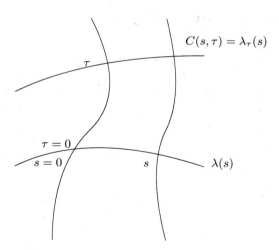

Fig. 16.1: Definition of $\lambda_\tau : s \to C(s, \tau)$.

Assign to each point $\lambda(s)$ the value of proper time (of that curve) zero. Identify the point $C(s, \tau)$ on the curve through the point $\lambda(s)$ and situated at a proper time τ from it along the curve. This determines a new curve $\lambda_\tau : s \to C(s, \tau)$ shifted by a proper time τ along the curves.

Define a local diffeomorphism ϕ_τ which maps points on the curves to points on the same curve a proper time τ away. Obviously, for any point p,

$$\frac{d}{d\tau} \phi_\tau(p) \Big|_{\tau=0} = N(p).$$

Under this mapping the curve $\lambda_{\tau'}$ is mapped onto $\lambda_{\tau'+\tau}$. The tangent vector

$$Z(s', \tau) = \frac{d}{ds} C(s' + s, \tau) \Big|_{s=0}$$

to λ_τ for different values of s, τ is mapped by the one-parameter group of diffeomorphisms whose orbits are the integral curves of N. Therefore, the Lie derivative $L_N Z = [N, Z] = 0$. As the torsion is zero ($T = 0 = D_X Y - D_Y X - [X, Y]$) this implies

$$D_Z N = D_N Z. \tag{16.33}$$

The deviation of nearby geodesics can now be calculated. We have

$$\begin{aligned} D_N(D_N Z) &= D_N(D_Z N) \\ &= D_Z(D_N N) - D_{[Z,N]} N + R(N, Z)N \end{aligned}$$

but $D_N N = 0$ and $[Z, N] = 0$ therefore

$$D_N(D_N Z) = R(N, Z)N. \tag{16.34}$$

16.4.2 Jacobi Equation for a Congruence

In the previous section we considered the trajectories of dust particles which fall freely in a gravitational field and constitute a congruence of time-like geodesics. If we have a fluid instead of dust, then infinitesimal elements of fluid follow time-like curves but not geodesics because the stresses act on the elements. Therefore we study a congruence of time-like curves not necessarily time-like geodesics.

We can construct the field N and Z as previously. Only, now $\dot{N} \equiv D_N N$ need not be zero as the congruence curves are not geodesics. From the geometry of construction we still have

$$\langle N, N \rangle = -1, \qquad [N, Z] = 0, \qquad D_N Z = D_Z N.$$

It is useful to project any vector field X perpendicular to the congruence. We can define the projection operator

$$h(X) = X + N\langle N, X \rangle$$

which gives

$$\langle N, h(X) \rangle = 0.$$

In particular $h(N) = 0$.

First Deviation

Take the covariant derivative (along N) of orthogonal part $h(Z)$ of Z. This derivative will have components both along the trajectory as well as orthogonal to it. The orthogonal part $h(D_N(h(Z)))$ of the covariant derivative (along N) of the orthogonal part of Z changes according to the first deviation equation:

$$h(D_N(h(Z))) = D_{h(Z)}N \qquad \text{[First deviation equation]}. \qquad (16.35)$$

The proof is straightforward,

$$
\begin{aligned}
D_N(h(Z)) &= D_N(Z + N\langle N, Z\rangle) \\
&= D_N Z + f D_N N + N.N(f), \qquad f = \langle N, Z\rangle.
\end{aligned}
$$

Taking projection with h, using $h(N) = 0$, as well as $D_N Z = D_Z N$,

$$
\begin{aligned}
h(D_N(h(Z))) &= h(D_Z N) + \dot{N}\langle N, Z\rangle \\
&= D_Z N + \dot{N}\langle N, Z\rangle
\end{aligned}
$$

because $D_Z N$ being orthogonal to N is already horizontal. The right-hand side can actually be written as

$$
\begin{aligned}
D_{h(Z)}N &= D_{Z+N\langle N,Z\rangle}N \\
&= D_Z N + D_N N\langle N, Z\rangle \\
&= D_Z N + \dot{N}\langle N, Z\rangle.
\end{aligned}
$$

This proves the first deviation equation.

Second Deviation: Jacobi Equation

Now we calculate the second deviation, that is one more derivative followed by taking the orthogonal projection. We have

$$h(D_N(h(D_N(h(Z))))) = h(D_N D_{h(Z)}N).$$

We write, using the definition of Riemann curvature tensor,

$$D_N D_{h(Z)}N = D_{h(Z)} D_N N + R(N, h(Z))N + D_{[N,h(Z)]}N.$$

Now

$$
\begin{aligned}
[N, h(Z)] &= [N, Z + fN], \qquad f = \langle N, Z\rangle \\
&= [N, Z] + f[N, N] + N\, N(f) \\
&= N\, N(f)
\end{aligned}
$$

as $[N, Z] = 0$ by construction and therefore

$$D_{[N,h(Z)]}N = N(f)D_N N = N(f)\dot{N}.$$

We can simplify it further by noting that

$$
\begin{aligned}
N(f) &= N(\langle N, Z \rangle) \\
&= \langle D_N N, Z \rangle + \langle N, D_N Z \rangle \\
&= \langle D_N N, Z \rangle + \langle N, D_Z N \rangle \\
&= \langle \dot{N}, Z \rangle;
\end{aligned}
$$

the last line follows because $D_Z N$ is orthogonal to N. Thus we get for the second deviation an equation called the **Jacobi equation**,

$$
h(D_N(h(D_N(h(Z))))) = h(D_{h(Z)} \dot{N}) + h(R(N, h(Z))N) + \dot{N}\langle \dot{N}, Z \rangle.
\tag{16.36}
$$

16.4.3 Jacobi Equation in Components

As explained above, it is the separation of nearby curves orthogonal to N which actually measures how fast they are converging. To be useful we must be able to define actual numbers which measure this deviation. For this purpose we choose three unit vectors, $E_i, i = 1, 2, 3$ orthogonal to N and to each other at some point on a chosen curve. The set

$$
\{E_a\}, a = 0, 1, 2, 3 \qquad E_0 = N
$$

is an orthonormal basis

$$
\langle E_a, E_b \rangle = \eta_{ab}.
$$

Then we propagate the three vectors along the curve by Fermi-Walker transport so that they remain orthogonal to N and to each other. This construction thus makes $\{E_a\} = N, E_i$ a basis along the curve, and the "horizontal part" $h(Z)$ of the deviation vector can be expanded in the three orthogonal vectors

$$
h(Z) = Z^i E_i.
$$

The rate of change of quantities like Z^i measures the deviation. If τ is the proper time along the curve (we are taking $c = 1$ so that $\langle N, N \rangle$ is equal to -1 and not $-c^2$)

$$
\begin{aligned}
D_N(h(Z)) &= D_N(Z^i E_i) \\
&= \frac{dZ^i}{d\tau} E_i + Z^i D_N E_i.
\end{aligned}
$$

Because E_i change along the trajectory according to Fermi-Walker transport we have

$$
\begin{aligned}
0 = D_N^{FW} E_i &= D_N E_i - N\langle \dot{N}, E_i \rangle + \dot{N}\langle N, E_i \rangle \\
&= D_N E_i - N\langle \dot{N}, E_i \rangle.
\end{aligned}
$$

Substituting $D_N E_i = N \langle \dot{N}, E_i \rangle$ gives

$$h(D_N(h(Z))) = \frac{dZ^i}{d\tau} E_i$$

because $h(N) = 0$.

On the other hand

$$D_{h(Z)} N = Z^i D_{E_i} N$$

therefore the first deviation equation $h(D_N(h(Z))) = D_{h(Z)} N$ becomes

$$\frac{dZ^i}{d\tau} E_i = Z^i D_{E_i} N.$$

Taking the inner product with E_j and using orthogonality $\langle E_i, E_j \rangle = \delta_{ij}$,

$$\frac{dZ^j}{d\tau} = \langle E_j, D_{E_i} N \rangle Z^i. \tag{16.37}$$

Similarly, the second deviation equation can be simplified.

$$\begin{aligned} D_N(h(D_N(h(Z)))) &= D_N\left(\frac{dZ^i}{d\tau} E_i\right) \\ &= \frac{d^2 Z^i}{d\tau^2} E_i + \frac{dZ^i}{d\tau} D_N E_i. \end{aligned}$$

Again operating with h gives $h(N) = 0$ and

$$h(D_N(h(D_N(h(Z))))) = \frac{d^2 Z^i}{d\tau^2} E_i.$$

The second deviation equation therefore becomes taking the inner product with E_j,

$$\frac{d^2 Z^j}{d\tau^2} = [\langle E_j, R(N, E_i)N \rangle + \langle E_j, D_{E_i}\dot{N} \rangle + \langle E_j, \dot{N} \rangle \langle \dot{N}, E_i \rangle] Z^i. \tag{16.38}$$

16.4.4   Raychaudhuri Equation

Let us go back to a congruence of time-like geodesics. Then $\dot{N} = 0$ and the components Z^i of the three-dimesional deviation vector satisfy the first and second deviation equations

$$\frac{dZ^j}{d\tau} = \langle E_j, D_{E_i} N \rangle Z^i \tag{16.39}$$

$$\frac{d^2 Z^j}{d\tau^2} = \langle E_j, R(N, E_i)N \rangle Z^i. \tag{16.40}$$

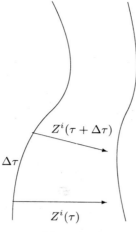

Fig. 16.2.

Let us define the τ dependent matrix

$$B_{ji}(\tau) \equiv \langle E_j, D_{E_i} N \rangle, \tag{16.41}$$

then the meaning of

$$\frac{dZ^j}{d\tau} = B_{ji} Z^i$$

can be understood as follows. Under an infinitesimal proper time $\Delta\tau$ the change in Z^j is linearly proportional to Z^i's again.

$$\Delta Z^j = B_{ji} Z^i \Delta\tau.$$

The coefficients $B_{ji}(\tau)$ are in general nine numbers at each τ and they determine the nature of change taking place in spatial separation of two neighbouring geodesics. If an observer 'falls freely' (like a dust particle) along one geodesic labelled by parameter s (as just discussed in section 16.4.1) then $Z^i\Delta s$ would be the spatial three-dimensional vector representing the position of another dust particle falling freely on a neighbouring geodesic with parameter $s + \Delta s$. We can think of a cloud of dust particles in the neighbourhood of the falling observer to see what gravity does to freely falling matter.

A general 3×3 matrix B_{ji} can be written as a sum of its symmetric part (six independent components) and the antisymmetric part (three). The symmetric part again can be written as a sum of a matrix proportional to the identity matrix (one) and a symmetric matrix which is also traceless (five).

$$B_{ji} = \frac{\theta}{3}\delta_{ji} + \sigma_{ji} + \omega_{ji} \tag{16.42}$$

where

$$\theta \ = \ \sum_i B_{ii}, \tag{16.43}$$

$$\sigma_{ji} \ = \ (B_{ji} + B_{ij})/2 - (\theta/3)\delta_{ij}, \tag{16.44}$$

$$\omega_{ji} \ = \ (B_{ji} - B_{ij})/2. \tag{16.45}$$

Each of these three types of matrices generate a physically visualizable type of change: if there was only an antisymmetric part $(B_{ji} = \omega_{ji})$, the change in Z^j would be an infinitesimal rotation because then the new position of the dust particles is obtained by an infinitesimal orthogonal transformation:

$$Z^j(\tau + \Delta\tau) = (\delta_{ji} + \omega_{ji}\Delta\tau)Z^i.$$

If there is only a symmetric traceless part $(B_{ji} = \sigma_{ji})$, then the change

$$Z^j(\tau + \Delta\tau) = (\delta_{ji} + \sigma_{ji}\Delta\tau)Z^i$$

will be a distortion of the cloud but no change of volume because the determinant

$$\det(\delta_{ji} + \sigma_{ji}\Delta\tau) = 1 + \sum_i \sigma_{ii} + O(\Delta\tau^2) \approx 1.$$

Such a change or distortion is called **shear**. If the dust cloud were a spherical ball at one time, it will become an ellipsoid of the same volume a little later.

Lastly, if the matrix were proportional to unit matrix $B_{ji} = (\theta/3)\delta_{ji}$, then the change is

$$Z^j(\tau + \Delta\tau) = (1 + \theta\Delta\tau)Z^j.$$

This is obviously **expansion** (or contraction if θ is negative) by the same proportionality factor $(1 + \theta\Delta\tau)$ for all points. θ is called the **expansion parameter**. This parameter determines how geodesics expand or contract and not just how they rearrange themeselves, because only in this case does the volume of the dust cloud change.

The second deviation equation tells us how these parameters of expansion, shear or rotation themselves change along the geodesics. Substituting the first deviation in the second we get

$$
\begin{aligned}
\langle E_j, R(N, E_i)N\rangle Z^i \ &= \ \frac{d}{d\tau}\left(B_{jk}Z^k\right) \\
&= \ \frac{dB_{jk}}{d\tau}Z^k + B_{jk}\frac{dZ^k}{d\tau} \\
&= \ \frac{dB_{ji}}{d\tau}Z^i + B_{jk}B_{ki}Z^i.
\end{aligned}
$$

Therefore we must have

$$\frac{dB_{ji}}{d\tau} + B_{jk}B_{ki} = \langle E_j, R(N, E_i)N \rangle.$$

Taking the trace of this equation (that is multiplying by δ_{ji} and summing over both indices) and using symmetry, antisymmery and traceless properties

$$\left(\frac{\theta}{3}\delta_{ik} + \sigma_{ik} + \omega_{ik}\right)\left(\frac{\theta}{3}\delta_{ki} + \sigma_{ki} + \omega_{ki}\right) = \frac{\theta^2}{3} + \sigma_{ij}\sigma_{ij} - \omega_{ij}\omega_{ij},$$

we get

$$\frac{d\theta}{d\tau} + \frac{\theta^2}{3} + \sigma_{ij}\sigma_{ij} - \omega_{ij}\omega_{ij} = \delta_{ji}\langle E_j, R(N, E_i)N \rangle.$$

We can simplify the right-hand side as follows: $E_0 \equiv N$ and E_i form an orthonormal basis $E_a, a = 0, 1, 2, 3$. Expand them in the coordinate basis

$$E_a = e_a{}^\mu \frac{\partial}{\partial x^\mu}, \qquad e_0{}^\mu = N^\mu$$

and we have

$$\langle E_a, E_b \rangle = \eta_{ab} = e_a{}^\mu e_b{}^\nu g_{\mu\nu}.$$

The basis $\{\varepsilon^a\}$ dual to $\{E_a\}$ is related to the dual basis of forms $\{dx^\mu\}$ by ("inverse-transpose rule" of Chapter 5)

$$dx^\mu = \varepsilon^a e_a{}^\mu,$$

therefore

$$\begin{aligned} g^{\mu\nu} = \langle dx^\mu, dx^\nu \rangle &= \eta^{ab} e_a{}^\mu e_b{}^\nu \\ &= -N^\mu N^\nu + \sum_i e_i{}^\mu e_i{}^\nu. \end{aligned}$$

Thus substituting $\sum_i e_i{}^\mu e_i{}^\mu$ from this relation,

$$\begin{aligned} \delta_{ji}\langle E_j, R(N, E_i)N \rangle &= -\sum_i R(E_i, N; E_i, N) \\ &= -\sum_i e_i{}^\mu e_i{}^\nu N^\sigma N^\tau R_{\mu\sigma\nu\tau} \\ &= -N^\sigma N^\tau R_{\sigma\tau} \end{aligned}$$

where the term with four factors of components of N does not survive because of the antisymmetry properies of the covariant Riemann tensor.

This is how we get the justly famous **Raychaudhuri equation**

$$\frac{d\theta}{d\tau} = -\frac{\theta^2}{3} - \sigma_{ij}\sigma_{ij} + \omega_{ij}\omega_{ij} - N^\mu N^\nu R_{\mu\nu}. \qquad (16.46)$$

Suppose ω_{ij} are zero, and we can ensure that $N^\mu N^\nu R_{\mu\nu}$ is positive; then the Raychaudhuri equation says that regardless of the initial value of θ it can only decrease along the geodesic in the forward time direction of N. This means that if θ was positive, expansion by factor $(1+\Delta\tau\theta)$, then the expansion will decrease. And if there was contraction before $\theta < 0$ there will be even more severe contraction. This 'focussing' property of gravitation is compatible our perception of gravitation being a attractive force for all matter and radiation.

The condition $N^\mu N^\nu R_{\mu\nu} > 0$ is called the **strong energy condition**. After taking the trace in the Einstein equation $R_{\mu\nu} - g_{\mu\nu}R/2 = (8\pi G/c^4)T_{\mu\nu}$, we find $-R = (8\pi G/c^4)T$ where $T = g^{\mu\nu}T_{\mu\nu}$. This allows us to write

$$R_{\mu\nu} = \frac{8\pi G}{c^4}\left(T_{\mu\nu} - \frac{1}{2}g_{\mu\nu}T\right).$$

Therefore the strong energy condition becomes

$$\left(N^\mu N^\nu T_{\mu\nu} + \frac{1}{2}T\right) > 0. \tag{16.47}$$

For the freely falling observer whose time axis is given by N, $N^\mu N^\nu T_{\mu\nu}$ is just the rest-mass density of dust (and is equal to T), the condition is certainly satisfied. The condition is also expected to be satisfied for all forms of matter and radiation.

Penrose and Hawking were able to prove with very general arguments using the Raychaudhuri equation that the kind of singularities that appear in the Schwarzschild or Kerr solutions or in the FRW metric at the beginning $t = 0$ of the universe are a generic feature of the general theory of relativity and not a pecularity of solutions with a very high degree of symmetry. This means that black holes cannot be wished away by arguing that exact spherical collapse requires initial conditions impossible in the actual world. Gravitational collapse of sufficiently large mass will give rise to a black hole after transients in the form of gravitational waves and radiation have eventually died down.

Similarly, the singularity at the very beginning of the universe is a real one, and we seem to have reached a physical phenomenon which cannot be explained by classical, non-quantum physics. Quantum gravitational phenomena will become important when lengths of the order of Planck length (equal to $\sqrt{\hbar G/c^3}$) are reached long before the mathematical limit of a singularity. The quest for a quantum theory of gravity has been going on for the last few decades.

It seems we are not there yet.

Index